NEGOTIATING CLIMATE CHANGE IN CRISIS

Negotiating Climate Change in Crisis

Edited by Steffen Böhm and Sian Sullivan

OpenBook
Publishers

ISBN Paperback: 9781800642607
ISBN Hardback: 9781800642614
ISBN Digital (PDF): 9781800642621
ISBN Digital ebook (epub): 9781800642638
ISBN Digital ebook (mobi): 9781800642645
ISBN XML: 9781800642652
DOI: 10.11647/OBP.0265

Cover image: Photo by Thijs Stoop on Unsplash available at: https://unsplash.com/photos/A_AQxGz9z5I
Cover design by Anna Gatti

Contents

List of Images and Videos

Introduction

Chapter 3

Chapter 8

Chapter 11

Chapter 12

Chapter 13

Chapter 14

Chapter 17

Acknowledgements

We are extremely grateful for a research grant from Bath Spa University to cover the production costs of this open access volume. We would also like to thank Alessandra Tosi, Melissa Purkiss and the team at Open Book Publishers, and two anonymous reviewers, for their enthusiasm about publishing this collection in time for the 26th Conference of the Parties of the United Nations Framework Convention on Climate Change, November 2021.

This work by eminent scholars from around the world offers a provocative and deeply insightful analysis of "the politics of paralysis and self-destruction" that have long hindered effective and equitable climate policy over the past 20 years. The book is very timely, and I hope will help to increase the sense of urgency for a deal that will save the planet and billions of poor people around the world that bear a disproportionate impact of climate change.

Prof Chukwumerije Okereke, Director Center of Climate Change and Development, Alex-Ekwueme Federal University, Ndufu-Alike, Nigeria

Every person of good will must work to make COP26 a decisive moment in the history of the world. This rich and eclectic collection offers a range of critical perspectives into the destructive role played by the coal, oil and gas sectors, and other crucial issues that will underpin the conference, as well as providing ideas for how we can yet secure a world of future flourishing.

David Ritter, CEO of Greenpeace Australia Pacific

This book presents perspectives from the Global South, highlighting voices from communities and sharing their daily lived experiences of climate change. These voices are often missing from international platforms such as COPs. The contributions included in the book are valuable for countries such as Namibia and others where the impacts of climate change are severe. Namibia strongly advocates for knowledge production regarding climate change and its impact on livelihoods, the coping mechanisms of vulnerable communities and their capacity to adapt.

Hon. Heather Mwiza Sibungo, Deputy Minister of Environment, Forestry and Tourism, Namibia

Bringing a wide range of social scientists together, this volume provides a much needed critical analysis of why meaningful action on climate change has been so elusive and what we can do about it. It asks the difficult questions and provides some promising answers.

Prof. Harriet Bulkeley, Durham University

One of the key challenges in responding to climate change is taking into account multiple perspectives across various theories and methods of research. Negotiating Climate Change in Crisis is a clear and thoughtfully organized anthology that rises to that challenge. Unlike some social science writing, this book is readily accessible to any generally interested reader. The contributors succeed in presenting a comprehensive, state-of-the-art report about social science perspectives on the climate crisis.

Sam Mickey, University of San Francisco

The Authors

Eeva-Lotta Apajalahti is Research Coordinator at the Helsinki Institute of Sustainability Sciences, HELSUS, in the University of Helsinki. She received her DSc (2018) from Aalto University School of Business by focusing on the role of large energy companies in energy system transition. Her current research focuses on energy transitions in cities, connections between sustainable consumption and production, energy citizenship and energy communities. She is one of the coordinators of the Finnish Expert Panel for Sustainable Development.

Ian Bailey is Professor of Environmental Politics at the University of Plymouth, UK. His research interests include the politics of designing and regulating carbon markets, social attitudes to onshore and offshore renewable energy, and debates on environmental and social justice within sustainability transitions. He has worked as an expert reviewer for the IPCC Fifth and Sixth Assessment Reports and has advised the UK Environmental Audit Committee, Cabinet Office, European Commission, and World Bank on aspects of climate mitigation policy. He is currently a member of the Devon Climate Emergency Net Zero Task Force.

Oliver Belcher is Assistant Professor in International Relations & Security at the School of Government and International Affairs, Durham University. He has written extensively on late-modern warfare, including the U.S. military as a climate actor.

Giovanni Bettini is a Senior Lecturer in International Development and Climate Politics at Lancaster University. His research investigates how environmental change—in its planetary but uneven character, and entangled with a series of contemporary 'crises' and historical legacies— is generating new spaces, modes of governance, subjectivities and forms of resistance. He has published extensively on the links between climate change and human mobility, and more recently on the role of 'the digital' in reshaping adaptation, resilience, and justice.

Patrick Bigger is an Honorary Research Fellow at Lancaster University. In addition to work on US military operations and their climate impacts, his research focuses on private investment for biodiversity conservation and climate adaptation.

Ingrid Boas is an Associate Professor at the Environmental Policy Group of Wageningen University. Ingrid conducts research in the fields of environmental change, mobilities, and governance. Her PhD (University of Kent, 2014) examined the securitisation of climate migration, and was funded by the UK Economic Social Research Council. In 2016, she was awarded a personal grant with the Netherlands Scientific Organization to study environmental mobility in the digital age. Ingrid's work has appeared in *Global Environmental Politics*, *Environmental Politics*, *Geoforum*, the *Journal of Ethnic and Migration Studies*, *Nature Climate Change*, and in a monograph on climate migration and security with Routledge (2015).

Steffen Böhm is Professor in Organisation & Sustainability at University of Exeter Business School. His research focuses on the political economy and ecology of the sustainability transition. He has published five books: *Repositioning Organization Theory* (Palgrave, 2006), *Against Automobility* (Wiley-Blackwell, 2006), *Upsetting the Offset: The Political Economy of Carbon Markets* (Mayfly), *The Atmosphere Business* (Mayfly, 2009), and *Ecocultures: Blueprints for Sustainable Communities* (Routledge, 2015). The book *Climate Activism* (with Annika Skoglund) is forthcoming with Cambridge. More details at www.steffenboehm.net.

Patrick Bond is an Honorary Professor of Geography at Wits University, both in South Africa. His doctorate (1993) was in Geography under David Harvey's supervision. Among his political ecology books are *Unsustainable South Africa: Environment, Development and Social Protest* (Merlin Press, 2002); *Looting Africa: The Economics of Exploitation* (Zed Books, 2006); and *Politics of Climate Justice: Paralysis Above, Movement Below* (University of KwaZulu-Natal Press, 2012).

Sarah Bracking is Professor of Climate and Society at King's College London, UK. She is editor of *Corruption and Development* (Palgrave, 2007); author of *Money and Power* (Pluto, 2009) and *The Financialisation of Power* (Routledge, 2016); and co-editor with Sian Sullivan, Philip Woodhouse, and Aurora Fredrikson of *Valuing Development, Environment and Conservation: Creating Values that Matter* (Routledge, 2019). She is currently researching climate and development finance, climate insurance and the wider political economy of development in southern Africa.

Mirjam de Bruijn is Professor in Contemporary History and Anthropology of Africa at Leiden University and the African Studies Centre Leiden (ASCL). Her specific fields of interest are mobility, youth, social (in)security, and Information and Communication Technologies (ICTs). She is an interdisciplinary scholar with a basis in social anthropology. In 2016 she founded the organisation Voice4thought (www.Voice4Thought.org) and is director of the project Voice4Thought Académie, based in Mali. Mirjam teaches 'innovative methods and methodology' in the BA and MA African Studies programmes. Her recent publications are on ICTs and society, radicalisation, and youth.

Ute Dieckmann is an anthropologist at the University of Cologne and currently German Principal Investigator for *Etosha-Kunene Histories* (www.etosha-kunene-histories.net), supported by the German Research Foundation and the UK's Arts and Humanities Research Council. For many years, she has worked at the Legal Assistance Centre in Windhoek, doing research with and advocacy for marginalised and indigenous communities in Namibia.

David Durand-Delacre is a PhD candidate in the Department of Geography at the University of Cambridge. Combining media analysis and interviews with academic researchers, journalists, NGOs, and policymakers, his thesis investigates the emergence and evolution of debates at the intersection of climate change and migration in France. David previously worked as an analyst for the United Nations Sustainable Development Network. In this role, he collected and analysed indicators for two editions of the network's flagship Sustainable Development Report (2015 and 2016). From 2015 to 2018, he also volunteered and then became President of Réfugiés Bienvenue, a Paris-based NGO providing housing to homeless asylum seekers.

Alexander Dunlap is a Post-doctoral Research Fellow at the Centre for Development and the Environment, University of Oslo. His work has critically examined police-military transformations, market-based conservation, wind energy development, and extractive projects more generally in both Latin America and Europe. He has published in *Anarchist Studies, Geopolitics, Journal of Peasant Studies, Capitalism Nature Socialism, Political Geography, Journal of Political Ecology, Environment and Planning E,* and a recent article in *Globalizations*.

James G. Dyke is a Senior Lecturer in Global Systems at the University of Exeter where he serves as the Assistant Director for the Global Systems Institute and programme director for the MSc Global Sustainability Solutions. He has previously held visiting positions at the Earth-Life Systems Institute at the Tokyo Institute of Technology, and the School of Geography at the University of Southampton. He is an environmental columnist for UK newspaper the *i*. His book *Fire, Storm & Flood: The Violence of Climate Change* (2021) is published by Head of Zeus / Bloomsbury. More details can be found at jamesgdyke.info.

Carol Farbotko is an Adjunct Fellow in Geography at the University of the Sunshine Coast, Australia. She is a cultural geographer with research interests in climate mobilities and the politics of climate risk. Much of her work is focused on climate change mobilities in the Pacific Islands.

Isabelle Fremeaux is a popular educator, action researcher and deserter of the neoliberal academy where for a decade she was Senior Lecturer in Media and Cultural Studies at Birkbeck, University of London. With Jay Jordan, Isabelle coordinates *The Laboratory of Insurrectionary Imagination,* bringing artists and activists together to design tools and acts of disobedience. Co-authors of the film and book *Paths Through Utopias* (La Découverte, 2011), they live and work on the zad of Notre-dame-des-landes, where an international airport project was abandoned after 50 years of struggle.

Eeva Furman is the Director of the Environmental Policy Center at the Finnish Environment Institute. Her background is in marine biology, and for the last twenty years she has worked with environmental governance. Her core interests are science-policy-society, the co-creation of sustainability transformations, and active citizenship. Her background is in biodiversity governance, ecosystem service management and sustainable development, and she engages actively internationally. She co-authored The *Global Sustainable Development Report* (GSDR) with fourteen other scientists—the report was handed to the heads of the UN's member states in the General Assembly held in September 2019. She is the Chair of the Finnish Expert Panel for Sustainable Development.

Sharon Gardham received her MA in Environmental Humanities from Bath Spa University and has a degree in History and Heritage Management from the University of Gloucestershire. She is currently undertaking PhD research at Bath Spa involving a multi and trans-disciplinary examination of the Cotswold Commons.

Giovanna Gioli is a Senior Lecturer in Human Geography at Bath Spa University. She has held research and teaching posts at several international universities, as well as a lectureship at the University of Edinburgh. She has also worked for various development organisations in South Asia, including the International Centre for Integrated Mountain Development (ICIMOD) in Kathmandu, Nepal.

Minna Halme is Professor of Sustainability Management at Aalto University School of Business. Her research focuses on co-creation of sustainable innovations in the context of grand challenges. She is Associate Editor of *Organization & Environment*, and on the editorial boards of several journals in the field of sustainability and management. She is a member of a number of national industry and public policy boards, including Finland's Expert Panel for Sustainable Development, and sits on the advisory boards of Finland's largest retailer SOK and the Central Chamber of Commerce. She is a co-founder of Aalto University's cross-disciplinary Creative Sustainability Master's programme, Aalto Sustainability Hub, and Aalto Global Impact.

Mike Hannis is Senior Lecturer in Ethics, Politics and Environment at Bath Spa University, UK. His academic publications include *Freedom and Environment: Autonomy, Human Flourishing and the Political Philosophy of*

Sustainability (Routledge, 2016). He lives off-grid in Somerset, and is an editor and feature writer for *The Land* magazine.

Paul G. Harris (www.paulgharris.net) is the Chair Professor of Global and Environmental Studies at the Education University of Hong Kong and a Senior Research Fellow in the Earth System Governance global research alliance. He is author/editor of twenty-five books on climate change and global environmental politics, policy and justice. His most recent books include, as author, *Pathologies of Climate Governance: International Relations, National Politics and Human Nature* (Cambridge University Press, 2021) and, as editor, *A Research Agenda for Climate Justice* (Edward Elgar, 2019).

M. Timm Hoffman is the Leslie Hill Chair of Plant Conservation in the Department of Biological Sciences at the University of Cape Town and is the Director of the Plant Conservation Unit. He uses repeat photography to understand long-term changes in the vegetation of southern Africa.

Mike Hulme is Professor of Human Geography at the University of Cambridge. His work illuminates the numerous ways in which the idea of climate change is deployed in public, political, religious, and scientific discourse, exploring both its historical, cultural and scientific origins and its contemporary meanings. He is the author of nine books on climate change, including most recently *Climate Change: Key Ideas in Geography* (Routledge, 2021), *Contemporary Climate Change Debates* (Routledge, 2020) and *Why We Disagree About Climate Change* (Cambridge University Press, 2009). From 2000 to 2007 Hulme was the Founding Director of the Tyndall Centre for Climate Change Research.

Saleemul Huq is an expert in the field of climate change, environment and development. He has worked extensively in the inter-linkages between climate change mitigation, adaptation and sustainable development, from the perspective of developing countries, particularly the least developed countries (LDCs). He was a lead author of the chapter on 'Adaptation and Sustainable Development' in the Third Assessment Report of the Intergovernmental Panel on Climate Change (IPCC), and was one of the coordinating lead-authors of 'Inter-relationships between adaptation and mitigation' in the IPCC's Fourth Assessment Report (2007). He has taught at Imperial College, the University of Dhaka, and the United Nations University. He was the founder, and is currently the Chairman, of the Bangladesh Centre for Advanced Studies (BCAS), a leading research and policy institute in Bangladesh.

Elodie Hut is a PhD candidate at the Hugo Observatory (University of Liège, Belgium), where she previously worked as a research assistant, conducting research for the MIGRADAPT project (on migration as a potential adaptation strategy to environmental changes). Prior to this, Elodie successively worked at the UNHCR and the IOM in South Africa, for GIZ in Senegal, and in a disaster risk reduction consultancy firm in South Africa. Elodie holds a Master's degree in Humanitarian Action and Law from the Institute of International Humanitarian Studies of Aix-en-Provence, as well a second Master's in International Relations from Sciences Po Aix.

Jouni J. K. Jaakkola is Professor of Public Health at the University of Oulu and Research Professor at the Finnish Meteorological Institute. He has broad long-term interests in global health. From the early 1990s he has pursued an international academic career working in Norway, Russia, the US, Sweden, and the UK. In 2008 he established the Center for Environmental and Respiratory Health Research (CERH) at the University of Oulu. In 2014 CERH was designated as a WHO Collaborating Centre in Global Change, Environment and Public Health. His professional mission is to conduct research on topics which help to solve emerging global public health problems. He is a member of the Expert Panel for Sustainable Development.

Jay Jordan (formerly John Jordan) is an art activist described as a "magician of rebellion" by the press and a "Domestic Extremist" by the UK police. Co-founder of "Reclaim the Streets" (1995–2000) and the *Clandestine Insurgent Rebel Clown Army*, Jay has co-authored *We Are Everywhere: The Irresistible Rise of Global Anticapitalism* (Verso, 2003) and *A User's Guide to Demanding the Impossible* (with Gavin Grindon, Minor Compositions, 2011). With Isabelle Fremeaux, Jay coordinates *The Laboratory of Insurrectionary Imagination*, bringing artists and activists together to design tools and acts of disobedience. Co-authors of the film and book *Paths Through Utopias* (La Découverte, 2011), they live and work on the zad of Notre-dame-des-landes, where an international airport project was abandoned after 50 years of struggle.

Rami Kaplan is a political and organizational sociologist at Tel Aviv University. He studies various aspects of global corporate capitalism, including its historical emergence and spread, corporate power and social responsibility, global diffusion of organizational practices, transnational business elite networks, global environmental politics, and the spread of populist rationality. His comparative research spans the USA, the UK, Germany, Venezuela, the Philippines, Israel, and the supra-national level.

Cara Kennelly holds a Master's by research in Carbon Accounting from Lancaster University and has supported a variety of carbon emissions research projects. She is currently the Sustainability Manager for VINCI Facilitie.

Mizan R. Khan is Deputy Director at the International Centre for Climate Change & Development (ICCCAD) and Programme Director of the LDC Universities' Consortium on Climate Change (LUCCC) at ICCCAD, Independent University Bangladesh, Dhaka. He served at North South University (NSU), Dhaka as Chair of the Department of Environmental Science & Management (DESM) from 2003–2009 and was Director of External Affairs at NSU in 2015. He was also an Adjunct Professor at the Natural Resources Institute (NRI), University of Manitoba, Canada, from 2009–2013.

Wolfgang Knorr is a Senior Researcher in the Department of Physical Geography and Ecosystem Science, Lund University. A physicist by training, he has published extensively on a broad range of climate and climate impacts research, including the global carbon cycle, climate impacts on terrestrial ecosystems, fire ecology, plant physiology, soil science, land surface-atmosphere feedbacks, and forestry for climate mitigation. He led a research group at the Max-Planck-Institutes for Meteorology and Biogeochemistry, served as the Deputy Leader of a major UK Earth system modeling research programme, and served for many years as editor for the American Physical Union, and as advisor to the European Commission and the European Space Agency.

Bruce Lankford is Emeritus Professor of Water and Irrigation Policy at the University of East Anglia, UK. He has more than thirty-five years of experience in agriculture, irrigation, and water resources management. His research interests are irrigation policy in Sub-Saharan Africa, serious games in natural resource management, irrigation efficiency and the paracommons, river basin management and water allocation, and irrigated catchment resilience.

Selma Lendelvo holds a PhD in Conservation Biology from the University of Namibia. She is Director of Grants Management and Resource Mobilisation at the University of Namibia (UNAM). She has been extensively involved in research with UNAM since 2001, mainly focusing on the field of environmental management and sustainable natural resources management across Namibia and abroad. She is interested in Natural Resources Management and Land Reform including wildlife management, community-based tourism, conservancy management, environmental management, rural development and gender. She is currently also the Namibian Principal Investigator for *Etosha-Kunene Histories* (www.etosha-kunene-histories.net).

David L. Levy is Professor of Management at the University of Massachusetts, Boston, and was a co-founder of the Sustainable Solutions Lab there. David, an Aspen Institute Faculty Pioneer Award Winner, conducts research on corporate and societal responses to climate change. His work explores strategic contestation over the governance and finance of controversial issues engaging business, governments, and NGOs, such as climate change and sustainability standards. David has spoken and published widely on these topics, for both academic and practitioner audiences.

Samuel Lietaer is an environmental social scientist working on the subjective dimensions of human interactions with environmental change, with a focus on marginal regions of low-income countries. He is currently working for the MIGRADAPT research project in Senegal, alongside which he is writing a PhD thesis at the Centre d'Etudes du Développement Durable (CEDD) of the Université Libre de Bruxelles (ULB). Both the project and his PhD dig further into translocal mechanisms of (political) remittances serving as adaptation

xxiv   *Negotiating Climate Change in Crisis*

strategies to environmental changes in Senegalese home communities. Previously, Samuel obtained his Master's in both Political Science and Law. He worked as a climate policy officer at 11.11.11, the overarching development NGO in Belgium. He also conducted research for the Belgian Development Cooperation in the field of climate-compatible private-sector development using perception approaches.

Lassi Linnanen is Professor of Environmental Economics and Management at LUT University, Lahti, Finland. Before joining academia, he was the CEO and co-founder of Gaia Group, a leading Finnish energy and environmental consultancy. He has also engaged in active management of over ten spin-off companies with business ideas around sustainable technology and management. He is a former member of the Finnish Climate Change Panel (2016–2019) and current Vice Chairman of the Finnish Expert Panel for Sustainable Development (2019-).

Jari Lyytimäki (PhD) works as a Senior Researcher at the Finnish Environment Institute, Environmental Policy Centre. He is also an Adjunct Professor at the University of Helsinki, Finland. His research interests cover various fields of environmental studies, including climate and energy issues, sustainability indicators, media analysis, risk communication and societal utilisation of scientific results. He is one of the coordinators of the Finnish Expert Panel for Sustainable Development.

Geoff Mann is Professor of Geography and Director of the Center for Global Political Economy at Simon Fraser University (Canada).

Shahrin Mannan is a Senior Research Officer at the International Centre for Climate Change and Development (ICCCAD) where, under the Locally Led Adaptation and Resilience programme, she currently manages three different projects related to community resilience, water, and livelihood security. She also manages ICCCAD's recently launched small grants programme and is the lead for its gender programme. She is actively involved in training and mentoring national and international students, grassroot representatives both online and in-person, and organised a side event at the COP25 in Madrid, presenting work on "Strengthening Climate Actions: Need for Decentralizing Climate Finance". By training, she is an urban planner, having graduated from the Bangladesh University of Engineering and Technology (BUET) and later completing her Master's in Development Studies at the University of Dhaka.

Mikko Mönkkönen is a Professor in Applied Ecology at the University of Jyväskylä. He leads the Boreal Ecosystems Research Group (BERG), a multidisciplinary research team studying the environmental and social impacts of natural resource use in boreal forests. This work combines socioeconomic scenario analyses, life-cycle analyses, environmental impact assessment using ecosystem service and biodiversity models, and forest planning optimisation tools. He is a member of Finland's Expert Panel for Sustainable Development.

Sarah Louise Nash is a political scientist working on climate change politics and policy, especially at the intersection of climate change and human mobilities, and is currently a postdoctoral researcher at the University of Natural Resources and Life Sciences (BOKU) in Vienna. She holds a Marie Skłodowska-Curie Individual Fellowship for her project "Climate Diplomacy and Uneven Policy Responses on Climate Change and Human Mobility" (CLIMACY). Her first book, *Negotiating Migration in the Context of Climate Change. International Policy and Discourse*, was published in 2019 by Bristol University Press. Sarah holds a PhD in Political Science from the University of Hamburg.

Benjamin Neimark is a Senior Lecturer at Lancaster University, UK. He is a human geographer and political ecologist who focuses on the green economy, resource extraction, high-value commodity chains, smallholders, agrarian change and development. Although he has a geographic focus on Sub-Saharan Africa and Madagascar, he also conducts research on the US military as a climate actor.

Peter Newell is a Professor of International Relations at the University of Sussex and co-founder and Research Director of the Rapid Transition Alliance. He has undertaken research, advocacy and consultancy work on different aspects of climate change for over twenty-five years. He sits on the board of directors of Greenpeace UK, is a board member of the Brussels-based NGO Carbon Market Watch and a member of the advisory board of the Greenhouse think-tank. His single and co-authored books include *Climate for Change* (Cambridge University

Press, 2000), *Governing Climate Change* (Routledge, 2010), *Climate Capitalism* (Cambridge University Press, 2010), *Transnational Climate Change Governance* (Cambridge University Press, 2014), and *Power Shift: The Global Political Economy of Energy Transitions* (Cambridge University Press, 2021).

Romie Nghitevelekwa holds a PhD in Anthropology from the University of Freiburg, Germany. She lectures in the subject areas of sociology of development, sociology of the environment, and rural sociology at the University of Namibia. Her research focuses on land reform, land rights, security of land tenure, land markets, land use and land-use change in rural areas, community conservation, rural socio-economic, and territorial restructuring. She is the author of *Securing Land Rights: Communal Land Reform in Namibia* (University of Namibia Press, 2020).

Peter North is Professor of Alternative Economies in the Department of Geography and Planning at the University of Liverpool. His research focuses on the politics of climate change and ecologically-focused social movements engaged in struggles about the implications of anthropogenic climate change and resource constraints for both humans and the wider ecosystems upon which we depend, and, using 'diverse economies' frameworks, understanding social and solidarity economies as tools for constructing more convivial, liveable, and sustainable worlds.

Daniel Nyberg is a Professor of Management at the University of Newcastle, Australia. His research explores the political activities of corporations, with a particular focus on how corporations engage both internally and externally with the climate catastrophe.

Matthew Paterson is Professor of International Politics at the University of Manchester and Research Director of the Sustainable Consumption Institute. His research has focused for thirty years on the political economy, global governance, and cultural politics of climate change. His latest work is *In Search of Climate Politics* (Cambridge University Press, 2021).

Mechtilde Pinto holds a Bachelor's degree in Tourism Management and is currently pursuing her Master's degree at the University of Namibia. Her research focus areas and interests are in natural resource management, community-based conservation, and community based-tourism. With Selma Lendelvo and Sian Sullivan, she is a co-author of 'A perfect storm? COVID-19 and community-based conservation in Namibia' (*Namibian Journal of Environment*, 2020).

Rick Rohde is a retired Research Fellow at the University of Edinburgh and Research Associate at the University of Cape Town. His interests include the environmental history of cultural landscapes, and historical and political ecology, primarily in Namibia, South Africa and Scotland. Visual ethnography and outsider photography are abiding interests that have found expression in projects in Namibia and South Africa.

Patrick Sakdapolrak is Professor at the Department of Geography and Regional Research, University of Vienna, Austria, where he leads the Working Group for Population Geography and Demography. He is also a Research Scholar at the International Institute of Applied Systems Analysis (IIASA) in Laxenburg, Austria. His research field is at the interface of population dynamics, environmental change, and development processes, with a focus on the topics of migration and displacement as well as health and disease, mainly in South- and Southeast Asia and East Africa.

Arto O. Salonen is Professor of Social Pedagogy and Sustainable Well-Being at the University of Eastern Finland, Faculty of Social Sciences and Business Studies, and a member of the Expert Panel for Sustainable Development. He is also an Adjunct Professor at the following universities: the University of Helsinki (education), the Finnish National Defence University (sustainable development), and the University of Eastern Finland (eco-social wellbeing). The title of his doctoral dissertation was *Sustainable Development and its Promotion in a Welfare Society in a Global Age*. His recent empirical research is on sustainable food, mobility and consumption, as well as active citizenship.

Rebecca Sandover is a Lecturer in Human Geography at The University of Exeter who undertakes Engaged Research focused on Sustainable Food Networks. Using a knowledge co-production approach, she has in recent years been investigating action toward the formation of sustainable food networks in the South West UK. Her research is particularly focused on building local food partnerships with local authorities, boosting access to sustainable local food, addressing food insecurity and addressing food-policy-related issues of health and wellbeing. She has been recently researching Public Participation in Climate Change policy making, exploring stakeholders' perceptions of the Devon Climate Emergency's Climate Assembly.

Katriina Siivonen is Vice Director and University Lecturer in Futures Studies at Finland Futures Research Centre, University of Turku (UTU), and Adjunct Professor in Cultural Heritage Studies (UTU). She holds several academic and societal board positions, including being Vice Chair of the Expert Panel for Sustainable Development in Finland, and she chairs the Advisory Board of the implementation of the UNESCO Convention for the Safeguarding of Intangible Cultural Heritage in Finland. She is an expert in qualitative research, participatory methodology, identities, heritage futures, and cultural sustainability transformation, and leads research on these themes.

Katriina Soini is Adjunct Professor and a Principal Research Scientist and Research Manager at the Natural Resources Institute Finland (Luke) in the field of Resilient Society. She has a background in Human Geography and her research has focused broadly on sustainability and sustainable governance in the rural context. Recently she has been working on sustainability transition/transformations and methods and theories of sustainability science. Soini has been leading a COST Action IS1007 on Culture and Sustainability and she is the editor of the Routledge Studies in Culture and Sustainable Development series. She is one of the coordinators of the Finnish Expert Panel for Sustainable Development.

Harald Sterly is a Senior Researcher at the Population Geography and Demography Working Group at the Department of Geography and Regional Research, University of Vienna, Austria. He works on the nexus of environmental migration with a special focus on the spatial and social aspects of migration, urbanisation, and technological change, and how the latter contribute to changes in vulnerable groups' scope for agency and their vulnerability and resilience.

Sian Sullivan is Professor of Environment and Culture at Bath Spa University and UK Principal Investigator for *Etosha-Kunene Histories* (www.etosha-kunene-histories.net). She is an environmental anthropologist, cultural geographer and political ecologist working to recognise diversity in perceptions and representations of the natural world, amidst contemporary concern about climate change and species decline. Over the last few years she has led an Arts and Humanities Research Council project called *Future Pasts* (www.futurepasts.net) focusing on understandings of sustainability in the conservation and cultural landscapes of west Namibia. She has also researched the 'financialisation of nature'—see The Natural Capital Myth (www.the-natural-capital-myth.net).

Tuuli Toivonen is a geographer and a Professor of Geoinformatics at the University of Helsinki. Her research focuses on spatial analyses and novel use of big and open data to support sustainable spatial planning. She is particularly interested in active and healthy mobility and the use of green spaces in urban areas and beyond. She is a member of a number of university and national policy boards, including Finland's Expert Panel for Sustainable Development. She leads the Geography BSc and MSc study programmes and the multidisciplinary Digital Geography Lab at the University of Helsinki.

Anne Tolvanen is a Professor in the ecology and multiple use of forests and the Programme Director of LandClimate research programme in the Natural Resources Institute Finland (Luke). Her work concentrates on the reconciliation of land uses to mitigate climate change and safeguard biodiversity in a sustainable and controllable manner. Her group develops models and tools that are used in the planning and management of peatland and forest ecosystems. She represents Luke in numerous domestic and international boards and working groups related to natural resources use, including the Finnish Expert Panel for Sustainable Development and the Society of Ecological Restoration (SER).

Basundhara Tripathy Furlong is a PhD candidate at Wageningen University and Research in the Netherlands. She is an anthropologist by training and her research interests include climate change, human mobility, gender, resilience, the anthropology-development nexus, and environment. She obtained her MSc in Social Anthropology from the

University of Oxford in 2012 and completed her undergraduate studies at the University of Delhi in Sociology (Hons.). She was a Lecturer at the University of Liberal Arts Bangladesh and carried out empirical research in India and Bangladesh.

Kees van der Geest (PhD) is Head of the "Environment and Migration: Interactions and Choices" (EMIC) Section at the United Nations University Institute for Environment and Human Security (UNU-EHS). As a human geographer he applies a people-centred perspective to study the impacts of climate change, human mobility, environmental risk, adaptation, livelihood resilience, and rural development. His work has contributed substantially to expanding the empirical evidence base on migration-environment linkages and impacts of climate change beyond adaptation ("loss and damage"). Kees has extensive fieldwork experience in the Global South, for example in Ghana, Burkina Faso, Vietnam, Bangladesh, Nepal, Marshall Islands and Bolivia.

Joel Wainwright is Professor of Geography at Ohio State University (USA).

Robert Watson is Emeritus Professor at the University of East Anglia. Robert is a physical chemist specialising in atmospheric science issues and a leading authority on the science of climate change due to human activity. His career spans research and advisory roles, including key roles with NASA, as a science policy adviser to US President Bill Clinton, and at the World Bank. For the UK government, Robert was Chief Scientific Adviser to the Department for Environment, Food and Rural Affairs. He has served as Chair of the Intergovernmental Panel on Climate Change (IPCC) and the Inter Intergovernmental Panel on Biodiversity Ecosystem Services (IPBES).

Lorraine Whitmarsh is a Professor in Environmental Psychology, specialising in perceptions and behaviour in relation to climate change, energy and transport, based in the Department of Psychology, University of Bath. She is Director of the ESRC-funded UK Centre for Climate Change and Social Transformations (CAST). She regularly advises governmental and other organisations on low-carbon behaviour change and climate change communication, and is Lead Author for IPCC's Working Group II Sixth Assessment Report. Her research projects have included studies of energy efficiency behaviours, waste reduction and carrier bag reuse, perceptions of smart technologies and electric vehicles, low-carbon lifestyles, and responses to climate change.

Christopher Wright is a Professor of Organisational Studies at the University of Sydney Business School and key researcher at the Sydney Environment Institute. His research explores organisational responses to climate change, with a particular focus on corporate environmentalism, risk, identity, and future imaginings. He is the author (with Daniel Nyberg) of the book *Climate Change, Capitalism and Corporations: Processes of Creative Self-Destruction* (Cambridge University Press, 2015).

Introduction:
Climate Crisis?
What Climate Crisis?

Steffen Böhm and Sian Sullivan

(At Least) Five Decades of Knowing and (Not) Acting[1]

In all the talk about the Paris Agreement, reached at the twenty-first Conference of Parties (COP21) of the United Nations Framework Convention on Climate Change in Paris in 2015, it is sometimes forgotten that the world's political leaders have held negotiations about climate change at the highest possible level for at least three decades. Many have known about climate change for a lot longer.

It was in the 1860s that the Irish scientist John Tyndall first established a link between CO_2 and what then became known as the 'greenhouse effect', which was further evidenced by the Swedish scientist Svante Arrhenius (Pain 2009). In 1938, the British scientist and engineer Guy Stewart Callendar "documented a significant upward trend in temperatures for the first four decades of the 20th century and noted the systematic retreat of glaciers" (Plass et al. 2010: online). In 1956, the American scientist Gilbert Plass (1956) published a seminal paper called 'Carbon Dioxide Theory of Climatic Change', creating a clear link between increases in the concentration of CO_2 in the atmosphere and global temperature rises.

---

1   The first part of this introduction draws on an earlier blog article by Böhm, published as 'The Paris Climate Talks and other Events of Carbon Fetishism', https://www. versobooks.com/blogs/2372-steffen-bohm-the-paris-climate-talks-and-other-events-of-carbon-fetishism.

    https://doi.org/10.11647/OBP.0265.29

This scientific knowledge has thus been 'out there' for a very long time, and was also not unnoticed in the political arena. As early as 1965, the US President's Science Advisory Committee

> told President Lyndon Johnson that greenhouse warming was a matter of real concern. There could be 'marked changes in climate,' they reported, 'not controllable through local or even national efforts.' CO_2 needed attention as a possibly dangerous 'pollutant' (Weart 2021: online).

In the 'mother country' of fossil fuel burning, the United Kingdom, politicians became increasingly aware of climate change in the 1960s. In 1969, the House of Lords (the upper chamber in the UK parliamentary system) discussed railway policy and the hereditary peer Jestyn Philipps asked the following question:

> [m]y Lords, can my noble friend say whether he and British Railways have taken account of the fact that what were abnormal temperatures last summer may not be abnormal if we continue to discharge CO2 into the air by the burning of various fossil carbons, so increasing the greenhouse effect? (Carbon Brief 2019a: online).

Public opinion, particularly in the highly industrialised, most polluting countries, had shifted markedly towards an awareness of environmental issues in the 1960s and 1970s. In 1962, Rachel Carson's influential and path-breaking book, *Silent Spring*, was published, becoming a bestseller worldwide. Anti-pollution, conservation and environmental protection movements sprang up everywhere. The first 'Earth Day' was held in the United States in 1970, becoming global in 1990 and marking the emergence of environmentalism as a serious social movement and political force (as also discussed by Hulme, this volume).[2] The world's first green political parties were founded in 1972, in the Australian state of Tasmania and in New Zealand. The German Green Party, which subsequently became one of the most successful national green parties worldwide, was founded in 1979. Climate change was written on the banners of these environmental activists from the start.

The rise of environmental consciousness from the 1960s onwards also made the bosses of fossil fuel companies take note. We now know that the corporate leaders of ExxonMobil, one of the biggest oil companies of the world, had known about climate change and the unsustainability of

---

2   See https://www.earthday.org/history/.

their business models since at least 1977 (Hall 2015), as also clarified by Wright and Nyberg, this volume. During the 1980s, Exxon and Shell had extensive internal discussions and memos on climate change (Franta 2018). We are constantly told that companies are always listening to what their customers want. Well, already in the 1970s it became clear that an increasing number of customers were worried about the degradation of nature and climate change in particular. Corporate leaders would have been aware of this shift in public consciousness and attention. Given that what companies hate most are business risks, and that climate change is the biggest risk to an oil and gas company's business model, it would be logical to assume that these companies were making climate change risk assessments from these decades.

The 1980s saw the rapid expansion of environmentalism worldwide. In 1987, the World Commission on Environment and Development (WCED) published *Our Common Future*, which became known as the 'Brundtland Report', named after the Commission's chairwoman Gro Harlem Brundtland. While the Report had a wider remit, focusing on a whole range of environmental issues, it clearly stated that there is

> the serious probability of climate change generated by the 'greenhouse effect' of gases emitted to the atmosphere, the most important of which is carbon dioxide (CO_2) produced from the combustion of fossil fuels (World Commission on Environment and Development 1987: 145–46).

It went on to say:

> [a]fter reviewing the latest evidence on the greenhouse effect in October 1985 at a meeting in Villach, Austria, organized by the WMO, UNEP, and ICSU, scientists from 29 industrialized and developing countries concluded that climate change must be considered a 'plausible and serious probability' [...] They estimated that if present trends continue, the combined concentration of CO_2 and other greenhouse gases in the atmosphere would be equivalent to a doubling of CO_2 from pre-industrial levels, possibly as early as the 2030s, and could lead to a rise in global mean temperatures 'greater than any in man's [sic] history'. Current modelling studies and 'experiments' show a rise in globally averaged surface temperatures, for an effective CO_2 doubling, of somewhere between 1.5°C and 4.5°C, with the warming becoming more pronounced at higher latitudes during winter than at the equator [...]. An important concern is that a global temperature rise of 1.5–4.5°C, with perhaps a two to three times greater warming at the poles, would lead to a sea level rise of 25–140 centimetres. A rise in the upper part of this

range would inundate low-lying coastal cities and agricultural areas, and many countries could expect their economic, social, and political structures to be severely disrupted (World Commission on Environment and Development 1987: 148).

The Brundtland Report and the continued gathering of scientific evidence catapulted climate change to the top of the political agenda of many countries at the end of the 1980s. On 23 June 1988—more than thirty years ago!—Dr James Hansen, then director of NASA's Institute for Space Studies, stated in a landmark testimony before the US Senate Energy and Natural Resources Committee, that

> [g]lobal warming has reached a level such that we can ascribe with a high degree of confidence a cause-and-effect relationship between the greenhouse effect and observed warming...In my opinion, the greenhouse effect has been detected, and it is changing our climate now (Brulle 2018: online).

Climate change was no longer only a concern for tree-hugging activists—if it ever was confined in that way. Now, NASA scientists and the top political class in the richest countries of the world were not only informed about climate change but were actively talking about what to do about it.

This recognition of the urgency of climate change, and the high risk of not doing anything to turn it around or address its predicted impacts, contributed to the Rio de Janeiro Earth Summit (Rio Summit) in 1992, which brought together leaders from government, business and NGOs from across the world, including most heads of state. At the Rio Summit, the United Nations Framework Convention on Climate Change (UNFCCC)—an official, international environmental treaty with binding obligations—was signed, coming into force in 1994.

The so-called Conference of the Parties (COP) is the UNFCCC's main decision-making body and meets annually. At COP3 in 1997, the Kyoto Protocol was signed, the first international agreement to curb global greenhouse gas emissions. At COP21, in 2015, the landmark Paris Agreement was reached to commit states across the world to keep global warming below 1.5 degrees Celsius.

Alongside these agreements under the UNFCCC, the United Nations has also included 'Climate Action' as one of seventeen global Sustainable Development Goals (SDGs) adopted in 2015, framing

SDG13 specifically as a call that governments "[t]ake urgent action to combat climate change and its impacts".[3]

Having arrived at 2021, however, scientific evidence for ongoing global temperature rises alongside industrial combustion of fossil fuels is now overwhelming. The simple graphic shown in Figure 1 communicates clearly where we are in terms of global temperature rises since 1850.

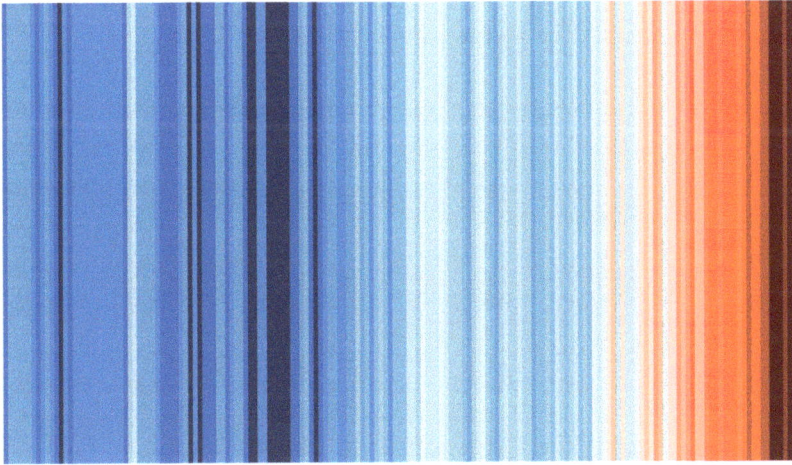

Fig. 1. Annual average temperatures for the world, 1850–2020, based on data by the UK Met Office, Graphics and lead scientist Ed Hawkins, National Centre for Atmospheric Science, University of Reading. Creative Commons, https://showyourstripes.info/.

Climate scientists now agree

> that 2011–2020 was the warmest decade on record, in a persistent long-term climate change trend. The warmest six years have all been since 2015, with 2016, 2019 and 2020 being the top three. The differences in average global temperatures among the three warmest years—2016, 2019 and 2020—are indistinguishably small. The average global temperature in 2020 was about 14.9°C, 1.2 ($\pm$ 0.1) °C above the pre-industrial (1850–1900) level (WMO 2021: online).

In other words, we have already seen a 1.2 degrees Celsius temperature rise globally, which makes it all but certain that we will fail to meet the

---

3   See https://www.un.org/sustainabledevelopment/climate-change/.

1.5 degrees commitment made by the UNFCCC's COP21 in Paris in 2015 – as confirmed in the recently published first instalment of the Sixth Assessment Report (AR6) of the Intergovernmental Panel on Climate Change (IPCC 2021). If current trends persist, and even if countries meet their Paris Agreement obligations, many climate scientists now warn that we are heading towards at least 3 degrees Celsius change compared to pre-industrial levels (UN 2019).

Common but Differentiated Responsibilities?

The UNFCCC treaty agreed in Rio in 1992 clearly acknowledged that the rich, highly developed and industrialised countries have a historical responsibility to take a lead in combating climate change, given that countries such as the UK, the US, France, Germany, etc. have been pumping greenhouse gases at scale into the atmosphere for at least three hundred years. The treaty says in Article 3, Principle 1:

> The Parties should protect the climate system for the benefit of present and future generations of humankind, on the basis of equity and in accordance with their common but differentiated responsibilities and respective capabilities. Accordingly, the developed country Parties should take the lead in combating climate change and the adverse effects thereof (UN 1992: 9).

The Kyoto Protocol, the first landmark, international agreement to commit to Greenhouse Gas (GHG) reductions, repeated this commitment:

> [a]ll Parties, taking into account their common but differentiated responsibilities and their specific national and regional development priorities, objectives and circumstances (UN 1998: 9).

The key phrase here is 'common but differentiated responsibilities', acknowledging that climate change is a global 'commons' problem (as further discussed in Lankford's chapter, this volume), but that different countries have different responsibilities in relation to their contributions to this problem. Carbon emissions do not respect national borders: if a large, coal-fired power station is built in one country, it ultimately affects the climate on the whole planet. In saying that responsibilities are 'differentiated', Global South countries are acknowledged to have not been the cause of climate change, and to have not emitted massive amounts of GHG for very long, with the inference that they cannot be expected to sort out the mess that Global North countries—the rich, industrialised nations with their expansionary colonising histories— have caused.

A glance at global history reveals how closely energy and GHG emissions have been linked to both economic growth and colonial expansion. The Netherlands was the first country to develop a taste for exponential industrial growth back in the sixteenth and seventeenth centuries, which would have been unthinkable without the availability of cheap domestic peat, as well as timber from Norwegian and Baltic forests (Moore 2010). One reason that Britain took over Holland's imperial leadership was due to its vast reserves of coal mined at great profit through the use of cheap labour, with the burning of coal taking off at the end of the eighteenth century, and growing exponentially in the nineteenth century (Malm 2016). Then came oil and gas, which have helped make the United States of America the global imperial master from the early twentieth century onwards (Foster 2006).

There is thus more than 250 years of fossil fuel burning by the Global North to account for. Many climate justice activists advocate for some form of reparations to be paid by the North to the poorest countries of the planet, particularly those that are already struggling to adapt to a rapidly changing climate, whether in the form of rising sea levels, increasing drought (as considered in the chapter by Lendelvo and colleagues, this volume), failed harvests, or bigger and more forceful weather events such as storms. The fact that approximately 80% of historical carbon emissions have to be attributed to the rich world (Centre for Global Development 2015), and are already causing havoc in many countries around the world, cannot simply be wished away.

In the Paris Agreement of 2015, 'common and differentiated responsibilities' were again mentioned repeatedly, for example in Article 4, Paragraph 19:

> [a]ll Parties should strive to formulate and communicate long-term low greenhouse gas emission development strategies, mindful of Article 2 taking into account their common but differentiated responsibilities and respective capabilities, in the light of different national circumstances (UN 2015: 6).

Yet, this principle has gradually been pushed into the background, and the discourse of 'differentiation' is now almost a fringe occurrence. The rapid rises of emissions, particularly in China and India, are often cited as reasons for why these fast-industrialising countries now also have to curb their emissions. Clearly, they have their own responsibilities and they need to be held to account: China, in particular, is now the largest GHG emitter by far in the world. Let us bear in mind, however, that India's carbon emissions per capita are still about a seventh of the figure for the United States (Carbon Brief 2019b), and China's rapidly rising emissions are to a great extent driven by export-driven industries, producing consumer goods for the rest of the world, particularly the Global North (Yang, Yuantao et al. 2020). If we add up historical *per capita* emissions over the past three hundred years, then China's carbon emissions—with its vast population—lag far behind those countries that industrialised first (Centre for Global Development 2015).

Western European countries like to portray themselves in green, responsible colours, highlighting that their carbon emissions are significantly lower than the Kyoto baseline of 1990. The UK, for example, which is hosting COP26 in Glasgow in 2021,[4] frequently and happily declares that "[i]n 2019, total UK greenhouse gas emissions were provisionally 45.2 per cent lower than in 1990" (Department for Business, Energy and Industrial Strategy 2020). What is conveniently forgotten is that the UK's apparent success in lowering GHG emissions is largely due to the early adoption of gas, which has lower emissions than coal and oil, in the early 1990s, i.e. before Kyoto. There are clearly carbon reduction successes in many Global North countries. The power generation sector in the UK, for example, has now phased out coal almost completely,[5] which, only four to five decades ago, would have been unthinkable. Renewable energy adoption rates are high in countries such as Germany. Global North governments have made efforts to put their countries on a decisive decarbonisation path with the UK being the first country to legislate for a net zero carbon emissions commitment by 2050. The UK's Office for National Statistics (ONS) claims there are already signs that there has been a decoupling

---

4   See https://ukcop26.org/.
5   Strangely and controversially, however, the UK government is currently considering to approve the establishment of a new coal mine in Cumbria; see https://www.bbc.co.uk/news/explainers-56023895.

of economic growth and GHG emissions in the country (Office for National Statistics 2019), apparently proving that the Environmental Kuznets Curve was right to predict that, as countries grow richer, their negative environmental impact will reduce (see, for example, Grossman and Krueger 1995). The ONS confirms that emissions peaked in 2007, and that the country is on the right path to meet its 2050 net zero commitments.

But there are four controversies that such statistics and resulting political posturing ignore or downplay.

First, most global GHG statistics are still based on the production principle: that is, carbon is counted in the countries where it is emitted (see discussion in Hannis's chapter, this volume). Countries such as the UK, however, are net importers of carbon emissions, as the ONS report rightly points out (Office for National Statistics 2019). If a consumption-based approach to carbon accounting is taken, the UK's national carbon emissions would be significantly higher than officially reported. How much higher is subject to which carbon accounting technique is used. This is also true for most Western European countries as well as the United States, which have seen increasing rates of deindustrialisation over the last two decades with not only jobs but also carbon emissions being offshored to countries of the Global South. In return the Global North receives cheap consumer goods whose embedded carbon emissions are not attributed to itself. Of course, some of the exponential growth in carbon emissions by India and China is also due to increases in home-grown consumption. China apparently now has the largest middle class in the world. If we take a consumption-based view, however, then even China's emissions per capita will not reach the US's current rate for a long time. India lags even further behind.

Second, there are three large sectors that are mostly and conveniently ignored by any carbon accounting techniques: the military, shipping and aviation. As Bigger et al. report in this volume, the US military 'bootprint' is higher than many middle-income countries. Calculating the carbon footprint of global military operations is nearly impossible, however, as governments do not report details of military fuel consumption, emissions and impacts. Some analysts estimate that the global 'bootprint' of the military could be as high as 6% of global emissions (Scientists for Global Responsibility 2020): bigger than Russia's entire share in 2019. The global shipping and aviation industries have also repeatedly evaded their climate responsibilities, given that their operations transgress national boundaries. Ships mostly operate outside national jurisdictions, and again an effective calculation of their carbon footprint is difficult. Estimates exist that put shipping and aviation on a combined 5.39% of global GHG emissions (Ritchie 2020; Saul 2020)—higher than the GHG emissions of high emitting countries such as Russia—and both global shipping and aviation emissions are rising fast.

The third area often conveniently forgotten in any national carbon accounting scheme concerns the financing of fossil fuel infrastructure. Large fossil fuel projects, such as the development of new oil and gas fields, or the building of pipelines and dams, requires finance that even national governments cannot muster. The largest banks of the world are also the largest financiers of fossil fuel developments. The Banking on Climate Chaos Report 2021 (Rainforest Action Network 2021) showed that the world's biggest sixty banks have provided $3.8tn of financing for fossil fuel developments, *since* the Paris Agreement was signed in 2015. JP Morgan Chase, which tops the table, has provided more than $300bn of finance alone. These global finance streams again distort the national pictures of GHG emissions. Without such finance, oil fields could not be developed nor coal-fired power stations built. But which country should be responsible and accountable for the carbon emissions caused by these new fossil fuel developments? These banks, which are normally headquartered in Global North countries, profit from these projects, creating economic growth activities in the countries they are based in and demonstrating the significant continuing global influence of the fossil fuel industry (as emphasised in Wright and Nyberg's chapter, this volume).

Fourth, but by no means least, we need to account for the fast-rising emissions associated with so-called 'green' industries, such as renewable energy. Biomass-burning power stations, such as the UK's Drax, electric vehicles, industrial-scale wind parks, large solar farms, nuclear power stations—these are all sold as 'low-carbon' solutions to the planet's climate change malaise. Yet, if the GHG emissions of the entire life cycle of these technologies are taken into account, their carbon footprint is significant, particularly in the context of their fast adoption around the world, not to mention their often-forgotten, grave social implications (Sullivan 2013a; Ramirez and Böhm 2021). Dunlap's chapter, this volume, thus proposes that industrial-scale renewable energy production should more accurately be labelled as 'Fossil Fuel+' to acknowledge the continued product-cycle dependence of these technologies on fossil-fuel based sources of energy.

Does the Environmental Kuznets Curve, i.e. the suggestion that beyond a certain degree of economic growth a society will reduce its environmental impacts, account for these four areas of contention? Are the rich, Global North countries really on a path of decarbonisation? Deepening the intractability of reducing CO_2 emissions whilst simultaneously maintaining an industrial growth pathway is the greenwashing that accompanies renewable energy production on an industrial scale. Perhaps only a Fossil Fuel Non-Proliferation Treaty, in which governments commit not to extract and exploit fossil fuels, will ultimately be the kind of governance mechanism that will prevent the exposure of the climate to future fossil fuel emissions—as proposed by Newell, this volume. On this point, it is encouraging to see a recent ruling by a Dutch court that oil giant Royal Dutch Shell must cut 45% of its 2019 greenhouse gas emissions by 2030 so as to contribute to national targets agreed under the Paris Agreement of COP21 (Farmer 2021).

Meanwhile, however, all the global GHG emissions curves go in the wrong direction.

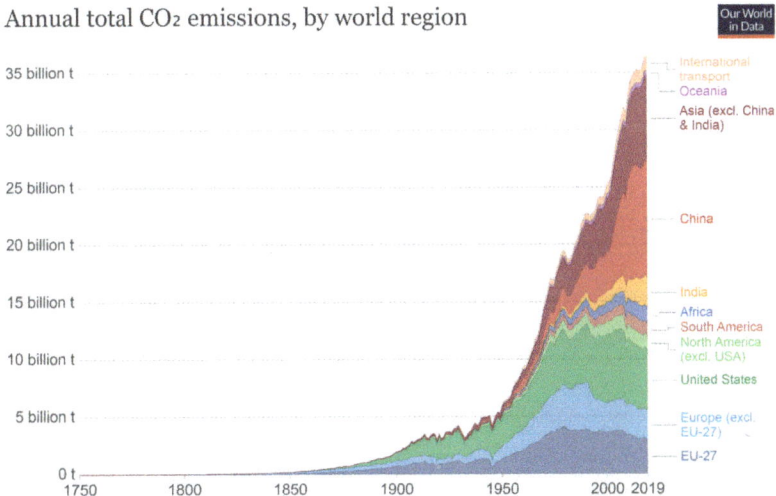

Fig. 2. Annual total CO_2 emissions, by world region, 1750–2019, Creative Commons, https://ourworldindata.org/grapher/annual-co-emissions-by-region.

Three Decades of Carbon Fetishism

For three decades now, there has been talk and action on climate change at the highest possible levels—in politics, business, finance and civil society. For three decades, climate change has shaped the consciousness of citizens, consumers, politicians, entrepreneurs, farmers—and particularly those land-based communities directly affected by climate change (see chapters by Dieckmann, Lendelvo et al., and Sullivan this volume). We have seen the rise of climate justice movements, such as Extinction Rebellion (XR)[6] and Fridays For Future[7] (as discussed by Gardham, this volume). Swedish environmental activist Greta Thunberg, whose activism began as a school strike in protest against the very limited political action *vis-à-vis* climate change, has been catapulted into a global phenomenon, speaking in front of the UN Assembly and the European Parliament. David Attenborough, the famous UK-based conservationist and broadcaster, has become an outspoken climate activist, producing advocacy films on the dangers of climate change shown around the world on platforms such as Netflix. Thousands of parliaments, local government authorities and other large public organisations around the world have declared Climate Emergencies. Not a day goes by without a large company making 'net zero' commitments (although see Dyke et al. and Bailey this volume for more detail regarding the effectiveness or otherwise of net zero policies).

Take, for example, the most recent, annual letter to the CEOs of companies invested in by BlackRock, the asset management company, written by Larry Fink, its founder, chairman and chief executive officer:

> I believe that the pandemic has presented such an existential crisis—such a stark reminder of our fragility—that it has driven us to confront the global threat of climate change more forcefully and to consider how, like the pandemic, it will alter our lives (Fink 2021: online).

Fink has committed BlackRock "to supporting the goal of net zero greenhouse gas emissions by 2050 or sooner" (BlackRock 2021: online). While there is very little detail on how this goal is to be achieved, it

---

6   https://extinctionrebellion.uk/.
7   https://fridaysforfuture.org/.

nevertheless is remarkable for Fink to become a corporate 'climate activist' (Skoglund and Böhm 2020) in a country, the United States, where a significant proportion of the population still believes climate change is a hoax. His commitment is part of a wider trend in the finance industry that is now, apparently, taking environmental, social and governance (ESG) criteria, such as climate change, very seriously (as also analysed by Kaplan and Levy, this volume). As the *Financial Times* says:

> Investment in companies that integrate environmental, social and governance factors continues to gain traction across public and private markets. Once considered a niche, the zeitgeist has gone past the notion of a 'seismic shift'. Instead, integration of ESG underpins most, if not all, debates about the future of the investment industry (Lampen 2021: online).

Sounds good, does it not? All this newly found commitment to tackle climate change as well as the wider environmental malaise we find ourselves in should be welcomed and celebrated. However, a heavy dose of scepticism and critical interrogation is also needed (as offered by Bracking, this volume), precisely because we have been here many times before.

Our outline above suggests that capitalism (and state socialism, for that matter) does not have a good track record in terms of environmental performance. For decades now, it has failed to adequately address the climate crisis (Böhm et al. 2012). Whatever has been tried has not worked. Global GHG emissions are still rising exponentially—as clarified in Figure 2. Most analyses indicate that the global COVID-19 pandemic will only temporarily halt emissions, with a massive rebound looming. This was certainly the case in 2008/09 when the last global crisis meant that GHG reduced slightly in most countries, only to continue on their path of exponential growth soon after.

Why is this? Why, despite all the talk and the good intentions by many, has the world not managed to reduce GHG emissions since the inception of the UNFCCC all those years ago? Why are emissions still rising fast? One answer lies in the carbon market instruments that have been invented over the past thirty years to deal with the climate crisis.

Already at COP3 in Kyoto in 1997, the rich world, led by the US, demanded 'flexibility' in terms of how the highly industrialised countries should be allowed to deal with cutting their carbon emissions. This demand resulted in a proliferation of market mechanisms, which, in proper capitalist market fashion, work by establishing property rights for carbon emissions, allowing carbon permits and credits to be traded globally (Böhm and Dhabi 2011; Böhm et al. 2015). The ensuing creation and 'primitive accumulation' of carbon units led to a number of Emissions Trading Schemes (ETS) being set up across the world, most prominently the EU-ETS, which came into force in 2005 (Lohmann 2009, 2014).

This market approach is in line with dominant neoliberal governance approaches that have been spearheaded by Anglo-Saxon countries, mainly the US and the UK, since the early 1980s, and which have since spread across the globe ultimately through finance conditionalities set by International Financial Institutions (Dunlap and Sullivan 2019). While states and their governments have not disappeared, the ideological approach by neoliberalism is to let market principles deal with most economic and increasingly social and environmental affairs, including climate change. But the success of ETSs around the world has been limited, to say the least, with only small percentage points of progress made towards reducing global GHG emissions.

A recent paper argues that despite low carbon prices in the EU-ETS, this continent-wide carbon trading scheme was responsible for a 3.8% cut in total EU-wide emissions between 2008 and 2016 (Bayer and Aklin 2020). This calculation can be challenged on various grounds, but even if such a reduction can be robustly attributed to the EU-ETS, such a modest success can hardly be a blueprint for the radical emissions cuts needed for any chance of limiting global climate chaos. Many academics, commentators and climate activists have argued that neoliberal, market-based approaches have been nothing more than a delaying tactic, allowing the big polluting companies and countries to continue to emit carbon at a massive scale, while offsetting their responsibilities through clever carbon accounting techniques (Lippert 2014) and the fantasies of "green success" they sustain (Watt 2021).

Carbon has now become a major commodity, traded on stock exchanges across the world. This financialisation of climate change is creating new carbon elites, benefitting from the world's quest to urgently curb global GHG emissions and seemingly decarbonise capital (Christophers 2019; Langley et al. 2021). Larry Fink and the other financiers now riding the ESG boom are sniffing gold. When he talks about "capturing opportunities created by the net zero transition" he has carbon pricing mechanisms in mind. Whilst these potentially financial(ising) instruments have been around since Kyoto in 1997, COP21 paved the way for a massive scaling up of voluntary trading of carbon units so as, again seemingly, to meet the urgent climate challenges facing global society today. For this reason,

> [a] host of top figures from business, finance and academia led by former Bank of England Governor Mark Carney have announced a global task force to accelerate the development of voluntary carbon markets across the private sector, ahead of anticipated surge in demand for CO2 offsets as the net zero transition gathers pace (Holder 2020: online).

Carney's 'taskforce' aims for "scaling voluntary carbon markets and allowing a global price for carbon to emerge", which is claimed will give companies the "right tools and incentives to reduce emissions at least cost" (ibid.). This new initiative conveniently ignores the problematic evidence and experience of operating carbon markets, and specifically voluntary offsetting schemes, over the past twenty years.

When in 2009, in the run-up to the COP15 climate change talks in Copenhagen, we published the book *Upsetting the Offset* (Böhm and Dabhi 2009), our intention was to show the negative and oft-ignored impacts of carbon offsetting on the ground in the Global South especially. Tamra Gilbertson (2009), for example, in her case of A. T. Biopower, showed how what used to be sustainable agricultural practices in Thailand were transformed into so-called carbon neutral operations that create profits and rents for local elites and international polluters, yet disadvantage local people and communities.

In 2014, one of us (Böhm) co-authored in the journal *Carbon Management* an analysis of evidence for 'Ten reasons why carbon markets will not bring about radical emissions reduction' (Pearse and Böhm 2014). The article argued that carbon markets do not work because they provide plenty of loopholes for the biggest emitters, often going hand-in-hand with a lack of political will to radically curb GHG emissions (see also Bryant et al. 2015). Lobbying by fossil fuel elites is rife and has resulted in dinosaur industries actually benefiting from the introduction of carbon markets, and there have also been many cases of corruption. Despite their stated commitments to further sustainable development, carbon offsetting schemes, which are normally implemented in Global South countries, often involve a whole range of negative social and environmental 'side-effects' not accounted for when 'net zero' or 'carbon neutral' claims are made by investors in this new imaginary of "carbon earth" (Sullivan 2010: 113; Ehrenstein and Muniesa 2013; Asiyanbi 2017). Carbon markets can also be regressive in terms of disproportionately affecting low-income households. What the carbon market approach amounts to is an almost blind belief in market mechanisms to solve the climate chaos we find ourselves in. But these carbon mechanisms are so complicated, technocratic, and obscure that they are really designed for use by corporate and financial elites only.

The reality and academic evidence presented in this 2014 paper has not changed. What has changed is that after almost ten years of bear climate markets, the new ESG activists, including Larry Fink, smell an opportunity. Whilst perhaps unintended, what the combined Greta Thunberg and David Attenborough effect has done is reignite the fantasy that carbon markets will solve climate chaos. Fink will always take BlackRock where the future money is. Carbon markets, in the guise of 'net zero' strategies, appear to offer this future.

Let us ask some questions in response. In the other, still ongoing, global crisis that is the COVID-19 pandemic, have governments around the world relied on markets to deal with the biggest health crisis the world has seen for a century? Have they created complicated COVID-credits that can be traded on global stock exchanges to determine the most efficient way of combating the virus? Have they allowed COVID-offsetting, so that some people or companies could buy themselves out of lockdown? No? No, they have not! (Although we note the increasing talk of creating and selling so-called 'Covid bonds'—see Postel-Vinay 2021). This is because there was no time. The pandemic was and is an emergency. Everybody had to go into lockdown, everybody was affected. The virus knew and knows no borders, and the countries that have most successfully dealt with the pandemic are those where there has been decisive government action. It is precisely this kind of political will that has been in short supply for combating climate change (as also observed by Halme and colleagues, this volume).

Carbon offsetting and trading schemes will undoubtedly see unprecedented growth in the coming years, if ESG activists and politicians have their way. Companies and countries will continue to ask for maximum 'flexibility' in meeting their climate commitments, which is code for saying: they want to offset their emissions without cleaning up their act at home.

What all this amounts to is what the critical geographer Erik Swyngedouw (2010) has called 'CO$_2$ fetishism'. For Swyngedouw, capitalism's attempt to deal with the climate crisis is a perfect example of 'post-politics', generating a lot of talk about what needs to change to make our existence on earth sustainable without changing much at all. What is important to bear in mind, though, is that this talk about change is not all there is. Swyngedouw (2010) also argues that capital attempts to materially reconfigure itself through the crisis of climate change, precisely by turning carbon (nature) into a commodity through putatively *decarbonising* capital, complementing observations of capitalism's reconfiguration through a 'financialisation of nature' deemed to effect nature's care (for example, Bracking 2012, 2019; Sullivan 2012, 2013b).

These innovations make sense as new layers of the commodification processes that have run through capitalism's history. With good reason Jason Moore (2015) calls capital an ecological regime, because it has always mixed human labour with nature's generative capacities so as to produce and reproduce natures in specific ways in particular times and places. Adopting this perspective, we can see how the battle of the twenty-first century is to further transform capital-as-ecology into capital-as-climate. Such abstractions involve immense economic, social, cultural and environmental forces, constituting a process of fetishisation in a Marxian sense, spurred on by the realities of the worsening climate in parallel with the profit and rent opportunities that emerge from this context for elites. Masked by the appearance of the carbon commodity that can be traded, then, are the socio-ecological contexts, calculative practices, and relations of exchange that permit a unit of carbon to assume a monetary value that can be traded.

Many environmentalists have become successful entrepreneurs and many investors are now riding on the ESG bandwagon. This is what capitalism does best. It commodifies, and a new commodity to be formed and traded is carbon. This is nothing else but what Marx (1976) called 'commodity fetishism', the process by which social relations appear as commodity relations. Without this most basic of all abstractions, capital could not appropriate people and things that exist outside its logic, in order to bring them into the workings of the capital machinery. Once appropriated, they are exploited either extensively (extending the time that capital has to work on its subjects) or intensively (squeezing more value from within a given time). And once the basic labour process cannot be more optimised and appropriation becomes too costly, financialisation kicks in. As Moore rightly argues (2015), capitalism is doing all of these three things simultaneously: appropriation, commodification and financialisation. 'Climate capitalism' is no different. It too expands through a unique combination of the exercise of state violence, business opportunity and cultural shifting.

This process of the creation of climate capitalism is not unchallenged of course. What the climate debate has shown is that this capitalisation process is a struggle of forces within capital and associated social actors themselves, as foregrounded in the chapters by Mannan and colleagues, North, Paterson and Bond, this volume. It is not something that is somehow masterminded by an evil force, placed in Washington, D.C. or London or even Beijing. It emerges out of the contradictions of capital, the outcomes of which are not predictable.

The latter observation also means that the commodity fetishisation of carbon is by no means an inevitable process. Given that capitalist processes can only deal with such a grave challenge as climate change through new layers of commodification that disenfranchise and dispossess people from other modes of production, conditions are also created that invite contestation and the expression of different concerns. The uneven, unequal and highly volatile process of climate capitalisation, which elites try to control, cannot help but engender resistance and greater consciousness of justice concerns (as considered in the chapter by Harris, this volume). Climate justice is not something that should somehow come after an acceptance of climate capitalism. A properly just response to climate change can only be brought about if we do not shy away from questioning the fundamental logic of carbon fetishism and the logic of the market that attempts to appropriate, commodify and financialise nature, and ourselves.

Purpose and Scope

We intend this book to fill a gap in the climate change debate, which is normally dominated by environmental, climate and natural science perspectives. This volume instead comprises twenty-eight short interventions by prominent social scientists and humanities scholars of climate change and societal responses, as well as emerging academic contributors and voices from climate activism, bearing in mind that these categories are not mutually exclusive. Contributors to the volume come from many parts of the world and share their perspectives both on what is important in climate change debates, and 'what is to be done' in terms of radical climate action.

The collection includes new essays as well as republished texts, organised around seven themes: paradigms; what counts?; extraction; dispatches from a climate change frontline country; governance; finance; and action(s). These themes, which we outline briefly below, emerged from our reading of the contributions submitted for the collection, rather than being established in advance. They thus seem salient as a representation of climate change concerns consolidating amongst social science and humanities scholars and activists.

Paradigms

The first section on **Paradigms** is comprised of contributions that address broad perspectives arising in climate change debates. These 'big-picture' essays introduce some key challenges of the climate crisis in terms of societal understanding and response.

The opening chapter in this section, titled 'One Earth, Many Futures, No Destination', is by geographer Mike Hulme, whose work is widely known for its critical engagement with how climate-change as an *idea* becomes deployed, mobilised and disagreed about in public, scientific and policy discourse: most recently in *Climate Change: Key Ideas in Geography* (Hulme 2021). Hulme observes that since the first Earth Day, more than half a century ago, it has become clear that it is easier to generate scientific insight into the ways human systems and behaviours are altering the planet than it is to redirect those human systems to lessen the planetary impacts predicted through scientific insight. Alongside other chapters in this volume, Hulme thus emphasises the significance of divergent human values for the possibility of making choices that redirect human systems such that their planetary impact is transformed.

The chapter that follows, by Minna Halme and colleagues on the Finnish Expert Panel on Sustainable Development, foregrounds how the COVID-19 pandemic has laid bare the vulnerability to crises such as climate change caused by a current global socio-economic system oriented towards short-term financial efficiency. 'From Efficiency to Resilience: Systemic Change towards Sustainability after the COVID-19 Pandemic' argues that sustainability at scale can only be effected through a replacement of the dominant efficiency paradigm with a resilience paradigm. Their chapter outlines key orientations towards sectoral planning and governance they consider necessary for society to walk a path that effects systemic transformation towards resilience, understood as a mutually supportive symbiosis of social and natural—socionatural—'systems'.

The section on paradigms closes with Sian Sullivan's chapter 'On Climate Change Ontologies and the Spirit(s) of Oil', which considers ontological dimensions at play in societal understandings of the causes and implications of climate change. She asks questions of how anthropogenic climate change is understood culturally, and of what responses may be promoted as appropriate for this systemic predicament. By gesturing to culturally-inflected differences in ways of seeing and knowing the world, she draws into focus the interplay of multiple realities in climate change understandings that may contribute to political disagreements around highly divergent values and worldviews.

What Counts?

In considering divergence in climate change understandings and responses, our first section leads clearly to the question of **What Counts?** in climate change management, as asked in the essays pursuing this second theme of the book. This section addresses issues of calculation and measurement, given that climate change debates are often about numbers, particularly in relation to emissions of GHG. A key aim of the contributions here is to look behind the numbers, providing a rationale for why processes of calculation and measurement should also always be seen as political and hence as open to contestation.

James Dyke and co-authors kick off the section with an essay provocatively titled 'Why Net Zero Policies Do More Harm than Good'. The policy idea and ideal of 'net zero carbon', i.e. that socio-economic activity should generate zero carbon emissions in aggregate, is important for a number of subsequent chapters in this book because it has been so definitive in proposing a route towards a 'solution' for managing CO_2 emissions globally. As Dyke and colleagues argue, however, 'net zero' discourse conceals and justifies the deployment of highly speculative technologies that pose 'fairytales' in terms of their CO_2 emissions reductions, whilst also nudging society on to a series of potentially dangerous technological pathways for which outcomes are unknown. They confirm instead that the only way to keep humanity safe is through immediate and sustained radical cuts to greenhouse gas emissions, acknowledging the simultaneous challenge of doing this in ways that are socially just so as to redress, rather than deepen, the inequalities bound up with uneven industrialisation endeavours.

In 'The Carbon Bootprint of the US Military and Prospects for a Safer Climate', Patrick Bigger and colleagues continue by asking searching questions of high-tech responses to CO_2 emissions reductions in one of the most impactful of global industries, namely the US military. They foreground how future military CO_2 emissions are locked into the US military's expansive and coupled global logistical networks, hardware, and interventionist foreign policy. Echoing Dyke et al.'s chapter, they argue that apparently well-intentioned calls to 'green' the military are insufficient to reign in military emissions, urging that for climate change management alone the scope of the US military must be dramatically scaled back as part of any serious initiative to maintain a safer climate.

David Durand-Delacre and his team of co-authors shift focus in the following chapter to modes of counting and their implications when directed towards the understanding and management of human dimensions of climate change, specifically human migration linked with climate change. 'Climate Migration is about People, not Numbers' examines the large numbers often invoked to underline alarming climate migration narratives, and outlines the serious methodological limitations linked with the production of these numbers. In arguing for a greater

diversity of knowledges of climate migration and emphasising the value
of qualitative and mixed methods in research, they also question the
usefulness of excessively inflated numbers to progressive agendas for
climate action, given the xenophobic fear of climate migration promoted
by such numbers. In short, they emphasise how decisions based on
meeting quantitative targets around migration reduction should be
refocused instead on peoples' needs, rights and freedoms, and that
understanding these dimensions of human experience requires a mode
of listening that does not reduce humans to statistical datapoints alone.

'We'll Always Have Paris' by Mike Hannis turns to the implications of
the provision for voluntary carbon trading in Article 6 of COP21's Paris
Agreement, considering how this can be robustly counted in practice so
as to ensure the elusive goal of global 'net zero' CO_2 emissions. Turning a
sceptical eye to theoretical carbon trading, fantastical Negative Emission
Technologies (NETs), and voluntary national 'contributions', Hannis
asks questions of how Nationally Determined Contributions (NDCs)
to global CO_2 emissions are calculated across the complexities posed by
voluntary carbon trading possibilities. In placing COP21's Article 6 in
the context of the internationalist spirit of 2015 that at least consolidated
the idea and impression of a globally coordinated effort, he additionally
asks what impact the resurgent nationalism of the years since may have
on NDC calculations. Will the new Democrat US presidency be able to
re-establish US climate leadership and move negotiations back towards
a position of constructive international engagement on climate issues
and the NDCs on which emissions reductions rest? How might COP26
provide a boost to morale in internationalist spirit around climate
change governance?

Bruce Lankford's closing chapter in this second section of the book
considers how CO_2 is conceptualised and counted in what he frames
as 'The Atmospheric Carbon Commons in Transition'. In his analysis
Lankford brings to bear the rich concept of "paracommons", as the
"commons of material salvages" currently arising from the context
of climate change crisis requiring that emitted CO_2 is systemically
retrieved or 'salvaged'. Drawing on analyses of resource commons, he
argues that the carbon/atmospheric commons can be framed in three
consecutive stages: a "sink-type atmospheric commons" occurring
prior to the 1980/90s; a "husbandry-type carbon commons" lasting
from the 1980/90s to 2030s; and an emergency "carbon paracommons"

post-2030s. The first stage sees the atmosphere treated as a dump or sink for carbon dioxide (CO_2) 'waste' resulting in rising CO_2 levels. The second stage sees climate change mitigation (e.g. carbon sequestration in forests) as an Ostromian-commons husbandry that attempts to reduce CO_2 emission rates but continues to result in levels remaining above 400 ppm. In the third stage being entered now, the "carbon paracommons" treats CO_2 and its 'salvaging' as a matter of urgency, requiring methods of permanent sequestration, non-use and transformation, amidst uncertainty as to how in practice these may be instituted and with what implications.

Extraction

Given that climate change is caused by the extraction and transformation of fossil fuels, the third theme of our book coalesces around **Extraction.** This section is comprised of contributions that critically reflect on our uneven addiction to fossil fuel extraction and ask whether the brave new world of the nascent renewable energy transition is set to be any more sustainable.

In 'The Mobilisation of Extractivism: The Social and Political Influence of the Fossil Fuel Industry', Christopher Wright and Daniel Nyberg highlight a key paradox: although the worsening climate crisis has led to growing social and political demands for meaningful climate action and the decarbonisation of economies, the modern global economy is defined by fossil fuel energy and its embedded legacy of two centuries of economic growth and development. They outline how the fossil fuel industry has defined the global economy and defended its position as the most powerful industry in the world. This context means that assumptions of corporate self-regulation as the logical response to the climate crisis will allow for the continuation of a 'business as usual' approach in which fossil fuel energy is maintained. They argue that an emphasis on corporate self-regulation deliberately ignores the urgent need for government regulation of carbon emissions. Worse, they foreground how current corporate responses to the climate crisis rely on a politics of 'predatory delay', wherein the fossil fuel industry cynically seeks to slow the process of decarbonisation to maximise their financial returns in the short term, whilst simultaneously appearing as concerned corporate citizens.

'End the "Green" Delusions: Industrial-Scale Renewable Energy is Fossil Fuel+', by Alexander Dunlap, matches this concern, arguing that industrial-scale renewable energy does little to remake exploitative relationships with the earth. In representing the renewal and expansion of the present capitalist order, particularly given the fossil fuels embedded in the making of these technologies, Dunlap considers instead that industrial-scale renewable energy production should be more accurately understood as "fossil fuel+". He urges a radical re-thinking of the socio-ecological reality of so-called renewable energy so as to create space for the step-change of strategies needed to mitigate and avoid climate and ecological catastrophe.

In 'I'm Sian, and I'm a Fossil Fuel Addict: On Paradox, Disavowal and (Im)Possibility in Changing Climate Change', Sian Sullivan draws on research with people who have known lives not determined by access to fossil fuels to face the reality of being completely personally dependent on fossil fuel extraction and the products made possible by fossil fuels. Her essay is an attempt to fully face the contradiction between maintaining hope for binding international climate agreements that have teeth, whilst being aware of her dependence on the fossil fuel extracting and emissions-spewing industrial juggernaut that permeates all our lives. Drawing critically on twelve-steps thinking and psychoanalytic literature, the chapter constitutes a reflection on fossil fuel addiction, highlighting the destructive paradox of not being able to live up to internalised but unreachable values regarding environmental care in a fossil-fuelled world.

Dispatches from a Climate Change Frontline Country— Namibia, Southern Africa

The fourth section of the volume changes tack to focus in on dimensions of climate change impacts and understanding for a single country at great risk of small systemic changes in climate parameters. In recognising that climate change debates are dominated and even colonised by perspectives from the Global North, **Dispatches from a Climate Change Frontline Country—Namibia, Southern Africa**, comprises contributions from and about one country with heightened potential vulnerability to small changes in climate due to its status as an arid or semi-arid 'Global South' country. These chapters are set against the background of amplified risk of climate change posed by present oil exploration in Namibia's north-east Kavango Region, bordering Botswana, and current frenzied excitement within the oil industry that Namibia might "become the biggest oil story of the decade" (Leigh 2021), although the essays here do not explicitly address this rapidly shifting situation. The section builds on long-term research collaborations in Namibia by one of us (Sullivan).[8]

Selma Lendelvo and colleagues in the opening chapter for this section draw attention to a specific set of climate change-related interactions that pose particular implications for women in some rural areas of Namibia. 'Gendered Climate Change-Induced Human-Wildlife Conflicts amidst COVID-19 in Erongo Region, Namibia' argues that the risks of climate change for drier countries have become more pronounced, with small increments in temperature changes considered to pose serious consequences for dry countries such as Namibia and neighbouring Botswana, both of which have experienced significant and sustained drought in recent years. Lendelvo and co-authors draw attention to one effect of such changes, namely the amplified intensity of interactions between people and wildlife, such as elephants, as drought in the drylands of west Namibia concentrates humans and wildlife around available water sources. They focus on some of the particular implications for women as a vulnerable social group that is bearing the brunt of climate change-induced 'human-wildlife conflict', and foreground how women here are adjusting to these pressures. In closing, Lendelvo and colleagues ask us to remember that climate change impacts are differentiated and that the most vulnerable social groups—women, the poor and others—tend not to be present at international round-table discussions such as COPs to share their experiences of dealing and living with the impacts of climate change in their daily lives. Their chapter is intended as a short communiqué to foreground the types of concerns

8   We gratefully acknowledge here two research grants that have supported this international collaboration with Namibian researchers and contexts: Future Pasts (www.futurepasts.net, supported by the UK's Arts and Humanities Research Council (AHRC), and Etosha-Kunene Histories (www.etosha-kunene-histories. net) supported by the AHRC and the German Research Foundation (DFG).

women in rural dryland communities might wish to voice if they were able to be present at COP26.

Rick Rohde and colleagues follow with 'Environmental Change in Namibia: Land-Use Impacts and Climate Change as Revealed by Repeat Photography', a chapter that draws attention to a specific environmental history methodology for understanding environmental change within a recent historical timeline. They demonstrate how repeat landscape photography can be used to explore and juxtapose different cultural and scientific understandings of environmental change and sustainability in west Namibia. Change in the landscape ecology of western and central Namibia over the last 140 years has been investigated using archival landscape photographs located and re-photographed or 'matched' with recent photographs. Each set of matched images for a site provides a powerful visual statement of change and/or stability that can assist with understanding present circumstances in specific places. Sometimes these image sets show trajectories of vegetation change that diverge from modelled climate change projections and scenarios, demonstrating the importance of drawing on multiple sources of information to contextualise, and perhaps complexify, projected and predicted environmental futures.

In 'On Climate and the Risk of Onto-Epistemological Chainsaw Massacres: A Study on Climate Change and Indigenous People in Namibia Revisited', Ute Dieckmann asks searching questions of what may be lost in the process of trying to translate indigenous environmental knowledges and experiences into internationally acknowledged scientific frameworks. She revisits a commissioned World Bank Trust Fund study on climate change and indigenous people in Namibia in which she was involved, to highlight the predicament of short-term 'participatory' research with indigenous communities on climate change, and the ways in which imposed conceptual frameworks may act to subordinate indigenous peoples' ontologies to 'western' ontologies. She reflects that the 'compartmentalising' necessitated by such a methodology risks losing the most important aspects of indigenous ecological knowledge related to climate change, thereby perpetuating both climate and epistemic injustices.

Governance

The fifth theme on **Governance** steps back into a 'wide-angle' view, emphasising that the failure to address global climate change so far has been a failure of governance at global and national levels. Contributions in this section reflect on this governance failure and outline a series of practical solutions for moving forwards, beyond COP26.

The section opens with 'Towards a Fossil Fuel Treaty' by Peter Newell, a clarion call for a new approach for tackling climate change that focuses explicitly on fossil fuels. Like Wright and Nyberg, Newell also highlights the power of fossil fuel lobbies to delay effective climate action, urging that it is time to reign in the power these actors have over our collective fate, through international agreements and laws which effectively and fairly leave large swathes of remaining fossil fuels in the ground. He proposes a *Fossil Fuel Non-Proliferation Treaty* (FF-NPT) based, like the Nuclear Non-Proliferation Treaty, on the three pillars of non-proliferation, disarmament and peaceful use, as an instrument of international governance that could fulfil this purpose.

In 'How Governments React to Climate Change: An Interview with the Political Theorists Joel Wainwright and Geoff Mann' by Isaac Chotiner, we republish an interview with leading theorists of climate change governance that considers how to approach the global politics of climate change. Their conversation foregrounds several different potential futures for our warming planet. They argue that a more forceful international order, or "Climate Leviathan", is emerging, but warn that this configuration remains unlikely to mitigate catastrophic warming.

'Inside Out COPs: Turning Climate Negotiations Upside Down' by Shahrin Mannan and colleagues highlights the complexities of negotiations by the Conference of Parties (COP), observing that COP25, the longest in history, did not achieve its intended outcomes. Government negotiators failed to agree on core issues meaning that implementation of the Paris Agreement (COP21) has been pushed further away. COP negotiations tend to be dragged into overtime and appear inefficient, a perspective not helped by the arcane language of the adopted texts. The chapter advocates for the entire negotiating process to be rethought: through the alternative concept of 'inside out' COPs, wherein actions on the ground to implement the Paris Agreement are given greater prominence than political negotiations around a patchwork of compromises for implementation. They affirm that many different actors, including civil society, private companies, cities, universities, indigenous communities, young people and others pressing for action, should be placed centre-stage in devising and delivering outcomes that may be more real than those spun in COP Agreement texts run through with constructed ambiguities.

Shifting towards governance by national governments, in 'Local Net Zero Emissions Plans: How Can National Governments Help?' Ian Bailey clarifies the support needed from national governments in order that local government bodies can act on urban and regional initiatives that catalyse capacity building, knowledge exchange, and practical action on climate change and other sustainability issues. Bailey affirms that while local government initiatives equally should not be viewed as a substitute for robust international and national action on climate change, they can provide important arenas for mobilising local actors, formulating policies and developing institutions to complement national strategies. This chapter examines three main areas where support from central governments for local climate change responses is needed: the creation of supportive national policy environments; ensuring local governments are enabled to exercise their delegated powers to influence emissions; and the provision of finance to support emissions-reduction activities.

In the final chapter for this section on governance, Paul Harris foregrounds relationships between global climate governance and climate justice. 'Reversing the Failures of Climate Governance: Radical Action for Climate Justice' again recognises that global governance of climate change has failed. Emissions of the greenhouse gas pollution that causes climate change are still *increasing* globally, and little has been done to help the most vulnerable communities adapt to the inevitable, potentially existential, impacts. Harris argues that radical action is needed to avert and cope with the most dangerous consequences of climate change but will require focused attention on identifying the most vital sources of failure in climate governance and overcoming them. He suggests that much, if not most, of the failure of climate governance can be attributed to a lack of multiple kinds of climate justice—a lack of ecological and environmental justice, a lack of social and distributive justice, and a lack of international and global justice—and that averting climate catastrophe will require governance practices that embrace and implement all forms of climate justice.

Finance

Extending the theme of governance, our sixth section on **Finance** consists of two incisive essays regarding the complex roles of finance in climate change governance. Financing climate change mitigation and adaptation will be arguably a much bigger undertaking than what has been witnessed so far in terms of the global response to the COVID-19 pandemic. The financial resources needed to combat climate change are considered to be immense and the climate finance industry has been growing steadily over the past twenty years. As clarified by the chapters in this section, however, this growth does not in itself demonstrate success in terms of climate change management. The question of how to design and institute sustainable climate finance futures that are also equitable remains. Is it possible to do this in a capitalist global economy that tends towards the concentration of financial(ised) assets, and the fetishised concealment of multiple contradictions?

Sarah Bracking's opening chapter in this section grapples head-on with these contradictions. 'Climate Finance and the Promise of Fake Solutions to Climate Change' illuminates how promises of money from global institutions and governments have financialised people's hopes and expectations of government action to adapt to climate change and slow the emission of greenhouse gases. Bracking asserts that the cultural power of money in our understanding of the world means that climate finance has had the particular effect of signifying action while delivering very little. She argues that moving forwards with the actual material changes to energy, infrastructure, production and income distribution lying at the heart of an effective response to climate change will require acceptance that largely fictional promises of money that "can change things" are a phantasmagorical expression of meaning acting as a "firewall" that prevents real change. The essay traces the small disbursement figures for the main pots of climate finance. In doing so, Bracking offers a stringent critique of the obfuscating power of the language of finance and its propositions in the financing of climate change governance.

Rami Kaplan and David Levy's chapter on 'The Promise and Peril of Financialised Climate Governance' emphasises the rise of investor-driven, "financialised governance" of corporate practices in relation to the natural environment. Investors and investment managers are demonstrating greater concern that the value of assets, from stock markets to real estate, is increasingly subject to climate risks. Financialised climate governance (FCG) puts investors and fund managers at the centre of efforts to limit greenhouse gas emissions, which suggests both the promise and peril of this advanced form of "climate capitalism". They describe these developments and point towards the peril that relying on investors and business self-interest is unlikely to result in the rapid structural shifts needed for full decarbonisation.

Action(s)

The volume's final section on **Action(s)** extends the practical and affirmative suggestions made elsewhere in the book to foreground specific proposals for negotiating responses that are both effective and equitable in addressing and averting the climate crisis.

Peter North's opening question 'What Is to Be Done to Save the Planet?' is a good place to start. The essay reviews the impacts of radical social movement activity on the climate based on observations over the past fifteen years or so. It considers experiences of grassroots prefiguration and experimentation such as transition initiatives, experiments for eco-localisation, and small business networks, contrasting these initiatives with more antagonistic, direct, action-based movements such as the climate camps, mobilisations around the COPs and Extinction Rebellion. The intervention concludes by discussing the perceived efficacy of these varied movements, suggesting a need for more strategic action to effect system change via a 'Green New Deal'.

In 'Climate Politics between Conflict and Complexity', Matthew Paterson similarly foregrounds how climate politics needs both moments of sharp, highly politicising, even over-simplifying moves to keep pressure up, but at the same time a sort of patient, careful attention to the complexity of socio-technical systems to work out how to generate radical shifts in infrastructure and practice. He observes that these different logics are in tension: the post-political/agonistic logic can reduce to slogans and deflect from how society may become decarbonised in practical terms; the observance of how to effect change through complex socio-technical systems can culminate in technocratic projects. Paterson's chapter navigates the question of how to keep both these logics and their affirmative engagements alive in political praxis linked with climate change.

Rebecca Sandover's chapter on 'Sustainable Foodscapes: Hybrid Food Networks Creating Food Change' connects food production practices specifically with climate change governance, asserting that food matters, from modes of production to global supply chains, to what we eat and how we address food waste. Considering that Agriculture, Forestry and Other Land Use (AFOLU) activities account for some 21–37% of total net anthropogenic GHG emissions it is clear that food practices shape not only climate and ecological breakdown but also human health and well-being including within our food producing communities, and in issues associated with unequal access to food, food justice and animal welfare. Sandover's chapter foregrounds in particular how place-based community groups have been self-organising and connecting with different national organisations whose campaigns overlap to form hybrid food networks, in the midst of a present food policy vacuum in England. It explores the dynamic potential of these hybrid networks in working towards place-based sustainable food solutions through a case study of Devon.

In 'Telling the "Truth": Communication of the Climate Protest Agenda in the UK Legacy Media', Sharon Gardham draws on the results of a thematic discourse analysis of UK media coverage of climate strike actions that took place in 2019, reflecting on the importance for the wider adoption of climate protest messages of how protester claims-making and identity are framed. Gardham's chapter revisits a key question for the organisers of such protests regarding how they can overcome the potential conflict between ensuring their actions pass the test of newsworthiness required to ensure media attention, without failing the tests of claims-making legitimisation necessary for an issue to become accepted as a societal problem that requires urgent resolution.

Picking up themes considered elsewhere in the volume regarding climate justice, Patrick Bond, in 'Climate Justice Advocacy: Strategic choices for Glasgow and Beyond', urges "non-reformist reforms" in climate action. He critiques the lack of ambition and action in the main UN processes, but also critically analyses the 'Glasgow Agreement' promoted by leading civil society activist groups. Drawing parallels with South Africa's resistance strategies to defeat apartheid, he calls for climate justice movements to similarly not cave in to the internal logic of the climate governance system, instead confronting its core dynamics by delegitimising the system of oppression on which it is overlain. What we need, instead, he argues, is to give confidence to critical ideas and social forces that can question climate capitalism wholesale.

In the penultimate chapter of the book, Lorraine Whitmarsh reviews 'Public Engagement with Radical Climate Change Action', arguing that it is a mistake to understand people's role only in terms of their actions as consumers of apparently low-carbon products and urging that it is critical that people are also engaged as political, social and professional actors to achieve the scale of societal transformation needed. Whitmarsh discusses the varied roles the public can play in decision-making and in taking rapid and radical climate action, their current levels of engagement with climate change, and how to foster further public action. She lands her chapter on a positive note: affirming that we have a unique opportunity now to build back society post-COVID-19 in a way that might lock in low-carbon habits created during the pandemic, and that builds on the growing social mandate for bold policy action to support sustainable lifestyles.

We close this collection with a republished intervention by artist-activists Isabelle Fremeaux and Jay Jordan. Their text clarifies—with poetic and gritty integrity—their choice to publicly refuse participation in an event (*Agir Pour le Vivant* / *Action for the Living*, in Arles in France, August 2020) on the grounds of the dissonance between the event's intentions and its sponsorship by a series of fossil fuel extractors and financiers. 'Five Questions whilst Walking' invites consideration of the sorts of choices that need to be made if we are collectively to walk away from the forces propelling global climate crisis.

***

Clearly, no book could be completely comprehensive regarding climate change, the multifaceted challenges it poses, and societal responses to these challenges. This book is rather posed as an intervention, collecting together indicative contributions regarding what social scientists, humanities scholars and climate activists around the world think needs to be done in terms of both understanding why climate action has failed to dramatically reduce emissions to date, and proposing some routes towards radical climate change action now. That is, the book is intended to provide an affirmative set of ideas about what is to be done and how it can be done, to bring about radical climate change governance so that we have a chance of avoiding runaway climate change.

We are publishing this collection of essays in the months leading up to the high-profile and eagerly awaited COP26 UN climate change conference, due to take place in Glasgow (Scotland, UK) in November 2021. At this conference, all the major stakeholders of the global climate change negotiation process will be present, including heads of state, large national government delegations, policy advisers, NGO and social movement activists, multinational corporations, industry associations, and inter-governmental institutions. There will be significant media interest in COP26, reaching millions of people around the world, linked, for example, with the re-entering of the agreement by the US, the newly entwined crises of COVID-19 and climate change, and negotiations around the form and content of Green New Deal proposals. We hope that this collection of essays will contribute to this discussion.

Despite more than thirty years of high-level, global talks on climate change, we are still seeing emissions rising dramatically around the world. Whatever we have done on this planet in terms of climate mitigation over the past thirty plus years has not worked. Given that most climate scientists believe we are soon running out of time, the authors contributing to this volume ask what has gone wrong and what now needs to be done. We hope the essays collated here will help us move more radically and urgently in the direction needed.

References

Asiyanbi AP, 'Financialisation in the Green Economy: Material Connections, Markets-in-the-making and Foucauldian Organising Actions', *Environment and Planning A*, 50(2) (2017), 531–48, https://doi.org/10.1177/0308518X17708787.

Bayer, Patrick, and Michaël Aklin, 'The European Union Emissions Trading System Reduced CO2 Emissions Despite Low Prices', *Proceedings of the National Academy of Sciences*, 117(16) (2020), 8804–12, https://doi.org/10.1073/pnas.1918128117.

BlackRock, 'Net Zero, A Fiduciary Approach' (BlackRock.com, 2021), https://www.blackrock.com/corporate/investor-relations/blackrock-client-letter.

Böhm, Steffen, and Siddharta Dabhi, *Upsetting the Offset: The Political Economy of Carbon Markets* (London: MayFlyBooks, 2009), http://mayflybooks.org/?p=206.

Böhm, Steffen, and Siddhartha Dhabi, 'Commentary: Fault Lines in Climate Policy: What Role for Carbon Markets?', *Climate Policy*, 11(6) (2011), 1389–92, https://doi.org/10.1080/14693062.2011.618770.

Böhm, Steffen, Maria Ceci Misoczky, and Sandra Moog, 'Greening Capitalism? A Marxist Critique of Carbon Markets', *Organization Studies*, 33(11) (2012), 1617–38, https://doi.org/10.1177/0170840612463326.

Böhm, Steffen, Vinicius Brei, and Siddhartha Dabhi, 'EDF Energy's Green CSR Claims Examined: The Follies of Global Carbon Commodity Chains', *Global Networks*, 15(s1) (2015), 87–107, https://doi.org/10.1111/glob.12089.

Bracking, Sarah, 'Financialisation, Climate Finance, and the Calculative Challenges of Managing Environmental Change', *Antipode*, 51 (2019), 709–29, https://doi.org/10.1111/anti.12510.

Bracking, Sarah, 'How do Investors Value Environmental Harm/care? Private Equity Funds, Development Finance Institutions and the Partial Financialization of Nature-based Industries', *Development and Change*, 43(1) (2012), 271–93, https://doi.org/10.1111/j.1467-7660.2011.01756.x.

Brulle, Robert, '30 Years Ago Global Warming Became Front-page News—And Both Republicans and Democrats Took it Seriously' (TheConversation.com, 2018), https://theconversation.com/30-years-ago-global-warming-became-front-page-news-and-both-republicans-and-democrats-took-it-seriously-97658.

Bryant, Gareth, Siddhartha Dabhi, and Steffen Böhm', "Fixing' the Climate Crisis: Capital, States, and Carbon Offsetting in India', *Environment and Planning A*, 47(10) (2015), 2047–63, https://doi.org/10.1068/a130213p.

Carbon Brief, 'Analysis: The UK Politicians Who Talk the Most About Climate Change' (Carbonbrief.org, 2019a), https://www.carbonbrief.org/analysis-the-uk-politicians-who-talk-the-most-about-climate-change.

Carbon Brief, 'The Carbon Brief Profile: India' (Carbonbrief.org, 2019b), https://www.carbonbrief.org/the-carbon-brief-profile-india.

Carson, Rachel, *Silent Spring* (Boston: Houghton Mifflin, 1962).

Centre for Global Development, 'Developed Countries Are Responsible for 79 Percent of Historical Carbon Emissions' (Cgdev.org, 2015), https://www.cgdev.org/media/who-caused-climate-change-historically.

Christophers, Brett, 'Environmental Beta or How Institutional Investors Think About Climate Change and Fossil Fuel Risk', *Annals of the American*

Association of Geographers, 109(3) (2019), 754–74, https://doi.org/10.1080/2 4694452.2018.1489213.

Department for Business, Energy and Industrial Strategy, '2019 UK Greenhouse Gas Emissions, Provisional Figures' (Gov.uk, 2020), https://assets. publishing.service.gov.uk/government/uploads/system/uploads/ attachment_data/file/875485/2019_UK_greenhouse_gas_emissions_ provisional_figures_statistical_release.pdf.

Dunlap, Alexander and Sian Sullivan, 'A Faultline in Neoliberal Environmental Governance Scholarship? Or, Why Accumulation-by-Alienation Matters', *Environment and Planning E: Nature and Space*, 5(2) (2019), 552–79, https:// doi.org/10.1177/2514848619874691.

Ehrenstein, V. and Fabian Muniesa, 'The Conditional Sink: Counterfactual Display in the Valuation of a Carbon Offsetting Restoration Project', *Valuation Studies*, 1(2) (2013), 161–88, https://doi.org/10.3384/vs.2001-5992.1312161.

Farmer, Matthew, 'Court Orders Royal Dutch Shell to Cut Emissions by 45%' (Offshore-technology.com, 2021), https:// www.offshore-technology.com/news/company-news/ shell-emissions-dutch-court-netherlands-ruling-paris-agreement/.

Fink, Larry, 'Larry Fink's 2021 Letter to CEOs' (BlackRock.com, 2021), https:// www.blackrock.com/corporate/investor-relations/larry-fink-ceo-letter.

Foster, John Bellamy, 'The New Geopolitics of Empire', *Monthly Review*, 57(8) (2006), 1, https://doi.org/10.14452/mr-057-08-2006-01_1.

Franta, Benjamin, 'Shell and Exxon's Secret 1980s Climate Change Warnings' (TheGuardian.com, 2018), https://www.theguardian. com/environment/climate-consensus-97-per-cent/2018/sep/19/ shell-and-exxons-secret-1980s-climate-change-warnings.

Grossman, Gene M., and Alan B. Krueger, 'Economic Growth and the Environment', *The Quarterly Journal of Economics*, 110(2) (1995), 353–77, https://doi.org/10.2307/2118443.

Hall, Shannon, 'Exxon Knew About Climate Change Almost 40 Years Ago', *Scientific American*, 26 (2015), https://www.scientificamerican.com/article/ exxon-knew-about-climate-change-almost-40-years-ago/.

Holder, Michael, 'Former Bank of England Governor Mark Carney Leads Task Force to Scale Global Market for CO2 Offsets' (Greenbiz.com, 2020), https:// www.greenbiz.com/article/former-bank-england-governor-mark-carney- leads-task-force-scale-global-market-co2-offsets.

Hulme, Mike, *Climate Change: Key Ideas in Geography* (London: Routledge, 2021).

IPCC, 'Climate Change Widespread, Rapid, and Intensifying' (Ipcc. ch, 2021), https://www.ipcc.ch/site/assets/uploads/2021/08/ IPCC_WGI-AR6-Press-Release_en.pdf.

Lampen, Katherine, 'ESG Trend is Changing Asset Management' (FTadviser. com, 2021), https://www.ftadviser.com/investments/2021/04/08/ esg-trend-is-changing-asset-management.

Langley, Paul, Gavin Bridge, Harriet Bulkeley and Bregje van Veelen, 'Decarbonizing Capital: Investment, Divestment and the Qualification of Carbon Assets', *Economy and Society*, 50(3), 494–516, https://doi.org/10.108 0/03085147.2021.1860335.

Leigh, Alex, 'Why Namibia Could Become The Biggest Oil Story of the Decade' (Oilprice.com, 2021), https://oilprice.com/Energy/Energy-General/Why-Namibia-Could-Become-The-Biggest-Oil-Story-of-the-Decade.amp.html.

Lippert, Ingmar, 'Environment as Datascape: Enacting Emission Realities in Corporate Carbon Accounting', *Geoforum*, 66 (2014), 126–35, https://doi. org/10.1016/j.geoforum.2014.09.009.

Lohmann, Larry, 'Performative Equations and Neoliberal Commodification: The Case of Climate', in *Nature™ Inc.: Environmental Conservation in the Neoliberal Age*, ed. by Büscher, Bram, Wolfram Dressler, and Rob Fletcher (Tucson: Arizona University Press, 2014), pp. 158–80.

Lohmann, Larry, 'Toward a Different Debate in Environmental Accounting: The Cases of Carbon and Cost-Benefit', *Account, Organizations and Society*, 34 (2009), 499–534, https://doi.org/10.1016/j.aos.2008.03.002.

Malm, Andreas, *Fossil Capital: The Rise of Steam Power and the Roots of Global Warming* (London: Verso, 2016).

Marx, Karl, *Capital, Vol. 1* (Harmondsworth: Penguin, 1976).

Moore, Jason W., *Capitalism in the Web of Life: Ecology and the Accumulation of Capital* (London: Verso, 2015).

Moore, Jason W., 'Amsterdam is Standing on Norway, Part I: The Alchemy of Capital, Empire and Nature in the Diaspora of Silver, 1545–1648', *Journal of Agrarian Change*, 10(1) (2010) 33–68, https://doi. org/10.1111/j.1471-0366.2009.00256.x.

Office for National Statistics, 'The Decoupling of Economic Growth from Carbon Emissions: UK Evidence' (Ons.gov.uk, 2019), https://www.ons.gov.uk/economy/nationalaccounts/ uksectoraccounts/compendium/economicreview/october2019/ thedecouplingofeconomicgrowthfromcarbonemissionsukevidence.

Pain, Stephanie, 'Before it was Famous: 150 Years of the Greenhouse Effect, *New Scientist*, 202(2708) (2009), 46–47, https://doi.org/10.1016/ S0262-4079(09)61329-4.

Pearse, Rebecca, and Steffen Böhm, 'Ten Reasons Why Carbon Markets Will Not Bring About Radical Emissions Eeduction', *Carbon Management*, 5(4) (2014), 325–37, https://doi.org/10.1080/17583004.2014.990679.

Plass, Gilbert N, 'The Carbon Dioxide Theory of Climatic Change', *Tellus*, 8(2) (1956), 140–54, https://doi.org/10.1111/j.2153-3490.1956.tb01206.x.

Plass, Gilbert N., James Rodger Fleming, and Gavin Schmidt, 'Carbon Dioxide and the Climate', *American Scientist*, 98(1) (2010), 58–62, https://www.americanscientist.org/article/carbon-dioxide-and-the-climate.

Postel-Vinay, Natacha '"Covid Bonds" Are of Limited Appeal Right Now, But They May Yet be Useful to the Government' (LSE.ac.uk, 2021), https://blogs.lse.ac.uk/covid19/2021/02/24/covid-bonds-are-of-limited-appeal-right-now-but-they-may-yet-be-useful-to-the-government.

Rainforest Action Network, 'Banking on Climate Chaos 2021' (RAN.org, 2021), https://www.ran.org/bankingonclimatechaos2021.

Ramirez, Jacobo, and Steffen Böhm, 'Transactional Colonialism in Wind Energy Investments: Energy Injustices Against Vulnerable People in the Isthmus of Tehuantepec', *Energy Research & Social Science*, 78 (2021), 102135, https://doi.org/10.1016/j.erss.2021.102135.

Ritchie, Hannah, 'Climate Change and Flying: What Share of Global CO2 Emissions Come From Aviation?' (Ourworldindata.org, 2020), https://ourworldindata.org/co2-emissions-from-aviation.

Saul, Jonathan, 'Shipping's Share of Global Carbon Emissions Increases' (Reuters.com, 2020), https://www.reuters.com/article/us-shipping-environment-imo-idUSKCN2502AY.

Scientists for Global Responsibility, 'The Carbon Boot-print of the Military' (SGR.org.uk, 2020), https://www.sgr.org.uk/resources/carbon-boot-print-military-0.

Skoglund, Annika, and Steffen Böhm, 'Prefigurative Partaking: Employees' Environmental Activism in an Energy Utility', *Organization Studies*, 41(9) (2020), 1257–83, https://doi.org/10.1177/0170840619847716.

Sullivan, Sian, *Financialisation, Biodiversity Conservation and Equity: Some Currents and Concerns* (Penang Malaysia: Third World Network Environment and Development Series 16, 2012), http://twn.my/title/end/pdf/end16.pdf.

Sullivan, Sian, 'After the Green Rush? Biodiversity Offsets, Uranium Power and the "Calculus of Casualties" in Greening Growth', *Human Geography*, 6(1) (2013a): 80–101, https://doi.org/10.1177/194277861300600106.

Sullivan, Sian, 'Banking Nature? The Spectacular Financialisation of Environmental Conservation', *Antipode*, 45(1) (2013b), 198–217, https://doi.org/10.1111/j.1467-8330.2012.00989.x.

Sullivan, Sian, '"Ecosystem Service Commodities"—A New Imperial Ecology? Implications For Animist Immanent Ecologies, With Deleuze and Guattari', *New Formations: A Journal of Culture/Theory/Politics*, 69 (2010), 111–28, https://doi.org/10.3898/NEWF.69.06.2010.

Swyngedouw, Erik 'Apocalypse Forever? Post-political Populism and the Spectre of Climate Change', *Theory Culture and Society*, 27(2–3) (2010), 213–232, https://doi.org/10.1177/0263276409358728.

UN, 'UN Emissions Report: World on Course for More Than 3 Degree Spike, Even if Climate Commitments are Met' (UN.org, 2019), https://news.un.org/en/story/2019/11/1052171.

UN, 'Kyoto Protocol to the United Framework Convention on Climate Change' (UNFCCC.int, 1998), https://unfccc.int/resource/docs/convkp/kpeng.pdf.

UN, 'Paris Agreement' (UNFCCC.int, 2015), https://unfccc.int/sites/default/files/english_paris_agreement.pdf.

UN, 'United National Framework Convention on Climate Change' (UNFCCC.int, 1992), https://unfccc.int/files/essential_background/background_publications_htmlpdf/application/pdf/conveng.pdf.

Watt, Robert, 'The Fantasy of Carbon Offsetting', *Environmental Politics*, online first https://doi.org/10.1080/09644016.2021.1877063.

Weart, Spencer, 'The Discovery of Global Warming' (History.aip.org, 2021), https://history.aip.org/climate/Govt.htm.

WMO, '2020 Was One of Three Warmest Years on Record' (WMO.int, 2021), https://public.wmo.int/en/media/press-release/2020-was-one-of-three-warmest-years-record.

World Commission on Environment and Development, 'Our Common Future' (1987), 1–91, http://www.un-documents.net/ocf-07.htm.

Yang, Yuantao, et al., 'Mapping Global Carbon Footprint in China', *Nature Communications*, 11(1) (2020), 1–8, https://doi.org/10.1038/s41467-020-15883-9.

I

PARADIGMS

1. One Earth, Many Futures, No Destination

Mike Hulme

Since the first Earth Day, more than half a century ago, it has become clear that it is easier to generate scientific insight into the ways human systems and behaviours are altering the planet, than it is to redirect those human systems to lessen their planetary impact. At the heart of this conundrum are divergent human values.

Earth Day 1970[1]

More than fifty years ago, twenty million Americans gathered in public streets, squares and parks across America to demonstrate their concern about the state of the planet. The first Earth Day rode the tide of late 1960s radicalism and protest in Western democracies and sought to "force the environmental issue into the political dialogue of the nation" (Lewis 1990: 10). Although it succeeded in doing so, and continues to do so more widely today in a very different world, it is questionable whether the larger ambitions of 1970 Earth Day to bring about a more sustainable civilisation have been met, not least with respect to a changing climate.

There is a paradox here. In the half century since 1970 it has been relatively easy for science to bring forward knowledge about the

---

1   This article was first published as Mike Hulme, 'One Earth, Many Futures, No Destination', *One Earth*, 2(4) (2020), 309–11, Copyright Elsevier. It has been lightly edited for this book volume.

https://doi.org/10.11647/OBP.0265.01

dynamics of the Earth system and identify the dangers of unmitigated climate change, knowledge that has now gained widespread public and political attention. And yet it has been manifestly harder to use such knowledge to orchestrate and deliver systematic change in the human sphere to mitigate future climatic risks.

In this essay I seek to analyse what is sometimes referred to as this 'knowledge-action gap' in three steps. First, I explain why facts alone can never be sufficient to drive policy and, second, I show that the facts of climate change can be consistent with different stories—sometimes radically different stories—that embody people's beliefs about the past, present and future. Third, this then explains why what I call 'climate solutionism' is the wrong framework within which to operate. I conclude by suggesting a focus less on the destination—i.e., 'stopping climate change'—and more on enhancing the political conditions of the journeying.

Why Facts Are Not Enough

As recently argued or observed, 'listening to the science' would appear to be the *sine qua non* of the new wave of climate protest movements (Schinko 2020; Kenis 2021). Making sure that "objective facts" are laid "on the table" is believed to put pressure on "obstructionist states" to deliver political change (Schinko 2020: 22). Or as the late Rajendra Pachauri asserted back in November 2014 at the launch of the IPCC's 5th Assessment Report, "all we need is the will to change, which we trust will be motivated by ... an understanding of the science of climate change". This 'science first' argument guides the consensus messaging campaign that seeks to emphasise above all else the "97% of scientists" who agree that human actions are changing the world's climate. It also leads cognitive psychologists such as Stephan Lewandowsky to develop climate science communication strategies based on "inoculation theory" (Cook et al. 2017). This theory asserts that people can be made immune to falsehoods by being exposed, ahead of time, to those falsehoods they are most likely to encounter on social media and elsewhere.

But facts are never enough. With regard to climate change, seeking merely to 'hit the numbers'—whichever one you choose: 2°C, 1.5°C,

350ppm, net zero—is not enough. It fuels what I have elsewhere called "climate reductionism" (Hulme 2011) and "climate deadline-ism" (Asayama et al. 2019) and encourages the type of "climate solutionism" of which I am critical (see below). 'Closed' timetables and emergency imperatives fail to respect the diverse moral horizons that characterise—and complicate—the difficult politics of climate change. Mere technique and technology crowds out wider explorations of human meaning and ethical purpose. Dan Sarewitz explains the flaw in this position:

> [...] our expectations for Enlightenment ideals of applied rationality are themselves irrational. We are asking science to do the impossible: to arrive at scientifically coherent and politically unifying understandings of problems that are inherently open, indeterminate and contested (Sarewitz 2017: para. 25).

Which Story?

Establishing scientific facts about climate change (or offering scientific projections of future change) does not on its own drive political change. Consensus messaging, for example, fails to work because risk is socially constructed and value driven. So, if, as Sarewitz says, climate change is "inherently open, indeterminate and contested", if in fact there is a surfeit of competing narratives each with different solutions to climate change, what should be our strategy? What are the wider resources beyond science—the motivational moral commitments that Jürgen Habermas refers to as "missing" in secularist societies (Habermas 2010)—that can enact and guide change? To illustrate what may be missing, I suggest below four different meta-narratives—guiding myths if you will, or ideologies—which are advocated by different voices to guide action in response to climate change. They differ from each other in various ways, sometimes profoundly. These future visions are rooted in different cultural values and often are antagonistic to each other (also see Dieckmann's and Sullivan's chapters, this volume). But they are similar in so far as they each require science and technology to be placed in a subservient role to their normative vision of how the world should be.

The first of these I group loosely under the label of 'eco-modernism'. The argument here is that modernity can, so to speak, both have its cake and eat it. Yes, climate change is an outcome of rapid and penetrating technological expansion and economic and population growth. But it is through adjusting and redirecting these very great achievements of modernity towards more just and ecologically sensitive ends that climate change can be arrested. Thus, for example, the *Ecomodernist Manifesto* claims that humans need to use all their "growing social, economic, and technological powers to make life better for people, stabilise the climate, and protect the natural world" (Asafu-Adjaye 2015: 6).

A second ideology—or motivational discourse—is that of 'ecological civilisation'. In essence, ecological civilisation is seen as the final goal of social, cultural and environmental reform within a given society. It argues that the changes to be wrought by climate change in the future can only be headed off through an entirely new form of civilisation, one based centrally on ecological principles. There are radically different techno and romantic versions of this envisioned future. The techno version of ecological civilisation has been embedded since 2012 in China's Communist Party's constitution. But it is very different from the romantic version espoused by deep ecologists and new cultural movements such as the Dark Mountain Project, which seek an unweaving of the core tenets of Western civilisation (Kingsnorth and Hine 2014).

A third narrative guiding political action in response to the challenges of climate change is the radical eco-socialist critique of capitalism. Following Naomi Klein's 2015 book *This Changes Everything: Climate vs Capitalism*, this has been articulated even more decisively by the new social movement Extinction Rebellion (XR) and in some versions of the Green New Deal (Pettifor 2019). XR have a clear belief that the only adequate response to climate change is the overturning of the social order and the capitalist economic system. The real enemy of a stable and benign climate is 'racialised capitalism' and its fetishing of economic growth and the centralisation of wealth and power that capitalism fuels. XR is rooted in what for many are the political extremisms of anarchism, eco-socialism and radical anti-capitalist environmentalism. The 'civil resistance model' espoused by XR is intended to achieve mass protest accompanied by law-breaking, leading eventually to the disruption of "business-as-usual" through the movement's calls to "tell the truth", "act now" and "go beyond politics"[2] (see discussion in Gardham, this volume).

A fourth guiding myth was given new focus in 2015 through the publication of Pope Francis' encyclical *Laudato 'Si: On Care for our Common Home* (Pope Francis 2015). Here, the facts of climate change 'reveal' an emaciation of the human spirit which is having adverse repercussions for the material world. Pope Francis is concerned first and foremost to offer a vision of human dignity, responsibility and purpose. He draws upon the rich traditions of Catholic theology and ethics, notably the idea of virtue ethics which is valorised above utilitarian and deontological modes of ethical reasoning.[3] *On Care for our Common Home* offers a powerful story, an inspirational account of divine goodness and healthy human living. It escapes the confines of a narrowly-drawn science and economics and shows the power, vitality and inspiration of a Christian worldview. Pope Francis draws attention to the centrality for the Christian faith of the idea of transformation, claiming "the ecological crisis is also a summons to profound interior conversion" (Pope Francis 2015: 158).

These ideologies offer different motivational commitments to tackling climate change and guide political action and public policy in different ways. For example, securing 'green growth' through a reformed capitalism is incommensurable with the eco-socialist ambition to dismantle the fetishism of growth upon which capitalism relies. Tackling climate change through inner spiritual transformation sits uneasily with the techno-modern vision of an ecological civilisation espoused by China's Communist Party. The Dark Mountain Project wants 'less' modernity; eco-modernists want 'more'. These meta-narratives illustrate why providing a coordinated global roadmap for climate action to deliver the 2°C target, in which all the pieces dovetail neatly into a single jigsaw, is not achievable.

---

2   Editors' note: see https://extinctionrebellion.uk/ and https://rebellion.global/.
3   Editors' note: virtue ethics focus on the morally virtuous dispositions of individuals that contribute to the flourishing of society more broadly, in contrast with a focus on actions designed for the purpose of generating broadly useful outcomes (utilitarian), or so as to be morally right in themselves, regardless of consequences (deontological).

Against Climate Solutionism

The belief that climate change can be solved can be traced back to its emergence in public life following the 1970 Earth Day as the latest in a series of environmental challenges facing the modern world. These challenges grew in scale from the merely local to the regional and then to the global. Climate change was in a line which can be traced back to Rachel Carson's intervention in the early 1960s about DDT (Dichlorodiphenyltrichloroethane) and chemical pesticides, and which then progressed through concerns about river and ocean pollution, smog, acid rain, the ozone hole and, eventually, in the late 1980s to the fully-developed awareness of the challenge of global climate change. Although inheriting this problem/solution framing, what 'solving' climate change actually means has always been harder to establish. It is not as simple as eradicating DDT, installing sulphur scrubbers on power stations or eliminating CFCs (Chlorofluorocarbons).

Uniting behind science, putting 'objective facts on the table' and thinking that solutions will flow naturally from them—what I mean by 'climate solutionism'—will not do. Science on its own offers no moral vision, no ethical stance, no political architecture for delivering the sort of world people desire. As Amanda Machin and Alexander Ruser have recently argued,

> [...] emblematic numbers and the production of political thresholds, targets and truths will not smooth out or settle down the political disputes over climate change. The reliance upon emblematic numbers may ignite a sense of urgency, but it may also fuel the suspicion of politicians, scientists and climate change policy (Machin and Ruser 2019: 223).

My examples above of different meta-narratives which give meaning to climate change show that the solutions to climate change are under-determined by the facts. In other words, climate change is a wicked problem (Hulme 2009), a problem that has no definitive formulation and no imaginable solution. Wicked problems are insoluble in the sense that solutions to one aspect of the problem reveal or create other, even more complex, problems which in turn demand further solutions. Proposed solutions to climate change can only ever be partial; they set in train secondary and tertiary consequences which always exceed what can be anticipated. This is the condition pointed to by the nomenclature of the

Anthropocene: namely, the modernist instinct for mastery, planning, optimisation and control is no longer an appropriate paradigm for living in the world of the twenty-first century.

Climate solutionism, driven by metrics, masks the contested politics and values diversity that lie behind different personal and collective choices—who wins, who loses, whose values count. It is a form of moral attenuation. Metrics are alluring because they simplify complex realities into 'objective' numbers and because they appear to short-circuit the need for difficult moral judgement. Metrification "may make a troubling situation more salient, without making it more soluble" (Muller 2018: 183). The circulation of ubiquitous carbon metrics operates as a facilitative and immanent mode of power. Morality by numbers also marginalises other modes of moral reasoning which cannot be reduced to calculation (also see chapters by Durand-Delacre et al. and Hannis, this volume). These other modes offer richer narrative contexts that enable the wisdom of different choices to be deliberated, interpreted and judged. Wise governance of climate—as indeed in the application of wisdom in everyday life—emerges best when rooted in larger and thicker stories about human purpose, identity, duty and responsibility.

No Destination

We have reached beyond a stage (if there ever was one) when steering the planet towards some long-term commonly agreed normative goal or benign state was feasible. At best, consensus messaging and inoculation theory may yield a thin veneer of agreement about the reality of human-caused climate change. But there is no trick that will force a convergence of human values. The stories people tell about themselves, their past, their futures, their place on the planet will continue to divide. Mobilising some new "solution science" (Doubleday and Connell 2020) resting on a putative cultural authority of science will not eradicate political conflict. We live on one Earth, but we imagine many futures and hence are not susceptible to alignment of our actions toward securing a common single destination.

We rather have to abandon the dream that a sustainable ecological equilibrium that works for everyone can be designed, implemented and reached. Securing a predetermined agreed destination, such as the 2°C target, is an illusion; delivering "Earth system management" (as proposed by Schellnhuber and Tóth 1999) is a chimera. What should be aimed for are less ambitious, more incremental and multi-scalar projects, that emerge from a humbler disposition toward the future and anticipating perverse outcomes. These interventions should be driven from the bottom up rather than by a top-down narrative of securing a singular global target. For example, there are many different local, culturally-sensitive policies that can be designed to progress toward securing one or more of the 169 UN Sustainable Development Targets. These interventions do not rely upon globally coordinated action, nor a commitment to one shared ideology, nor do they measure success according to just one index.

The corollary of this disposition is that investing in new participatory and agonistic forms of democracy (Mouffe 2006), where value-conflicts and political disagreements are acknowledged, voiced and worked with, is as important—perhaps more important—than investing in new scientific or technical knowledge. There is a balance to be struck between the twin dangers of, on the one hand, the crisis politics of emergency and, on the other, perpetually 'kicking the can down the road'. But good politics requires agonistic listening—the pursuit of what Nicholas Rescher (1993) calls 'acquiescence' in a decision—rather than consensual agreement. Have all interested parties been heard? Has their case been understood? Have their concerns been recognised? Over-emphasising the epistemic force of narrow science-based indicators—like global temperature or net zero emissions, or the emotional rhetoric of 'only 10 more years'—are poor substitutes or short-cuts for political forms of closure.

References

Asafu-Adjaye, John, et al., *An Ecomodernist Manifesto* (Ecomodernism.org, 2015), http://www.ecomodernism.org/.

Asayama, Shin, Rob Bellamy, Oliver Geden, Warren Pearce, and Mike Hulme, 'Why Setting a Climate Deadline is Dangerous', *Nature Climate Change*, 9(8) (2019), 570–72, https://doi.org/10.1038/s41558-019-0543-4.

Cook, Jon, Stefan Lewandowsky, and Ullrich Ecker, 'Neutralizing Misinformation Through Inoculation: Exposing Misleading Argumentation Techniques Reduces Their Influence', *PLoS ONE*, 12(5) (2017), e0175799, https://doi.org/10.1371/journal.pone.0175799.

Doubleday, Zoë, and Sean Connell, 'Shining a Bright Light on Solution Science in Ecology', *One Earth*, 2 (2020), 16–19, https://doi.org/10.1016/j.oneear.2019.12.009.

Habermas, Jürgen, *An Awareness of What is Missing: Faith and Reason in a Post-Secular Age* (Cambridge: Polity, 2010).

Hulme, Mike, *Why We Disagree About Climate Change: Understanding Controversy, Inaction and Opportunity* (Cambridge: Cambridge University Press, 2009), https://doi.org/10.1017/cbo9780511841200.

Hulme, Mike, 'Reducing the Future to Climate: A Story of Climate Determinism and Reductionism', *Osiris*, 26(1) (2011), 245–66, https://doi.org/10.1086/661274.

Kenis, Anneleen, 'Clashing Tactics, Clashing Generations: The Politics of the School Strikes for Climate in Belgium', *Politics & Governance*, 9(2) (2021), 135–45, https://doi.org/10.17645/pag.v9i2.3869.

Kingsnorth, Paul, and Dougal Hine, *'Uncivilisation: The Dark Mountain Manifesto'* (Dark-mountain.net, 2014), https://dark-mountain.net/product/uncivilisation-the-dark-mountain-manifesto/.

Klein, Naomi, *This Changes Everything: Capitalism vs. the Climate* (London: Penguin, 2015).

Lewis, Jack, 'The Spirit of the First Earth Day', *EPA Journal*, 16(1) (1990), 8–12.

Machin, Amanda, and Alexander Ruser, 'What Counts in the Politics of Climate Change? Science, Scepticism and Emblematic Numbers', in *Science, Numbers and Politics*, ed. by Markus J Prutsch (London: Palgrave MacMillan, 2019), pp. 203–25, https://doi.org/10.1007/978-3-030-11208-0_10.

Mouffe, Chantel, *On the Political* (Abingdon: Routledge, 2006), https://doi.org/10.4324/9780203870112.

Muller, Jerry, *The Tyranny of Metrics* (Princeton, NJ: Princeton University Press, 2018), https://doi.org/10.1515/9780691191263.

Pettifor, Ann, *The Green New Deal* (London: Verso, 2019).

Pope Francis, *Laudato Si' of the Holy Father Francis—On Care for our Common Future* (Rome: Vatican Press, 2015).

Rescher, Nicholas, *Pluralism: Against the Demand for Consensus* (Oxford: Oxford University Press, 1993).

Sarewitz, Dan, 'Stop Treating Science Denial Like a Disease' (Theguardian.com, 2017), https://www.theguardian.com/science/political-science/2017/aug/21/stop-treating-science-denial-like-a-disease.

Schellnhuber, John, and Ferenc Tóth, 'Earth System Analysis and Management', *Environmental Modeling & Assessment*, 4 (1999), 201–07, https://doi.org/10.1023/A:1019084805773.

Schinko, Thomas, 'Overcoming Political Climate-change Apathy in the Era of #FridaysForFuture', *One Earth*, 2 (2020), 20–23, https://doi.org/10.1016/j.oneear.2019.12.012.

2. From Efficiency to Resilience: Systemic Change towards Sustainability after the COVID-19 Pandemic

M. Halme, E. Furman, E.-L. Apajalahti, J. J. K. Jaakkola, L. Linnanen, J. Lyytimäki, M. Mönkkönen, A. O. Salonen, K. Soini, K. Siivonen, T. Toivonen and A.Tolvanen

The COVID-19 pandemic has revealed the vulnerability of current socio-economic systems and thrown into question the dominant global paradigm geared towards short-term financial efficiency. Although it has been acknowledged for several decades that this paradigm has detrimental impacts on the climate, the environment and global welfare, the pandemic has now offered a grim 'rehearsal round' for more serious crises that are to come with the accelerating climate emergency, loss of biodiversity and growing human inequalities. Along with worsening climate change, there are looming risks for mass migrations and armed conflicts as habitats capable of supporting human wellbeing become scarce, such as through the loss of potable water, an increasing lack of suitable land for agriculture, or the rise of unliveable temperatures. Although the COVID-19 pandemic has temporarily decreased some of the climate impacts, e.g. in the energy and transportation sectors, it has at the same time accelerated several global welfare problems. In this chapter, we claim that the way out of the crisis scenario is to replace

https://doi.org/10.11647/OBP.0265.02

the dominant efficiency paradigm with a resilience paradigm. Against the backbone of the key societal systems outlined in the *Global Sustainable Development Report* (GSDR 2019), we show how the pursuit of narrowly-defined efficiency hampers present and future sustainability, and chart some key actions on the path to transforming these systems towards resilience.

The Problem of Extreme Efficiency

The efficiency paradigm ruling global business has led to the dominance of global trade and supply networks, in which a British citizen is dependent on medicine manufactured only in China, or in which a citizen of the Nordic countries, in the barren midwinter, buys tulips grown in Kenya by Dutch companies, and Brazilian farmers depend on seeds supplied by multinational corporations. Efficiency has become a taken-for-granted organising principle for the global economy (Martin 2019), meaning we seldom pause to think about the 'costs'—widely defined—accompanying the efficiency of the current global economy. Many times, efficiency actually refers to low cost—cheap clothes, electronics, food—but often not to better products with lower overall costs. Efficiency often generates what in economics are called 'externalities'—uncosted costs or benefits for third parties, including 'the environment'—and has limited capacity to bring about a reduction in use of natural resources and accumulation of waste on a global scale (also see Lankford, this volume). Furthermore, gains in efficiency leading to lower prices are likely to be offset by increased consumption, which in turn has led to increased overall emissions and resource use (Heindl and Kanschik 2016; Alcott 2005), and compromised the resilience of economic and ecological systems (Martin 2019) (recognising that these 'systems' are also interconnected).

The Socio-Ecological Price of Efficiency

On the social side, the efficiency paradigm has led to the exploitation of those that have weak negotiating power in the (global) marketplace. Despite the benefits that international trade has brought to a number of people, trade also comes with externalities, such as salaries pushed below a living wage, human rights violations in supply chains, and increasing economic inequality (GSDR 2019, authored by Independent Group of Scientists appointed by the Secretary-General, Global Sustainable Development Report 2019; Shorrocks et al. 2016). The sharpening inequalities indicate that efficiency currently disproportionately benefits those in power: executives and shareholders of global firms or local elites in developing countries.

On the ecological front, efficiency as the organising principle externalises costs related to climate change, pollution, biodiversity loss, and dwindling natural resources. One of the key enablers of efficiency is incomprehensible and weak environmental legislation that allows these externalities to exist, creating possibilities for companies and consumers to avoid paying the costs of environmental damage such as carbon emissions that will be borne by society as a whole, and making the slow response to climate change "the biggest environmental market failure in human history" (Auffhammer 2018). This dynamic is exacerbated by global supply chains, in which a company headquartered in a country with stronger environmental legislation can take advantage of lax environmental laws in supplier countries.

Towards Resilience

The COVID-19 pandemic has made visible the vulnerabilities of current efficiency-based systems, and generated an urgent need to create more resilient societies. Resilience can be defined as a symbiosis of human and natural systems that can support one another to survive and transform through natural and manmade shocks (Walker et al. 2004; Elmqvist et al. 2019). This means that the processes of natural systems are sustained by supportive societal actions, and social systems are sustained by well-functioning natural systems. The *Global Sustainable Development Report* (GSDR 2019) proposes a universal framework for transforming six connected dimensions of societal organisation towards sustainability. In the rest of this essay, we provide a rough idea of how extreme efficiency hampers these six systems and how they could be organised so as to lock-in greater resilience.

1. Economy

Current global trade has been widely extolled as a prime example of efficiency, but its efficiency gains do not materialise at the whole system level. Mainstream business models are based on selling high volumes of easily breakable products and many externalities follow from low-cost sourcing in countries with lax regulations and old technologies. Further, global freight shipping, one of the cornerstones of global trade, comes with an ecological price: its CO_2 emissions would make it the ninth biggest country in the world (EU Edgar database) and its NOx emissions make thousands of people ill annually.

The COVID-19 pandemic has revealed the vulnerability of global supply chains: when one part of the chain is disabled, negative impacts are felt by many (O'Rourke 2014). When China closes down factories that manufacture up to 70% of ingredients of common drugs, or India limits exports of drugs like paracetamol or popular antibiotics, those in need of medicines in the Global South, but also in the US or Europe, suffer the consequences. Further, the efficiency quest has made us believe that labour is an expense that should be minimised. At its extreme, the low-wage trend has meant that employees cannot make a living with their wages, and need social benefits. As a result, the wider economy suffers when taxpayers end up paying employers' costs. In societies with no social benefits available, the low-wagers suffer from unfulfilled basic needs. In a resilient economy firms would focus on long-term productivity. Means to avoid the above negative impacts include curbing the excess concentration of ever larger firms, re-deploying smart trade barriers and reducing the widening wealth gaps that breed social unrest and populism (Edelman Trust Barometer 2020). Curbing the size of firms would leave room for smaller, often innovative, competitor firms and, through firm diversity and genuine competition, build resilience at the system level (Martin 2019). Reducing the dominance of large corporations would pave the way for a resilient economy where other stakeholders could bargain for institutions, which in turn could divide economic benefits more justly (Piketty 2013).

2. Food

The efficiency-driven agricultural system, based on large monoculture farming, commercial fertilisers, chemical pest control, fossil fuels and global logistics, comes with underlying problems of loss of fertile top soils and biodiversity, large-scale use of antibiotics in meat production and the subsequent threat of antibiotic resistance in humans, and the lack of affordable, healthy food. As a result, the number of people suffering from severe food insecurity is about 750 million, and about two billion people lack regular access to nutritious and sufficient food, whereas at the same time, about two billion people suffer from obesity and related illnesses, including thirty-eight million children (FAO, IFAD, UNICEF, WFP and WHO 2020). Resilience can be built instead by localising food production (as also argued by Sandover, this volume), switching to organic farming and agroforestry to provide alternatives to monocultures, increasing the organic matter content in soils, and carbon sequestration through the agroecological practices adapted to local conditions. Ensuring land property rights and other support for the 600–750 million smallholder farmers that are likely to be operating in 2030 will be a key component of a resilient food system (Thornton et al. 2018). The COVID-19 pandemic may have led to 83–132 million more undernourished people in 2020 (FAO, IFAD, UNICEF, WFP and WHO 2020). Acknowledging that modern agriculture and food production cannot escape the realities of ecological food chains is key for preventing the emergence and spread of vector-borne diseases. A shift toward plant-based diets adds resilience by reducing the high demand for land for livestock, the climate impacts of meat production, and the overuse of antibiotics, and also supports innovations against food loss in local production chains by enhancing the viability of local businesses.

3. Energy

The COVID-19 pandemic has demonstrated the weaknesses of the current centralised fossil fuel paradigm. During the first quarters of the pandemic, coal demand fell by 8% and oil demand fell by 5%, leading to serious financial crises for fossil fuel-based energy producers (Global Energy Review 2020). At the same time, the demand for renewable energy continued to grow due to a larger installed capacity and priority dispatch[1] (Global Energy Review 2020). Three persistent structural vulnerabilities were revealed in our fossil fuel-dependent economy and energy systems. First, declining system efficiency is a result of decreases in Energy Return on Investment (EROI) of fossil fuels. This means that, although oil deposits exist, extracting oil is becoming increasingly costly and difficult with larger environmental damage. Second, the rebound effects of improving energy efficiency have decreased emissions per unit, but the absolute amount of emissions continues to increase. Third, indirect energy use, i.e. energy embedded in products and services, continues to grow due to increasing consumption and global trade. Moving from a centralised fossil fuel-based structure toward distributed renewable energy systems will be key to enabling more resilient energy systems. Resilience provided by off-grid technologies and localising energy production and consumption (O'Brien and Hope 2010) will be critical for mitigating the poor infrastructure in large urban centres, extending the grid to rural areas and enhancing just, secure, and affordable energy for all. Furthermore, reducing consumption-based carbon footprints with new sufficiency measures will be important in order to reach climate targets (see Linnanen et al. 2020).

4. Urbanisation

COVID-19 has hit the 4.2 billion people living in cities around the world particularly hard. Dense urban structures have made urban areas hotspots of virus spread. This situation highlights the need to rethink urban structures from a new, more local perspective, embracing resilience over efficiency. Maximising urban efficiency from the viewpoint of infrastructure and economics easily leads to urban environments with fewer green areas, sparse service networks, long commutes and distant food production. Cities that have emphasised human scale in their planning are likely to be more resilient, not only during crises like pandemics but also when confronting disturbances from climate events. Furthermore, diversity in urban structure and flexible use of buildings and open areas are beneficial for cities and citizens in general (Jacobs 1961), because neighbourhoods with high social capital are able to provide a support network and social resilience. Well-functioning,

---

1   Editors' note: wherein the dispatch of energy from renewable generators is prioritised ahead of other generators.

locally connected administrations equipped with up-to-date data and analytical practices are also considered crucial for increasing the resilience of cities and their populations. Planning cities for people goes hand-in-hand with building more sustainable and resilient cities that are also better prepared for future crises.

5. Human Wellbeing and Capabilities

From the perspective of extreme efficiency seeking, the main roles for humans are top-performing professionals, cheap labour and consumers constantly buying new products and services. Individuals who do not meet these standards become framed as 'friction' in an otherwise efficient system. In organisational contexts, performance measurement, with its roots in industrial efficiency, has penetrated to all sectors, including healthcare and education. Each societal actor assesses their actions based on the efficiency and profitability of only their own sector. Maximisation of efficiency in the short run, however, leads to inefficiency in the long term, as well as to a lack of holistic wellbeing. The illusion of efficiency contributes to the crises of our time and risks reducing the capabilities of humanity in total. The COVID-19 pandemic has revealed the lack of resilience also in societies with substantial material wealth. It has widened inequalities that affect wellbeing, e.g. between different categories of labour: some people can work from home, others, often in low-paid jobs and more vulnerable positions based on their socio-economic income levels, cannot. This has resulted in situations where people working in low-paid jobs either lost their income completely or were exposed to the virus. Resilience can be strengthened by supporting the agency and diversity of human capabilities, sustaining cultural practices connected to identities, raising awareness about ecological problems connected to remediating practices, as well as by fostering global belonging and ecological citizenship (Duxbury et al. 2017). Instituting lower pay differentials and a basic income for all would also increase human wellbeing and create space for capability building. A holistic view of health and wellbeing is needed to complement specialised healthcare and contribute to a substantial shift from curative to preventive action and to increase preparedness, so as to improve the resilience of communities, societies and humanity in the face of grand challenges.

6. Global Environmental Commons

Economic growth has been largely enabled through intensive and wide-scale exploitation of resources in terrestrial and ocean ecosystems. Thus, global environmental commons provide 'source material' for the efficiency paradigm and are also where the consequent externalities are most visible. Despite relative efficiency gains, since 1970 global material extraction has more than tripled (Oberle et al. 2019) to fulfil the needs of the growing population and higher economic growth. Species loss, habitat destruction, pollution, the spread of invasive species, and climate change reflect the overexploitation of Earth's resources, which constitutes a threat to human health and wellbeing (Montanarella et al. 2018). By threatening the environmental commons, we are also enabling emerging zoonotic[2] diseases. The COVID-19 pandemic has emphasised how human wellbeing is intimately connected with the wellbeing of the natural environment. Increasing resilience in this system calls for active measures aimed at a putative 'no-net loss' in biodiversity and other environmental commons. Proposals include the conservation of large parts of the Earth (Wilson 2016), and the restoration of certain degraded habitats to fully compensate for the loss and degradation of habitats elsewhere (Moilanen and Kotiaho 2018). These bold conservation objectives, however, conflict with other demands for land use.[3] Thus, even though it is likely that increased resilience in the five other systems will have positive impacts on global environmental commons, resilience requires concerted cross-sectorial action, e.g. tackling the drivers of land-use change. For example, without the above outlined transformation of the food system, the protection of global biodiversity will be in conflict with affordable food provision (Leclère et al. 2020).

---

2 Zoonotic diseases are diseases that "pass from an animal or insect to a human. Some don't make the animal sick but will sicken a human" (see https://www.healthline.com/health/zoonosis).

3 Editors' note: such proposals also require caution since they can act to devalue land-use practices and modes of production by communities who may have sustained biodiversity over the long term.

Conclusions: Towards a Resilience Transformation

The COVID-19 pandemic has revealed the fundamental problems in current efficiency-driven, global socio-economic systems. A way out is to promote radical changes in the six key systems discussed in this chapter (GSDR 2019), so as to foster a transition towards resilience. Mitigating climate change is not simply a case of reducing emissions, but rather requires parallel changes in all of the six global systems discussed here. Despite posing a major threat to humanity, the COVID-19 crisis also paradoxically gives us hope that this kind of change is possible. First, the forced economic slowdown has demonstrated that considerable and rapid changes in emissions and pollution levels to reduce climate impacts are possible, but require considerable alteration and transformation of the current efficiency paradigm to make the impact durable. Second, and more importantly, the reactions to COVID-19 in many countries have shown that it is possible to change behaviours fast when the evidence shows that current paths are unsustainable. This may create new hope and invigorate our belief in the possibility of transformation through evidence-informed decisions. Simply put, the economy is governed not by natural laws, but by routines, conventions, rules, and policy decisions made by human individuals and communities that can be adjusted. This experience has shown the power of the crisis and supports the idea of declaring a climate emergency as a global climate crisis.

The COVID-19 pandemic shines a light on the co-benefits for humanity and nature that can be achieved by a series of interconnected activities aimed at resilience. Moving aspects of production processes closer to where consumption takes place reduces dependency on long supply networks. It would also provide a 'face' to production workers, making extreme forms of labour exploitation more difficult. A transition toward more plant-based diets is not only healthier, but also reduces CO_2 emissions and helps maintain biodiversity, as less space is needed for feeding livestock. Renewable-based distributed energy production creates more jobs and opportunities for income amongst local communities and households who produce wind and solar energy (although see Dunlap's critique of industrial-scale renewable energy, this volume). The resilience transformation could also be called 'the transformation to a globally informed, but more localised economy'.

Removing externalities, which are the main economic drivers of unsustainable development, requires more comprehensive and global environmental governance. Localised economies with globally coherent environmental governance would not harm the economy as a whole, but would rather give more opportunities and hope to those who have been losers in the extreme efficiency paradigm. To raise hope, it is this aspect of the resilience story that we must first start talking about, so that we can then start walking the talk.

References

Alcott, Blake, 'Jevons Paradox', *Ecological Economics*, 54 (2005), 9–21, https://doi.org/10.1016/j.ecolecon.2005.03.020.

Auffhammer, Maximilian, 'Quantifying Economic Damages from Climate Change', *Journal of Economic Perspectives*, 32 (2018), 33–52, https://doi.org/10.1257/jep.32.4.33.

Duxbury, Nancy, Anita Kangas, and Christiaan De Beukelaer, 'Cultural Policies for Sustainable Development: Four Strategic Paths', *International Journal of Cultural Policy*, 23 (2017), 214–30, https://doi.org/10.1080/10286632.2017.1280789.

Edelman Trust Barometer (Edelman.com, 2020), https://www.edelman.com/trustbarometer.

Elmqvist, Thomas, Erik Andersson, Niki Frantzeskaki, Timon McPhearson, Per Olsson, Owen Gaffney, Kazuhiko Takeuchi, and Carl Folke, 'Sustainability and Resilience for Transformation in the Urban Century', *Nature Sustainability*, 2 (2019), 267–73, https://doi.org/10.1038/s41893-019-0250-1.

EU Edgar database, *Top CO2 emitters* (Data.jrc.ec.europa.eu, 2021), https://data.jrc.ec.europa.eu/collection/edgar.

FAO, IFAD, UNICEF, WFP and WHO, *The State of Food Security and Nutrition in the World 2020. Transforming Food Systems for Affordable Healthy Diets* (Rome: FAO, 2020), https://doi.org/10.4060/ca9692en.

Global Energy Review, *The Impacts of the COVID-19 Crisis on Global Energy Demand and CO2 Emissions* (Iea.org, 2020), https://www.iea.org/reports/global-energy-review-2020.

GSDR, Independent Group of Scientists appointed by the Secretary-General, Global Sustainable Development Report 2019: The Future is Now; Science for Achieving Sustainable Development (Sustainabledevelopment.un.org, 2019), https://sustainabledevelopment.un.org/gsdr2019.

Heindl, Peter, and Philipp Kanschik, 'Ecological Sufficiency, Individual Liberties, and Distributive Justice: Implications for Policy Making', *Ecological Economics*, 126 (2016), 42–50, https://doi.org/10.1016/j.ecolecon.2016.03.019.

Jacobs, Jane, *The Death and Life of Great American Cities* (New York: Random House, 1961).

Leclère, David, Michael Obersteiner, and Lucy Barrett, 'Bending the Curve of Terrestrial Biodiversity Needs an Integrated Strategy', *Nature*, 585 (2020), 551–56, https://doi.org/10.1038/s41586-020-2705-y.

Linnanen, Lassi, Tina Nyfors, Tero Heinonen, Heikki Liimatainen, Ari Nissinen, Kristiina Regina, Merja Saarinen, Jyri Seppälä, and Riku Viri, *The Sufficiency Perspective in Climate Policy: How to Recompose Consumption*, Report of Finnish Climate Change Panel (Urn.fi, 2020), http://urn.fi/URN:NBN:fi-fe20201222102703.

Martin, Roger, 'The High Price of Efficiency', *Harvard Business Review*, 97 (2019), 42–55.

Moilanen, Atte, and Janne S. Kotiaho, 'Fifteen Operationally Important Decisions in the Planning of Biodiversity Offsets', *Biological Conservation*, 227 (2018), 112–20, https://doi.org/10.1016/j.biocon.2018.09.002.

Montanarella, Luca, Robert Scholes, and Anastasia Brainich (eds), *The Assessment Report on Land Degradation and Restoration* (Germany: Bonn, 2018), https://doi.org/10.5281/zenodo.3237392.

Oberle, Bruno, Stefan Bringezu, Steve Hatfield-Dodds, Stefanie Hellweg, Heinz Schandl, and Jessica Clement, *Global Resources Outlook 2019: Natural Resources for the Future We Want* (United Nations Environment Programme, 2019), http://hdl.handle.net/20.500.11822/27517.

O'Brien, Geoff and Alex Hope, 'Localism and Energy: Negotiating Approaches to Embedding Resilience in Energy Systems', *Energy Policy*, 38 (2010), 7550–58, https://doi.org/10.1016/j.enpol.2010.03.033.

O'Rourke, Dara, 'The Science of Sustainable Supply Chains', *Science*, 344 (2014), 1124–27, https://doi.org/10.1126/science.1248526.

Piketty, Thomas, 'About Capital in the Twenty-First Century', *American Economic Review*, 105 (2013), 48–53, https://doi.org/10.1257/aer.p20151060.

Shorrocks, Anthony F., James B. Davies, and Rodrigo Lluberas, *Global Wealth Databook 2016* (Credit.suisse.com, 2016), https://www.credit-suisse.com/gwr.

Thornton, Philip, Dhanush Dinesh, Laura Cramer, Ana Maria Loboguerrero, and Bruce Campbell, 'Agriculture in a Changing Climate: Keeping Our Cool in the Face of the Hothouse', *Outlook on Agriculture*, 47 (2018), 283–90, https://doi.org/10.1177/0030727018815332.

Walker, Brian, C. S. Holling, Stephen R. Carpenter, and Ann Kinzig, 'Resilience, Adaptability and Transformability in Social–ecological Systems', *Ecology and Society*, 9 (2004), 5, http://www.ecologyandsociety.org/vol9/iss2/art5/.

Wilson, Edward O., *Half-Earth: Our Planet's Fight for Life* (New York: Liveright, 2016).

3. On Climate Change Ontologies and the Spirit(s) of Oil

Sian Sullivan

The last major UNFCCC COP Agreement—the so-called Paris Agreement of COP21 in 2015—emphasised international cooperation through market-based instruments. International carbon trading was insisted on, so as to (seemingly) allow mitigation, rather than reduction/cessation, of emissions from industrial production. Repeated utterances of the positive impacts of carbon markets in terms of reducing emissions and speeding the transition to a low-carbon economy, however, were also met with equally repetitive and forceful claims that carbon markets have failed. The polarised disagreement between these positions and the numbers supporting them demonstrates that climate management and carbon markets are not merely technical problems that can be fixed by measurement, modelling and technocratic solutions. They are political problems representing highly divergent values and worldviews. This essay asks questions about how anthropogenic climate change is understood, and which responses are promoted as appropriate for this systemic predicament. It argues that ontological dimensions are at play here, arising from different ways of seeing and knowing the world.

https://doi.org/10.11647/OBP.0265.03

The Push and Pull of Climate 'Agreements'[1]

In building up to the 26th United Nations Conference of the Parties (COP26) on the Framework Convention on Climate Change (UNFCCC), it is worth recalling the intense debate, planning and redrafting of texts preceding the so-called Paris Agreement of COP21 in 2015. In the run-up to any UN COP (or 'Summit'), government negotiators engage in multiple redrafts of the deal to be agreed by the Convention deadline.[2] Their every edit is scrutinised by those with varying interests in the exact wording of the deal (Yeo 2015).

Market-based instruments (MBIs) play a key but controversial role in these negotiated texts regarding how climate change mitigation is to be achieved. The International Carbon Action Partnership (ICAP) thus submitted to the 2015 Conference Working Group a call for the Paris Agreement to support the use of market mechanisms to help countries achieve the targets laid down in their Intended Nationally Determined Contributions (INDCs) (ICAP 2015). Other organisations argued instead that carbon pricing, trading and markets fail to do what they are repetitively promised to do.

Many social movements and environmental NGOs campaign vigorously against the 'false solution' of carbon markets. They see market mechanisms as legitimising capitalist structures at the root of fossil fuel production and consumption, as well as of growing global

---

1   This essay develops a blog by the same title first published in November 2015 by the Sheffield Political Economy Research Institute for their series of blogs coinciding with the 2015 Paris UNFCCC COP—see http://speri. dept.shef.ac.uk/2015/11/19/speri-spotlight-on-the-un-climate-summit-part-2/ (longer version at https://the-natural-capital-myth.net/2015/11/19/on-climate-change-ontologies-and-the-spirits-of-oil/).

2   See, for example, texts produced by the UN Ad Hoc Working Group on the Durban Platform For Enhanced Action (ADP), including an eighty-six-page 'Negotiating Text' released on 12 February 2015, followed by a seventy-six-page 'Draft Agreement' released on 24 July, greatly reduced to a twenty-page 'Draft Agreement' by the Conference co-chairs, Dan Reifsnyder from the US and Ahmed Djoghlaf from Algeria, released on 5 October 2015: UNFCCC Negotiating Text (Unfccc. int, 2015), https://unfccc.int/files/bodies/awg/application/pdf/negotiating_text_12022015@2200.pdf; UNFCCC Draft Agreement (Unfccc.int, 2015), https://unfccc.int/resource/docs/2015/adp2/eng/4infnot.pdf; UNFCCC Draft Agreement (Unfccc.int, 2015), https://unfccc.int/resource/docs/2015/adp2/eng/8infnot.pdf.

inequities in wealth concentration,[3] emphasising poor outcomes of carbon pricing in realising deep decarbonisation (Rosenblum et al. 2020). Echoing campaigns at previous COPs, climate justice activists mobilise instead for much more ambitious international collaboration and cooperation, their activities at COP21 framed around setting out the minimal necessities for a liveable planet as "red lines" that must never be crossed (Hudson 2015). For COP26, carbon markets are set to again be a critical dimension of concern for activists seeking to "stop climate chaos" (Stop Climate Chaos Scotland 2020).

These contrary positions—the notion that pricing and trading carbon on markets is essential for reducing climate-forcing carbon emissions *versus* the notion that carbon markets make money for trading parties but fail to reduce carbon emissions—drive the push and pull of international climate negotiations. Polarised disagreement between these positions and the numbers used to support them demonstrates, however, that climate management and carbon markets are *not* only technical problems that can be fixed by measurement, modelling and technocratic solutions. As Hulme, Bigger et al., Durand-Delacre et al., Hannis, and Bracking also clarify in this volume for different dimensions of climate change measurement and management, they are political problems revealing highly divergent values and worldviews.

High Stakes / End Times?

It has become normal in pre-summit moments to assert that the stakes are high. How high they may be is connected with observations of a series of rapidly accelerating changes in socioeconomic and earth system indicators associated with global economic and human population growth since World War II (Steffen et al. 2015). These changes include marked increases in atmospheric methane and carbon dioxide levels, both of which correspond with higher climate temperatures. Methane and carbon dioxide are now at levels that constitute a data outlier whose prediction would have been improbable if simply extrapolating from levels over the previous 800,000 years (IGBP 2015). Given that

---

3   For figures, see 'Global Inequality' (Inequality.org, 2020), https://inequality.org/facts/global-inequality/.

the previous 800,000 years indicate that climate temperature is tightly coupled with levels of both atmospheric methane and carbon dioxide (IGBP 2015 and references therein), it is reasonable to assume that climate temperature levels will rise too. And since temperature is a factor in the geographical presence or absence of species, it is also reasonable to assume that significantly rising temperatures will have significant implications for species, not to mention for human cultural and economic activity. This is why there is a UN Framework Convention on Climate Change, and why people are so concerned about the probable impacts of actual and predicted climate change. As Naomi Klein (2015) asserts, "this changes everything".

Moreover, the connected and recursive feedback loops at play between atmospheric gases, climate temperatures and biocultural materialities suggest that the momentum of changing values is becoming greater in magnitude. These circumstances indicate the sorts of ratcheted up interactions that chaos and complex systems theories predict will generate significant but not necessarily predictable system shifts (Prigogine 1997), implying "a massive, imminent phase transition in human historical experience" (Danowski and Viveiros de Castro 2017: 18). The linking of COVID-19 with habitat changes linked in turn with climate change, might constitute one of these kinds of system shifts (The Lancet 2020). If this is indeed the case, then we are on the cusp of changes which contemporary calculative and forecasting practices may be unable to foretell with any degree of accuracy. The horizon of the future is increasingly murky, giving rise to a sense that we are *Living in the End Times*, as philosopher Slavoj Žižek (2011) has put it.

But crises are opportunities too (Klein 2008). Credit Suisse (2020) proclaims "Climate change—Decarbonizing the economy" to be a "Supertrend" for investment, its relevance underpinned by "the global COVID-19 pandemic". Economists, accountants and financiers tinker with methodologies for designing and embedding calculated and priced units of nature further into economic spreadsheets and capital asset reports, seeking 'solutions' to the impacts of these system changes that simultaneously sustain economic momentum (Asiyanbi 2017; Sullivan 2018). Climate change management and ecological health thus become further enmeshed with an economic machine that is itself an engine of volatility, leaving societal and environmental damage in its wake. Yes, the stakes are high.

What's Ontology Got to Do with It?

To put this differently, a shift in the complex dynamic system we call Earth is being generated by an expansionary economic culture based on particular practices of extraction, measurement, calculation, accounting and accumulation of 'value'. This 'culture' is itself built on recursivity (i.e. positive feedback). Capitalist values and production practices drive the accumulation and concentration of 'surplus' value and monetised assets, such that capital concentrates exponentially (Marx 1974[1867]; Luxemburg 2003[1913]). Notwithstanding the efficiency drive invoked by Halme and colleagues, this volume, the movements of commodity prices demonstrate trending and volatility, rather than unrisky 'market efficiency': they are characterised by an abundance of seemingly improbable or erratic price swings, rather than by a normal distribution around a mean (Mandelbrot and Hudson 2006).

Yet this hegemonic economic 'culture' is conventionally perceived to be efficient, rational, potentially equitable and predictable. In projecting its own image on to beyond-human natures, it misperceives the complex biophysical system within which it is embedded. Seeing only a complicated but predictable and accountable machine, its truth claim is that management may be perfected simply through better measurement and calculation of the carbon and 'natural capital' 'units' of which it is considered made (EU 2014).

Such measurement, however, selectively *determines* what becomes visible to markets, whilst disavowing the recursive and unpredictable nature of the interacting biophysical phenomena exceeding the balance sheets that thereby arise. In acting to consolidate forms of wealth that are amenable to such calculation, they may amplify, rather than reduce, system parameters that enhance volatility (Sullivan and Hannis 2017). Claims to pragmatism and superior expertise framed as beyond ideology (Helm 2015) additionally occlude different knowledges and values, effecting a climate management colonialism that denies the self-determination of cultural perspectives that think—with consistency and coherence—otherwise (Clastres 2010).

Through these multiple collisions of phenomena that are complex, organic and unpredictably emergent with thinking that is complicated, calculative and predictably additive, conditions for improbable catastrophic events are likely to be enhanced rather than reduced. These are, in part, *ontological concerns* arising from different ways of seeing and knowing the world: from different ways of understanding both the nature of nature, and the nature of appropriate forms of use, value and appreciation that humans negotiate with the beyond-human natures with which we co-exist and retain evolutionary kinship. Ontology is the study and naming of the fundamental, assumed, and known nature of reality (Sullivan 2017). It defines what entities exist, into what categories they can be sorted, and by what practices and modes of verification they can be known. Cultural and historical differences and agreements shape ontological perception and understanding, and imply the possibility of diverse, consistent and coherent explanations of causality regarding socioecological change and appropriate responses to this (Burmann 2017; also see Dieckmann, this volume), as considered further below.

On The Spirit(s) of Oil

Let us step for a moment towards the cosmology of Sápara ('Zapara') peoples of Pastaza in the upper Amazon Forest of Ecuador. I learned a little of their shared worldview through meeting, some years ago, Manari Ushigua, formerly Vice-President of the Confederation of all the Indigenous Nationalities of Ecuador (CONAIE) and later President of the Bi-National Sápara Federation of Ecuador and Peru (Ushigua and Tryon 2020).[4] Fewer than 600 Sápara live on land sustaining biological diversity with which Sápara culture, language, and cosmology have long been entangled. Only four individuals, Manari included, now speak the Sápara language,[5] which in 2001 was recognised by the UN's Educational, Scientific and Cultural Organization (UNESCO) as a unique "depository" of intangible cultural heritage and memory of the people and the region.[6]

---

4   Manari travelled to the UK through the support of the Pachamama Alliance (https://www.pachamama.org/) and the School of Movement Medicine (https://www.schoolofmovementmedicine.com/), with whom I was studying dance movement at the time.

5   Naku North, 'The Sapara History and Legend' (Nakunorth.com, 2020), https://nakunorth.com/sapara/.

6   See UNESCO, 'Oral heritage and cultural manifestations of the Zápara people' (Ich.unesco.org, 2020), https://ich.unesco.org/en/RL/oral-heritage-and-cultural-manifestations-of-the-zapara-people-00007.

For several years, Sápara have engaged in intense struggles to retain their land and the integrity of the forest that is their home, in the face of enormous pressure for the extraction of oil from beneath Sápara territory. Sápara legally own their land, and Ecuador has appeared to be a leading light on environmental issues due to its constitutional recognition of the "Rights of Nature" (Republic of Ecuador 2008). Nonetheless, the Ecuadorean government claims rights to below-ground fuels and minerals, meaning that huge areas of the Amazon are cut up into blocks franchised for prospecting—and potentially for extraction—to international oil corporations (as shown in Figure 3).

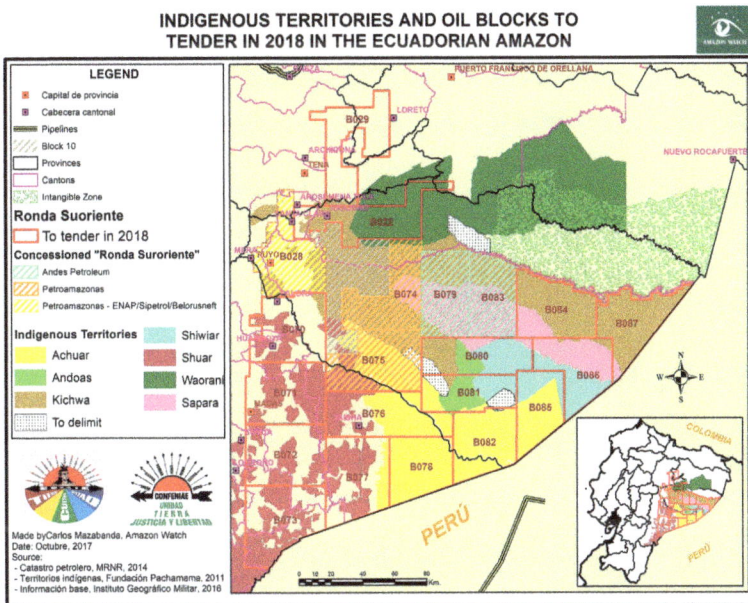

Fig. 3. Indigenous territories and tendered oil blocks in the Ecuadorian Amazon, 2018. ©Amazon Watch, public domain, https://amazonwatch.org/news/2017/1026-amazonian-indigenous-peoples-reject-ecuadors-plans-for-new-oil-tender

In October 2019, sustained resistance by Sàpara to oil extraction from these lands led to the extraordinary granting by Ecuador's Ministry of Energy and Non-Renewable Natural Resources of a *force majeure* request to the company concerned—Andes Petroleum Ltd Ecuador (Amazon Watch 2019). Although succeeding to halt oil extraction for the time being, we can see here how fossil fuel momentum unfurled even in the midst of more than two decades of climate change negotiations intent on managing and reducing carbon emissions. In this case, the normalisation of fossil fuel extractive rights continued, even though the area is considered by ecologists to be amongst the most biodiverse localities on the planet, its sustenance arguably simultaneously entwined with that of Sápara language and knowledge (see, for e.g., Gorenflo et al. 2012).

Sápara ontology, as spoken of by Manari Ushigua, affirms the presence of spirit beings deep in the earth associated with the oil found there. These spirits confer vitality to the oil, also nourishing different spirit beings around five metres below the surface of the soil, which in turn animate the roots of plants that burst through the surface of the soil to provide food and habitat to animals and humans dwelling above the earth's surface. In this spirited understanding of the connected nature of being—in which mineral, plant and animal-human entities are animate and mutually nourishing—extraction of the earth's potent below-ground materials disrupts the lifeforce of the connected entities above ground. This perspective affirms that the zone of life on earth referred to as the 'biosphere' by environmental scientists, is intractably entwined with fluids and minerals found deep in the earth. Above-ground socio-ecological health and diversity is connected with the spirited liveliness of intact below-ground fluids and minerals.

There are echoes of this spirited earth ontology in many other cultural contexts. U'wa of Colombia reportedly understand oil as the blood of a mothering earth, and in the late 1990s threatened collective suicide in protest against the affront of oil exploitation by US-based corporation Occidental Petroleum (Global Nonviolent Action Database 2011). American Indian Movement activist the late John Trudell (2000) describes another potent mineral—uranium—as a spirited "DNA" of the earth, from which industrial mining-refinement processes create a mutated form of power that ultimately is toxic to life.

These perspectives and the distinct, but diverse, 'indigenous paradigm' they invoke suggest that the effects of pulling fuel and minerals out of the earth may be more unpredictable, mysterious and far-reaching than the echoes of an Enlightenment mechanistic worldview are able to register. They give weight to an understanding that the holes puncturing earth through mining processes, coupled with changes in atmospheric composition caused by pumping mined elements into the layer of gases permitting life to thrive on earth, are causing sickness in the living, breathing body of the earth itself.

Compassion in an Apocalyptic Moment?

Of course, there is complexity here too. Manari flew to the UK using the substance whose exploitation threatens his people with extinction. We are all caught within the web of industrial-techno-capitalism in ways that make it impossible to fully shrug off our culpability in systemic planetary changes that many consider are drawing us towards broad spectrum catastrophe (discussed further in Sullivan, Chapter 11 in this volume). And seeking to learn from those living in the recent echo of colonialism's extractive impetus might be construed as one more colonising engagement, this time to capture and extract "indigenous knowledge" (see Dieckmann this volume).

These paradoxes constitute critical challenges for our times. To sit with compassion for our own accountability for the losses now occurring; whilst acting for the possibility of systemic change that prevents these losses. To face what can seem to be the impossibility of reorienting the global compass bearing away from financial profit and economic growth; whilst keeping hope alive for a systemic re-orientation towards equitable socio-ecological relationships in which a diversity of beings and cultures may flourish.

Placing indigenous realities at the heart of UNFCCC negotiations requires taking seriously perspectives and ontologies that view the nature of climate change differently. Perhaps it is for this reason that indigenous concerns have tended to be sidelined in the COPs, even though a widening of the circle of perspectives regarding this critical juncture for humankind is desperately needed to strengthen the legitimacy of these talks that affect us all. To echo Yukon leader Stanley James, commenting on the slow pace of negotiations at COP15 in Copenhagen in 2009: "we need to have the aboriginal people at the table with those government people ... then things will change, I think" (CBC News 2009).

References

Asiyanbi, Adeniyi P., 'Financialisation in the Green Economy: Material Connections, Markets-in-the-making and Foucauldian Organising Actions', *Environment and Planning A*, 50(3) (2017), 531–48, https://doi.org/10.1177/0308518X17708787.

Amazonwatch, 'Indigenous Opposition Forces Andes Petroleum Out of Controversial Rainforest Oil Block' (Amazonwatch.org, 2019), https://amazonwatch.org/news/2019/1106-indigenous-opposition-forces-andes-petroleum-out-of-controversial-rainforest-oil-block.

Burmann, Anders, 'The Political Ontology of Climate Change: Moral Meteorology, Climate Justice, and the Coloniality of Reality in the Bolivian Andes', *Journal of Political Ecology*, 14 (2017), 921–38, https://doi.org/10.2458/v24i1.20974.

CBC News, 'Indigenous Groups Push for Progress at Climate Summit' (Cbc.ca, 2009) https://www.cbc.ca/news/canada/north/indigenous-groups-push-for-progress-at-climate-summit-1.799492.

Clastres, Pierre, *Archaeology of Violence* (New York: Semiotext(e), 2010).

Credit Suisse, 'Supertrends' (Creditsuisse.com, 2020), https://www.credit-suisse.com/microsites/investment-outlook/en/supertrends.html.

Danowski, Déborah, and Eduardo Viveiros de Castro, *The Ends of the World*, trans. by Rodrigo Nunes (Oxford: Polity Press, 2017).

EU, 'The Economics of Nature' (Europa.eu, 2014), https://europa.eu/capacity4dev/articles/economics-nature.

Global Nonviolent Action Database, 'U'wa people block Occidental Petroleum (Colombia), 1995–2001' (Nvdatabase.swarthmore.edu, 2020), https://nvdatabase.swarthmore.edu/content/uwa-people-block-occidental-petroleum-colombia-1995-2001.

Gorenflo, Larry J., Suzanne Romaine, Russell A. Mittermeier, and Kristen Walker-Painemilla, 'Co-occurrence of Linguistic and Biological Diversity in Biodiversity Hotspots and High Biodiversity Wilderness Areas', *Proceedings of the National Academy of Sciences*, 109(21) (2012), 8032–37.

Helm, Dieter, *Natural Capital: Valuing the Planet* (London: Yale University Press, 2015).

Hudson, Drew, 'Red Lines at the Cop21 Conference Closing' (Environmental-action.org, 2015), https://environmental-action.org/blog/d12-red-lines-at-the-cop21-conference-closing/.

ICAP, 'ICAP Calls for a Paris Agreement Supporting Market Mechanisms' (Icapcarbonaction.com, 2015), https://icapcarbonaction.com/en/icap-unfccc-adp-submission-2015.

IGBP, 'Anthropocene' (Igbp.net, 2015), http://www.igbp.net/globalchange/ant hropocene.4.1b8ae20512db692f2a680009238.html.

Klein, Naomi, *The Shock Doctrine: The Rise of Disaster Capitalism* (London: Penguin, 2008).

Klein, Naomi, *This Changes Everything: Capitalism vs. the Climate* (London: Penguin, 2015).

Luxemburg, Rosa, *The Accumulation of Capital* (London: Routledge, 2003[1913]).

Mandelbrot, Benoit and Richard L Hudson, *The Misbehavior of Markets: A Fractal View of Financial Turbulence* (New York: Basic Books, 2006).

Marx, Karl, *Capital, Vol. 1*, ed. by Frederick Engels, trans. by Samuel Moore and Edward Aveling (London: Lawrence and Wishart, 1974[1867]).

Prigogine, Ilya, *The End of Certainty* (New York: The Free Press, 1997).

Republic of Ecuador 2008 'Constitution of the Republic of Ecuador' (Pdba. georgetown.edu, 2020), https://pdba.georgetown.edu/Constitutions/Ecuador/english08.html.

Rosenbloom, Daniel, Jochen Markard, Frank W. Geels, and Lea Fuenfschilling, 'Opinion: Why Carbon Pricing is Not Sufficient to Mitigate Climate Change—and How "Sustainability Transition Policy" Can Help', *PNAS*, 117(16) (2020), 8664–68, https://doi.org/10.1073/pnas.2004093117.

Steffen, Will, Wendy Broadgate, Lisa Deutsch, Owen Gaffney, and Cornelia Ludwig, 'The Trajectory of the Anthropocene: The Great Acceleration', *The Anthropocene Review*, 2(1) (2015), 81–98, https://doi. org/10.1177/2053019614564785.

Stop Climate Chaos Scotland, *Delivering Climate Justice at COP26 in Glasgow* (Stopclimatechaos.scot, 2020), https://www.stopclimatechaos.scot/wp-content/uploads/2020/10/Delivering-climate-justice-at-COP26.pdf.

Sullivan, Sian, 'What's Ontology Got to Do with it? On Nature and Knowledge in a Political Ecology of "The Green Economy"', *Journal of Political Ecology*, 24 (2017), 217–42, https://doi.org/10.2458/v24i1.20802.

Sullivan, Sian, 'Making Nature Investable: From Legibility to Leverageability in Fabricating "Nature" as "Natural Capital"', *Science and Technology Studies*, 31(3) (2018), 47–76, https://doi.org/10.23987/sts.58040.

Sullivan, Sian and Mike Hannis, '"Mathematics Maybe, but Not Money": On Balance Sheets, Numbers and Nature in Ecological Accounting', *Accounting, Auditing and Accountability Journal*, 30(7) (2017), 1459–80, https://doi. org/10.1108/AAAJ-07-2017-3010.

The Lancet, 'Climate and COVID-19: Converging Crises', *The Lancet*, 397 (2020), 71, https://doi.org/10.1016/S0140-6736(20)32579-4.

Trudell, John, 'They're Mining Us', *Descendants Now Ancestors* (ASITIS Productions, 2000).

Ushigua, Manari and Zoë Tryon, 'Of the Forest', trans. by Nick Caistor (Granta.com, 2020), https://granta.com/of-the-forest/.

Yeo, Yeo, 'New UN Climate Deal Text: What's In, What's Out' (Carbonbrief.org, 2015), https://www.carbonbrief.org/new-un-climate-deal-text-whats-in-whats-out.

Žižek, Slavoj, *Living in the End Times* (London: Verso, 2011).

II

WHAT COUNTS?

4. Why Net Zero Policies Do More Harm than Good

James G. Dyke, Wolfgang Knorr and Robert Watson

Although well-intentioned, net zero policies have licensed a reckless 'burn now, pay later' approach, in which continuing with climate impacts is justified via promises of future technological salvation. Net zero thinking continues a three-decades-long process of mitigation delay in which academia has at times played an underappreciated role.

Introduction

Sometimes realisation comes in a blinding flash. Blurred outlines snap into shape and suddenly it all makes sense. Underneath such revelations is typically a much slower-dawning process. Doubts at the back of the mind grow. The sense of confusion that things cannot be made to fit together increases until something clicks. Or perhaps snaps. Collectively we three authors of this article must have spent more than eighty years thinking about climate change. Why has it taken us so long to speak out about the obvious dangers of the concept of net zero? In our defence, the premise of net zero is deceptively simple—and we admit that it deceived us.

The threats of climate change are the direct result of there being too much carbon dioxide in the atmosphere. So it follows that we must stop emitting more and even remove some of it. This idea is central to the world's current plan to avoid catastrophe. In fact, there are many suggestions as to how to actually do this, from mass tree planting, to high-tech direct air capture devices that suck out carbon

https://doi.org/10.11647/OBP.0265.04

dioxide from the air (Hanna et al. 2021; also see Lankford's chapter, this volume). The current consensus is that if we deploy these and other so-called 'carbon dioxide removal' techniques at the same time as reducing our burning of fossil fuels, we can more rapidly halt global warming. Hopefully around the middle of this century we will achieve 'net zero'.

Net zero is the point at which any residual emissions of greenhouse gases are balanced by technologies removing them from the atmosphere. This is a great idea, in principle. For example, it is currently hard to see how all emissions from agriculture will be zeroed out in time. Consequently, there will need to be some drawdown of carbon dioxide in order to offset such emissions. Unfortunately, in practice it helps perpetuate a belief in technological salvation and diminishes the sense of urgency surrounding the need to curb emissions now.

We have now arrived at the painful realisation that the idea of net zero has licensed a recklessly cavalier 'burn now, pay later' approach which has seen carbon emissions continue to soar. It has also hastened the destruction of the natural world by increasing deforestation today, and greatly increases the risk of further devastation in the future. In fact, already in 2008, when the G8 were discussing a target of 50% reduction by 2050, one of us [Knorr] co-authored a paper which pointed out that in order to stabilise the climate, net zero will be a necessity in the long term and remaining emissions would have to be balanced out by a residual "artificial sink" (House et al. 2008). In the IPCC's projections, this 'artificial sink' grew out of proportions creating a fantasy world of planetary-scale carbon removal. The shocking revelation is that by bringing the need for net zero on to the table, we have also given licence for its abuse.

To understand how this has happened, how humanity has gambled its civilisation on no more than promises of future solutions, we must return to the late 1980s, when climate change awareness broke on to the international stage.

Steps towards Net Zero

On 22 June 1988, James Hansen was the administrator of NASA's Goddard Institute for Space Studies, a prestigious appointment but someone largely unknown outside of academia. By the afternoon of the 23 June he was well on the way to becoming the world's most famous climate scientist. This was as a direct result of his testimony to the US congress, when he forensically presented the evidence that the Earth's climate was warming and that humans were the primary cause: "[t]he greenhouse effect has been detected, and it is changing our climate now" (United States Congress Senate Committee on Energy and Natural Resources 1988).[1]

If we had acted on Hansen's testimony at the time, we would have been able to decarbonise our societies at a rate of around 2% a year in order to give us about a two-in-three chance of limiting warming to no more than 1.5°C. It would have been a huge challenge, but the main task at that time would have been to simply stop the accelerating use of fossil fuels while fairly sharing out future emissions. This would have required increasing the efficiency of fossil fuel use in transportation, buildings, and industry.

Four years later, there were glimmers of hope that this would be possible. During the 1992 Earth Summit in Rio, all nations agreed to stabilise concentrations of greenhouse gases to ensure that they did not produce dangerous interference with the climate (Grubb et al 2019). The 1997 Kyoto Summit attempted to start to put that goal into practice (UNFCCC 1998). But as the years passed, the initial task of keeping us safe became increasingly harder, given that climate forcing was increasing due to increasing carbon emissions from fossil fuel use along with increased emissions of methane, nitrous oxide, and fluorinated chemicals.

It was around that time that the first computer models linking greenhouse gas emissions to impacts on different sectors of the economy were developed. These hybrid climate-economic models are known as Integrated Assessment Models (Gambhir et al. 2019). They allowed modellers to link economic activity to the climate by, for example, exploring how changes in investments and technology could lead to changes in greenhouse gas emissions. They seemed like a miracle: you could try out policies on a computer screen before implementing them, saving humanity costly experimentation. They rapidly emerged to become key guidance for climate policy, a primacy they maintain to this

---

1   Editors' note: see https://www.sealevel.info/1988_Hansen_Senate_Testimony.html.

day. Unfortunately, they also removed the need for deep critical thinking. Such models represent society as a web of idealised, emotionless buyers and sellers and thus ignore complex social and political realities, or even the impacts of climate change itself. Their implicit promise is that market-based approaches will always work. This meant that discussions about policies were limited to those most convenient to politicians: incremental changes to legislation and taxes.

Around the time they were first developed, efforts were being made to secure US action on the climate by allowing it to count carbon sinks of the country's forests (Dessai 2001). The US argued that if it managed its forests well, it would be able to store a large amount of carbon in trees and soil which should be subtracted from its obligations to limit the burning of coal, oil and gas. In the end, the US largely got its way. Ironically, the concessions were all in vain, since the US Senate never ratified the agreement. Postulating a future with more trees could in effect offset the burning of coal, oil and gas now. As models could easily churn out numbers that saw atmospheric carbon dioxide go as low as one wanted, ever more sophisticated scenarios could be explored which reduced the perceived urgency to reduce fossil fuel use. By including carbon sinks in climate-economic models, a Pandora's box had been opened. It is here we find the genesis of today's net zero policies.

That said, most attention in the mid-1990s was focused on increasing energy efficiency and energy switching (such as the UK's move from coal to gas) and the potential of nuclear energy to deliver large amounts of seemingly carbon-free electricity. The hope was that such innovations would quickly reverse increases in fossil fuel emissions.

But by around the turn of the new millennium it was clear that such hopes were unfounded. Given their core assumption of incremental change, it was becoming more and more difficult for economic-climate models to find viable pathways to avoid dangerous climate change. In response, the models began to include more and more examples of carbon capture and storage, a technology that could remove the carbon dioxide from coal-fired power stations and then store the captured carbon deep underground indefinitely. This had been shown to be possible in principle: compressed carbon dioxide had been separated from fossil gas and then injected underground in a number of projects since the 1970s. These Enhanced Oil Recovery schemes were designed to force gases into oil wells in order to push oil towards drilling rigs and so allow more to be recovered—oil that would later be burnt, releasing even more carbon dioxide into the atmosphere.

Carbon capture and storage offered the twist that instead of using the carbon dioxide to extract more oil, the gas would be left underground and removed from the atmosphere. This promised breakthrough technology would allow climate-friendly coal and so the continued use of this fossil fuel. But long before the world would witness any such schemes, the hypothetical process had been included in climate-economic models. In the end, the mere prospect of carbon capture and storage gave policy makers a way out of making the much-needed immediate cuts to greenhouse gas emissions.

The Rise of Net Zero

When the international climate change community convened in Copenhagen in 2009, however, it was clear that carbon capture and storage was not going to be sufficient, for two reasons. First, it still did not exist. There were no carbon capture and storage facilities in operation on any coal fired power station and no prospect the technology was going to have any impact on rising emissions from increased coal use in the foreseeable future. The biggest barrier to implementation was essentially cost. The motivation to burn vast amounts of coal is to generate relatively cheap electricity. Retrofitting carbon scrubbers on existing power stations, building the infrastructure to pipe captured carbon, and developing suitable geological storage sites required huge sums of money. Consequently, the only application of carbon capture in actual operation then—and now—is to use the trapped gas in enhanced oil recovery schemes. Beyond a single demonstrator, there has never been any capture of carbon dioxide from a coal fired power station chimney with that captured carbon then being stored underground (Power Technology 2021).

Just as important, by 2009 it was becoming increasingly clear that while emissions reductions were both technically and economically feasible, there remained a serious lack of political action. The amount of carbon dioxide being pumped into the air each year meant humanity was rapidly running out of time. With hopes for a solution to the climate crisis fading again, another magic bullet was required. A technology was needed not only to slow down the increasing concentrations of carbon dioxide in the atmosphere, but actually reverse it. In response, the climate-economic modelling community—already able to include plant-based carbon sinks and geological carbon storage in their models—increasingly adopted the "solution" of combining the two.

So it was that Bioenergy Carbon Capture and Storage, or BECCS, rapidly emerged as the new saviour technology (Hickman 2016). By burning 'replaceable' biomass such as wood, crops, and agricultural waste instead of coal in power stations, and then capturing the carbon dioxide from the power station chimney and storing it underground, BECCS could produce electricity at the same time as removing carbon dioxide from the atmosphere. That is because as biomass such as trees grow, they suck in carbon dioxide from the atmosphere. By planting trees and other bioenergy crops and storing carbon dioxide released when they are burnt, more carbon could be removed from the atmosphere. With this new solution in hand the international community regrouped from repeated failures to mount another attempt at reining in our dangerous interference with the climate. The scene was set for the crucial 2015 climate conference in Paris.

A Parisian False Dawn

As its general secretary brought the twenty-first United Nations conference on climate change to an end, a great roar issued from the crowd. People leaped to their feet, strangers embraced, tears welled up in eyes bloodshot from lack of sleep. The emotions on display on 13 December 2015 were not just for the cameras. After weeks of gruelling high-level negotiations in Paris a breakthrough had finally been achieved.

Against all expectations, after decades of false starts and failures, the international community had finally agreed to do what it took to limit global warming to well below 2°C, preferably to 1.5°C, compared to pre-industrial levels. The Paris Agreement was a stunning victory for those most at risk from climate change. Rich industrialised nations will be increasingly impacted as global temperatures rise. But it is the low-lying island states such as the Maldives and the Marshall Islands that are at imminent existential risk. As a later UN special report made clear, if the Paris Agreement was unable to limit global warming to 1.5°C, the number of lives lost to more intense storms, fires, heatwaves, famines and floods would significantly increase (IPCC 2018) (on the Paris agreement, also see Hannis, this volume).

But dig a little deeper and you could find another emotion lurking within delegates on 13 December. Doubt. We struggle to name any climate scientist who at that time thought the Paris Agreement was feasible. We have since been told personally by some scientists that the Paris Agreement was "of course important for climate justice but unworkable" and "a complete shock, no one thought limiting to 1.5°C was possible".[2] Rather than being able to limit warming to 1.5°C, a senior academic involved in the IPCC concluded we were heading beyond 3°C by the end of this century. Relying on untested carbon dioxide removal mechanisms to achieve the Paris targets when we have the technologies to transition away from fossil fuels today is plain wrong and foolhardy. Instead of confronting our doubts, we scientists decided to construct ever more elaborate fantasy worlds in which we would be safe. The price to pay for our cowardice: having to keep our mouths shut about the ever-growing absurdity of the required planetary-scale carbon dioxide removal.

Taking centre stage was BECCS because at the time this was the only way climate-economic models could find scenarios that would be consistent with the Paris Agreement. Rather than stabilise, global emissions of carbon dioxide had increased some 60% since 1992. Alas, BECCS, just like all the previous solutions, was too good to be true. Across the scenarios produced by the Intergovernmental Panel on Climate Change (IPCC) with a 50% or better chance of limiting temperature increase to 1.5°C, BECCS would need to remove billions of tonnes of carbon dioxide each year. BECCS at this scale would require massive planting schemes for trees and bioenergy crops.

The Earth certainly needs more trees. Humanity has cut down some three trillion since we first started farming some 13,000 years ago. But rather than allow ecosystems to recover from human impacts and forests to regrow, BECCS generally refers to dedicated industrial-scale monoculture plantations regularly harvested for bioenergy, rather than carbon stored away in forest trunks, roots and soils.[3] Currently, the two most efficient biofuels are sugarcane for bioethanol and palm oil for

---

2   Personal communications.
3   Editors' note: whose conservation and financing is envisaged through the UN REDD+ programme (Reducing Emissions from Deforestation and Forest Degradation in Developing Countries), see https://redd.unfccc.int/.

biodiesel—both grown in the tropics (Chiriboga et al. 2020). Endless rows of such fast growing monoculture trees or other bioenergy crops harvested at frequent intervals devastate biodiversity.

It has been estimated that BECCS could demand an area of land approaching twice the size of India (Furman et al. 2020). How will that be achieved at the same time as feeding eight to ten billion people around the middle of the century or without destroying native vegetation and biodiversity? Large-scale monoculture tree plantations can adversely impact water availability for agriculture as well as drinking. Increasing forest cover in higher latitudes can have an overall warming effect because replacing grassland or fields with forests means the land surface becomes darker (Mykleby et al. 2017). This darker land absorbs more energy from the Sun and so temperatures rise. Focusing on developing vast plantations in poorer tropical nations comes with real risks of people being driven off their lands. The massive amount of offsetting needed for most net zero scenarios with the aim of staying within safe climate limits cannot be met by leaving nature alone. It demands fast growing, mostly alien species that are cut down often and regularly, thereby releasing carbon. We are already seeing the beginning of this in European forests. The consequences of net zero can look almost as scary as those of climate warming. As these impacts are becoming better understood, the sense of optimism around BECCS has diminished.

Pipe Dreams

Given the dawning realisation of how difficult Paris would be in the light of ever rising emissions and the limited potential of BECCS, a new buzzword emerged in policy circles: the "overshoot scenario" (Ricke et al 2017). Temperatures would be allowed to go beyond 1.5°C in the near term, but then be brought down with a range of carbon dioxide removal by the end of the century. This means that net zero actually means 'carbon negative'. Within a few decades, we will need to transform the global economy from one that currently pumps out forty billion tons of carbon dioxide into the atmosphere each year, to one that produces a net removal of tens of billions. Mass tree planting, for bioenergy or as an attempt at offsetting, had been the latest attempt to stall cuts in fossil fuel use. But the ever-increasing need for carbon removal was calling for more. This is why the idea of direct air capture, now being touted by some as the most promising technology out there, has taken hold. It is generally more benign to ecosystems because it requires significantly less land to operate than BECCS, including the land needed to power them using wind or solar panels.

Unfortunately, it is widely believed that because of its exorbitant costs and energy demand (Lebling et al. 2021), direct air capture—if it ever becomes feasible to be deployed at scale—will not be able to compete with BECCS with its voracious appetite for prime agricultural land (Hanssen et al. 2020).

It should now be getting clear where the journey is heading. As the mirage of each magical technical solution disappears, another equally unworkable alternative pops up to take its place. The next is already on the horizon—and it is even more ghastly. Once we realise net zero will not happen in time, or even at all, geoengineering—the deliberate and large-scale intervention in the Earth's climate system—will probably be invoked as the solution to limit temperature increases. One of the most researched geoengineering ideas is solar radiation management—the injection of millions of tons of sulphuric acid into the stratosphere that will reflect some of the Sun's energy away from the Earth (Reynolds 2019). It is a wild idea, but some academics and politicians are deadly serious about it, despite its significant risks. The US National Academies of Sciences, for example, has recommended allocating up to US$200 million over the next five years to explore how geoengineering could be deployed and regulated. Funding and research in this area is sure to significantly increase.

It is astonishing how the continual absence of any credible carbon removal technology never seems to affect net zero policies. Whatever is thrown at it, net zero carries on without a dent in the fender. The argument appears to be that net zero technologies will work because they have to work. But beyond fine words and glossy brochures there is nothing there. The emperor has no clothes.

Difficult Truths

In principle there is nothing wrong or dangerous about carbon dioxide removal proposals. In fact developing ways of reducing concentrations of carbon dioxide can feel tremendously exciting. You are using science and engineering to save humanity from disaster. What you are doing is important. There is also the realisation that carbon removal will be needed to mop up some of the emissions from sectors such as aviation and cement production. So there will be some small role for a number of different carbon dioxide removal approaches. The problems come when it is assumed that these can be deployed at a vast scale. This effectively serves as a blank cheque for the continued burning of fossil fuels and the acceleration of habitat destruction.

Carbon reduction technologies and geoengineering should be seen as a sort of ejector seat that could propel humanity away from rapid and catastrophic environmental change. Just like an ejector seat in a jet aircraft, it should only be used as the very last resort. But policymakers and businesses appear to be entirely serious about deploying highly speculative technologies as a way to land our civilisation at a sustainable destination when these are no more than fairytales. The only sure way to keep humanity safe is the immediate and sustained radical cuts to greenhouse gas emissions in a socially and economically just way.

Academics typically see themselves as serving society. Those working at the climate science and policy interface desperately wrestle with an increasingly difficult problem. Similarly, those that champion net zero as a way of breaking through the barriers holding back effective action on the climate also work with the very best of intentions. This was certainly the motivation of a key group of international academics and activists that can be seen as one of the important centres for the emergence of the net zero concept (Darby 2019). This important work was designed around ways to accelerate actual mitigation that would be required in order to limit warming to well below 2°C. The tragedy is that their collective efforts were never able to mount an effective challenge to a climate policy process that would only allow a narrow range of scenarios to be explored.

Most scientists feel distinctly uncomfortable stepping over the invisible line that separates their day job from wider social and political concerns. There are genuine fears that being seen as advocates for or against particular issues could threaten their perceived independence. Scientists inhabit a largely trusted profession. Trust is very hard to build and easy to destroy.

But there is another invisible line, the one that separates academic integrity and self-censorship. As scientists, we are taught to be sceptical, to subject hypotheses to rigorous tests and interrogation. But when it comes to perhaps the greatest challenge humanity faces, we often show a dangerous lack of critical analysis. In private, scientists express significant scepticism about the Paris Agreement, BECCS, offsetting, geoengineering and net zero. Apart from some notable exceptions, in public we quietly go about our work, apply for funding, publish papers and teach (Anderson 2015). The path to disastrous climate change is paved with feasibility studies and impact assessments. Rather than acknowledge the seriousness of our situation, we instead continue to participate in the fantasy of net zero.[4] What will we do when reality bites? What will we say to our friends and loved ones about our failure to speak out now?

The youth of today and future generation will look back in horror that our generation gambled with catastrophic changes in climate and biodiversity for the sake of cheap fossil fuel energy when cost effective and socially acceptable alternatives were available. We have the knowledge needed to act. The most recent IPCC and IPBES assessments clearly show we are failing to meet any of the agreed targets for limiting climate change or loss of biodiversity.

The time has come to voice our fears and be honest with wider society. Current net zero policies will not keep warming to within 1.5°C because they were never intended to. They were and still are driven by a need to protect business as usual for as long as possible, not the climate. If we want to keep people safe then large and sustained cuts to carbon emissions need to happen now. That is the very simple acid test that must be applied to all climate policies and it needs to be solidly on the negotiating table at COP26. The time for wishful thinking is over.

---

4   Editors' note: It is also alarming to observe the proliferation of this 'netting' fantasy into other areas of environmental management. In conceiving of so-called 'natural capital' in aggregate, for example (cf. Helm 2015), a fairytale can be sustained in which biodiversity will gain from its measurable harm in the course of development, so as to produce a 'no net loss' or 'net gain' in biodiversity 'units', even though losses have occurred (for a critical engagement with aggregate thinking in environmental governance, see Sullivan 2017).

References

Chiriboga, Gonzalo, Andrés De La Rosa, Camila Molina, Stefany Velarde, and Ghem Carvajal C., 'Energy return on investment (EROI) and life cycle analysis (LCA) of biofuels in Ecuador', *Heliyon*, 6(6) (2020), https://doi.org/10.1016/j.heliyon.2020.e04213.

Darby, Megan, 'Net zero: the story of the target that will shape our future' (Climate Change News, 2019), https://www.climatechangenews.com/2019/09/16/net-zero-story-target-will-shape-future.

Dessai, Suraje, 'Tyndall Working Paper: The climate regime from The Hague to Marrakech: Saving or sinking the Kyoto Protocol?' (2001), https://web.archive.org/web/20121031094826/http:/www.tyndall.ac.uk/content/climate-regime-hague-marrakech-saving-or-sinking-kyoto-protocol.

Fajardy, Mathilde, Alexandre Koeberle, Niall MacDowwell, and Andrea Fantuzz, 'BECCS deployment: a reality check. Grantham Institute Briefing Paper No. 28' (2018), https://www.imperial.ac.uk/media/imperial-college/grantham-institute/public/publications/briefing-papers/BECCS-deployment---a-reality-check.pdf.

Friedlingstein, Pierre, et al., 'Global Carbon Budget 2020', *Earth System Science Data*, 12(4) (2020) 3269–3340, https://doi.org/10.5194/essd-12-3269-2020.

Fuhrman, Jay, Haewon McJeon, Pralit Patel, Scott C. Doney, William M. Shobe, and Andres F. Clarens, 'Food–energy–water implications of negative emissions technologies in a + 1.5 C future', *Nature Climate Change*, 10(10) (2020) https://doi.org/10.1038/s41558-020-0876-z.

Gambhir, Ajay, Isabela Butnar, Pei-Hao Li, Pete Smith, and Neil Strachan, 'A review of criticisms of integrated assessment models and proposed approaches to address these, through the lens of BECCS', *Energies*, 12(9) (2019) https://doi.org/10.3390/en12091747.

Grubb, Michael, Matthias Koch, Koy Thomson, Abby Munson, and Francis Sullivan, *The Earth Summit Agreements: A Guide and Assessment: An Analysis of the Rio'92 UN Conference on Environment and Development* (Vol. 9) (Abingdon: Routledge, 2019), https://doi.org/10.4324/9780429273964-6.

Hanna, Ryan, Ahmed Abdulla, Yangyang Xu, and David G. Victor, 'Emergency deployment of direct air capture as a response to the climate crisis', *Nature Communications*, 12(1) (2021), https://doi.org/10.1038/s41467-020-20437-0.

Hanssen, S. V., V. Daioglou, Z. J. N. Steinmann, J. C. Doelman, D. P. Van Vuuren, and M. A. J. Huijbregts, 'The climate change mitigation potential of bioenergy with carbon capture and storage', *Nature Climate Change*, 10(11) (2020), https://doi.org/10.1038/s41558-020-0885-y.

Helm, Dieter, *Natural Capital: Valuing the Planet* (London: Yale University Press, 2015).

Hickman, Leo, 'Carbon Brief. Timeline: How BECCS became climate change's 'saviour' technology' (Carbonbrief.org, 2016), https://www.carbonbrief. org/beccs-the-story-of-climate-changes-saviour-technology.

House, Joanna I., Chris Huntingford, Wolfgang Knorr, Sarah E. Cornell, Peter M. Cox, Glen R. Harris, Chris D. Jones, Jason A. Lowe, and I. Colin Prentice, 'What do recent advances in quantifying climate and carbon cycle uncertainties mean for climate policy?', *Environmental Research Letters*, 3 (2008), https://doi.org/10.1088/1748-9326/3/4/044002.

Lebling, Katie, Noah McQueen, Max Pisciotta, and Jennifer Wilcox, 'Direct Air Capture: Resource Considerations and Costs for Carbon Removal' (World Resources Institute, 2021), https://www.wri.org/insights/ direct-air-capture-resource-considerations-and-costs-carbon-removal.

Masson-Delmotte, V. P., Zhai, H.-O. Pörtner, D. Roberts, J. Skea, P. R. Shukla, A. Pirani, W. Moufouma-Okia, C. Péan, R. Pidcock, S. Connors, J. B. R. Matthews, Y. Chen, X. Zhou, M. I. Gomis, E. Lonnoy, T. Maycock, M. Tignor, and T. Waterfield (eds), 'IPCC Summary for Policymakers', in *Global Warming of 1.5°C. An IPCC Special Report on the impacts of global warming of 1.5°C above pre-industrial levels and related global greenhouse gas emission pathways, in the context of strengthening the global response to the threat of climate change, sustainable development, and efforts to eradicate poverty* (IPPC, 2018), https://www.ipcc.ch/ site/assets/uploads/sites/2/2019/05/SR15_SPM_version_report_LR.pdf.

Mykleby, P. M., P. K. Snyder, and T. E. Twine, 'Quantifying the trade-off between carbon sequestration and albedo in midlatitude and high-latitude North American forests', *Geophysical Research Letters*, 44 (2017), https://doi. org/10.1002/2016GL071459.

Power Technology, 'SaskPower Boundary Dam and Integrated CCS' (Power-technology.com, 2021), https://www.power-technology.com/projects/ sask-power-boundary-dam/.

Reynolds, Jesse L., 'Solar geoengineering to reduce climate change: a review of governance proposals', *Proceedings of the Royal Society A*, 475(2229) (2019) https://doi.org/10.1098/rspa.2019.0255.

Ricke, K. L., R. J. Millar, and D. G. MacMartin, 'Constraints on global temperature target overshoot', *Scientific Reports*, 7(1) (2017) https://doi.org/10.1038/ s41598-017-14503-9.

Sullivan, Sian, 'On "natural capital", "fairy-tales" and ideology', *Development and Change*, 48(2) (2017), 397–423.

UNFCCC, 'Kyoto Protocol to the United Nations Framework Convention on Climate Change' (1988), https://doi.org/10.1017/9781316577226.067.

United States Congress, 'Greenhouse effect and global climate change: hearings before the Committee on Energy and Natural Resources, United States Senate, One Hundredth Congress, first session 1988. Senate. Committee on Energy and Natural Resources. U.S. G.P.O.' (1988), https://pulitzercenter. org/sites/default/files/june_23_1988_senate_hearing_1.pdf.

5. The Carbon Bootprint of the US Military and Prospects for a Safer Climate

Patrick Bigger, Cara Kennelly,
Oliver Belcher and Ben Neimark

The United States military is the largest institutional consumer of fossil fuels in the world, but until recently accurate data on its fuel consumption were not widely available. Using Freedom of Information Act requests, we compiled data on how much fuel the US military consumes and calculated its 'carbon bootprint.' We explain how the US military's expansive and coupled global logistical networks, hardware, and interventionist foreign policy paradigms help to 'lock-in' future military emissions. Even though they are well-intentioned, calls to 'green' the military are insufficient to rein in military emissions. Instead, the scope of the US military must be dramatically scaled back as part of any serious initiative to maintain a safer climate.

The US Military's Carbon Bootprint

The United States military's carbon bootprint is enormous. Like corporate supply chains, it relies on an extensive global network of container ships, pipelines, trucks, and cargo planes to supply its operations with everything from bombs to hydrocarbon fuels to humanitarian aid. This is no coincidence: historically, many of the parts of complex global logistics networks were developed by the US military (Cowen 2014), including the containerisation of freight (Levinson 2016) and even online shopping portals (Fryar 2012). We have traced these global logistical networks and conducted multiple Freedom of Information Act (FOIA) requests on recent US military fuel purchases to understand the extent and intensity of the climate impacts from fossil fuels by sprawling US military operations.

Calculations of greenhouse gas emissions usually focus on civilian energy and fuel use. Recent work also shows that the US military is one of the largest institutional polluters in history, consuming more liquid fuels and emitting more climate-changing gases than most medium-sized countries (Belcher et al. 2019; Crawford 2019).[1] If the US military were a country, its fuel usage alone would make it the forty-seventh largest emitter of greenhouse gases in the world, sitting between Peru and Portugal. In 2017, the US military bought about 269,230 barrels of oil a day and emitted more than 25,000 kilotons of carbon dioxide. It effectively takes the capacity of one medium-sized refinery operating at full tilt to keep up with military fuel demand. These US military fuels are sourced from, and consumed at, thousands of sites around the world: from Hampton Roads, VA, the largest naval installation in the world, now threatened by rising sea levels, to remote forward operating bases throughout Afghanistan in support of the nearly twenty-year-old war there.

Indeed, the US military operates more than 800 bases around the world through its 'lily-pad' network that renders all the globe a potential theatre of war. These bases all house energy-hungry equipment. Regarding specific branches, in 2017 alone the US Air Force purchased US\$4.9 billion worth of fuel, and the Navy US\$2.8 billion, followed by the Army at US\$947 million and the Marines at US\$36 million. These figures reflect the overwhelming amount of jet fuel (JP-8) purchased and consumed by both the Air Force and the Navy, which is both the highest among all fuel types in total volume burned, and amongst the most climate damaging in terms of emissions, since nitrogen oxide (NOx)

---

1   See too the important work by Scientists for Global Responsibility, who have tracked environmental effects of militaries more broadly, especially Stuart Parkison, 'The Carbon Boot-print of the Military' (Sgr.org, 2020), https://www.sgr.org.uk/resources/carbon-boot-print-military.

gases contained in the fuel have greater radiative forcing potential when combusted higher in the atmosphere (Fahey et al. 2016).

It is no coincidence that quantitative assessments of US military emissions tend to be absent in climate change studies, although there is a robust and growing literature on various intersections of militarism and global change (Dalby 2020), itself building on decades of scholarship on the environmental impacts of military intervention, training, and discourse more broadly (see Westing 2008). The absence of military emissions totals stems, in part, from the difficulty of accessing consistent data from the Pentagon and across US government departments. This difficulty is arguably an intended consequence of specific policy positions, given that the US insisted on an exemption for reporting military emissions in the 1997 Kyoto Protocol (Nelson 2015). This loophole was closed by the Paris Agreement of 2015, but reopened when the Republican Trump administration withdrew from the accord in 2017. Although Biden has enlisted the US military to focus on climate change as a recurring threat to US national security, we have seen very little movement in terms of transparency of DoD emission reporting out of the new administration.

We arrived at these volumes of fuel and associated CO_2 emissions through data retrieved from multiple FOIA requests to the US Defense Logistics Agency (DLA). The DLA is the massive, and often shadowy, bureaucratic agency tasked with managing the US military's supply chains, including its hydrocarbon fuel purchases and distribution. As has been well documented (Ali and Stone 2018), it is effectively impossible to audit the US military's budget, and the DLA has recently been embroiled in accounting scandals as the scope of wasteful, or outright reckless, spending throughout the 'War on Terror' has come into focus (Lindorff 2018). The Department of Defense (DoD) is by far the largest of all federal agencies relying on discretionary budget allocations. Despite protestation from a few lonely corners of Congress, the DoD effectively had a blank check for much of the twenty-first century (Lindorff 2018). Even at $8.7 billion, however, fuel comprises less than 2% of the total DoD spending of $523.9 billion in 2017, a figure that does not include other channels through which war is pursued, like the CIA drone programme.

The US military is particularly ideal to study. It is the third largest active military personnel, next to China and India. It boasts over thirteen thousand aircraft, and eleven aircraft carriers—the second closest are China, Italy and UK with two each (GlobalFirepower 2021). It operates one of the largest and most complex material supply chains, responsible for enormous built infrastructure (e.g. forward operating bases, roads and airports), yet its socio-environmental effects remain relatively unexamined in most major climate and environmental policy agreements. The US military is the largest single logistical operation in the world that is still exempt from having to report its carbon emissions (Neslen 2015). To put this another way, although the CO_2 emissions of the US military count very significantly in terms of their global contribution to total emissions, they did not count in global carbon emissions reporting until the US rejoined the Paris Agreement.

Threat Multipliers

While the US military continues to emit globally significant volumes of greenhouse gases, it has also long understood that it is not immune from the potential consequences of climate change—recognising this as a 'threat multiplier' that can exacerbate other risks on top of the possibility of environmental change itself producing new conflicts (Gilbert 2012). In forward-looking public documents, the US military envisions a dangerous future that returns to great-power geopolitics alongside the murky, diffuse, and emergent threats that may be called into being by environmental change (also see Durand-Delacre et al.'s critique in this volume of xenophobic discourses around migration 'floods' attributed to climate change). The military's response in this regard is somewhat tautological. Because climate change will produce new threats, the military will continue to build its interventionist capacity—as can be seen through the massive build-up of US forces across Sub-Saharan Africa (Turse 2018), in turn continuing to burn massive volumes of fuel and thus exacerbating the exact threats to which the military will respond (also see Chapter 11 by Sullivan, this volume). The very discourse of 'threat multipliers' threatens to bring into being the very situation it describes, putting vast swathes of the globe at more, rather than less, risk.

While US climate policy has been inexcusably slow and ineffective due to a politics mired in climate denialism and state capture by oil firms, the military has some degree of autonomy in defining and responding to threats. As far back as the 1990s, climate change has in fact been identified as one of those threats (White House 1991), and many, although not all, military bases have been preparing for climate change impacts such as sea level rise (Mathews 2019). Nor has the military ignored its own contribution to the problem, having dabbled in developing alternative energy sources such as biofuels (generating considerable pushback from lawmakers in oil-producing states in the process). Alternative energy sources comprise only a tiny fraction of military spending on fuels, however, and also may generate their own socio-environmental problems (as explored in more detail in Dunlap's chapter, this volume).

Turn Down the Furnace

The American military's climate policy remains contradictory. There have been attempts to 'green' aspects of its operations by increasing renewable electricity generation on bases (Gardner 2017), but it remains the single largest institutional consumer of hydrocarbons in the world (Bigger and Neimark 2017). It has also locked itself into hydrocarbon-based weapons systems for years to come, by depending on existing aircraft and warships for open-ended operations. The F-35 fighter, for example, a product of one of the most costly and delayed military acquisition programmes in history (Sullivan 2016), could hypothetically run on third generation biofuels were they available at the scale required to power the fleet. But these fuels are not currently, or for the foreseeable future, available at the scale needed (Banerjee et al 2019), plus the large-scale production of feedstock for biofuels already creates serious environmental (Cruzen et al. 2016) and social (Neville and Dauvergne 2016) problems (also see Dyke et al., this volume).

As these new fighters are rolled out and pilots perform regular training missions, despite the complete absence of air battles for the last thirty years, fossil fuels will thus continue to power the DoD's fleet of more than 6,500 airplanes, 6,700 helicopters, untold numbers of HumVees, APCs, base vehicles, non-nuclear ships, and diesel electricity generators that power a stunning number of bases around the world (Vine 2015). This massive volume of kit, if kept operational, represents a significant level of fossil-fuel lock-in (Unruh 2000; Urry 2003).

It is also worth considering the overarching role of the US military in producing and enforcing a fossil-fueled global economic system (Surprise 2020). At this point, it would be relatively uncontroversial to state that much ongoing US overseas intervention was, at least initially, predicated on securing access to, and the distribution of, fossil fuels from the Middle East. This observation has been confirmed (to whatever extent can be believed) in statements by former US President Trump, claiming that the US should "take" Iraqi oil as recompense for the cost of sixteen years of occupation (Borger 2016). Even leaving this adventurism aside, significant resources—both material and in terms of relationship maintenance—are devoted to maintaining the free flow of oil around the world, especially through key shipping routes. In this way, the US military not only locks-in its own fuel consumption, but also ensures oil supplies remain cheap, plentiful, mobile, and accessible.

Don't Just Green the Military. Shrink It.

While new spending initiatives like Biden's 2021 Infrastructure plan include significant (though still insufficient) outlays for decarbonisation and climate adaptation, the military's contribution to environmental change remains off-radar. Indeed, rather than scaling back military spending to pay for urgent climate-related spending, initial budget requests for military appropriations are actually *increasing* even as some US foreign adventures are supposedly coming to a close (Macias 2021). This includes vast outlays for new or retrofitted fuel-intensive vehicles, from tanks to new fleets of aircraft that will continue to demand liquid fossil fuels for decades to come. For any green initiative of national scope to be effective, the US military's carbon bootprint must be addressed in domestic policy and international climate treaties.

Action on climate change demands shutting down vast sections of US military machinery. There are few activities on Earth as environmentally catastrophic as waging war. Significant reductions to the Pentagon's budget and shrinking its capacity to wage war would reduce demand from the biggest consumer of liquid fuels in the world. This is critical in a world awash in cheap oil in which the US military continues to have vast resources for its acquisition. Indeed, we might speculate that the US military may function as a buyer of last resort for some fraction of global output (especially given political influence in procurement decisions), so as to delay the closure of marginal production and refining facilities (Surprise 2020).

It does no good in terms of anthropogenic climate change management to tinker around the edges of the US war machine's environmental impact. In considering alternatives, the money spent procuring and distributing fuel across the US empire could instead be spent as a peace dividend, helping to fund a Green New Deal that is international in outlook, and includes significant technology transfer and no-strings-attached funding for adaptation and clean energy to those countries most vulnerable to climate change, who bear little historic or contemporary responsibility for emissions (Belcher et al. 2020). There is no shortage of policy priorities that could use a funding bump. With Lai et al. (2017), we agree that any of these options would be better than continuing to wastefully fuel one of the largest military forces in history.

References

Ali, Idrees, and Mike Stone, 'The Department of Defense Fails its First-ever Audit.' (Reuters.com, 2018), https://uk.reuters.com/article/uk-usa-pentagon-audit/pentagon-fails-its-first-ever-audit-official-says-idUKKCN1NK2UL.

Banerjee, Sharmistha, Shuchi Kaushi and Rajesh Singh Tomar, 'Global Scenario of Biofuel Production: Past, Present and Future', in *Prospects of Renewable Bioprocessing in Future Energy Systems*, ed. by Ali Asghar Rastegari, Ajar Nath Yadav, and Arti Gupta (Cambridge: Springer, 2019), pp. 499–518.

Belcher, Oliver, Patrick Bigger, Benjamin Neimark, and Cara Kennelly, 'Hidden Carbon Costs of the Everywhere War: Logistics, Geopolitical Ecology, and the Carbon Boot-print of the US Military', *Transactions of the Institute of British Geographers*, 45(1) (2020), 65–80, https://doi.org/10.1111/tran.12319.

Bigger, Patrick, and Benjamin Neimark, 'Weaponizing Nature: The Geopolitical Ecology of the US Navy's Biofuel Program' *Political Geography*, 60 (2017), 13–22, https://doi.org/10.1016/j.polgeo.2017.03.007.

Borger, Julian, 'Trump's Plan to Seize Iraq's Oil: "It's Not Stealing, We're Reimbursing Ourselves"' (Theguardian.com, 2016), https://www.theguardian.com/us-news/2016/sep/21/donald-trump-iraq-war-oil-strategy-seizure-isis.

Colman, Zack, 'Pentagon: Climate Change Threatens Military Installations' (Politico.com, 2019), https://www.politico.com/story/2019/01/18/pentagon-military-installations-climate-1098095.

Cowen, Deborah, *The Deadly Life of Logistics: Mapping Violence in Global Trade* (Minneapolis: University of Minnesota Press, 2014).

Crawford, Neta, *Pentagon Fuel Use, Climate Change, and the Costs of War* (Watson Institute, Brown University, 2019), https://watson.brown.edu/costsofwar/files/cow/imce/papers/Pentagon%20Fuel%20Use%2C%20Climate%20Change%20and%20the%20Costs%20of%20War%20Revised%20November%202019%20Crawford.pdf.

Crutzen, Paul, Arvin R. Mosier, Keith A. Smith, and Wilfried Winiwarter. 'N$_2$O Release from Agro-biofuel Production Negates Global Warming Reduction by Replacing Fossil Fuels', in *Paul J. Crutzen: A Pioneer on Atmospheric Chemistry and Climate Change in the Anthropocene*, ed. by Paul Crutzen and Hans Blauch (New York: Springer, 2016), pp. 227–38.

Dalby, Simon, *Anthropocene Geopolitics: Globalization, Security, Sustainability* (Ottawa: University of Ottawa Press, 2020).

Fahey, David W. and David S. Lee, 'Aviation and Climate Change: A Scientific Perspective', *Carbon and Climate Law Review*, 97(2) (2016), 1–8, http://www.jstor.org/stable/44135212.

Fryar, Debra, *The History of PartNet* (Partnet.com, 2012), https://partnet.com/the-history-of-partnet/.

Gardner, Timothy, 'U.S. Military Marches Forward on Green Energy, Despite Trump' (Reuters.com, 2017), https://www.reuters.com/article/us-usa-military-green-energy-insight/u-s-military-marches-forward-on-green-energy-despite-trump-idUSKBN1683BL.

Gilbert, Emily, 'The Militarization of Climate Change', *ACME: An International E-Journal for Critical Geographies*, 11(1) (2012), 1–14.

Lai, K. K. Rebecca, Troy Griggs, Max Fisher, and Audrey Carlsen, 'Is America's Military Big Enough?' (Nytimes.com, 2017), https://www.nytimes.com/interactive/2017/03/22/us/is-americas-military-big-enough.html?auth=login-smartlock.

Levinson, Marc, *The Box: How the Shipping Container Made the World Smaller and the World Economy Bigger* (Princeton: Princeton University Press, 2016).

Lindorff, Dave, 'Exclusive: The Pentagon's Massive Accounting Fraud Exposed' (Thenation.com, 2018), https://www.thenation.com/article/archive/pentagon-audit-budget-fraud/.

Macias, Amanda, 'Here's the Firepower the Pentagon is Asking For in its $715 Billion Budget' (Cnbc.com, 2021), https://www.cnbc.com/2021/05/28/pentagon-asks-for-715-billion-in-2022-defense-budget.html.

Mathews, Mark, 'Not All Military Bases Plan for Warming, Watchdog Finds' (Scientificamerican.com, 2019), https://www.scientificamerican.com/ article/not-all-military-bases-plan-for-warming-watchdog-finds.

Nelson, Arthur, and Climate Desk, 'Why the U.S. Military Is Losing Its Carbon-Emissions Exemption' (Theatlantic.com, 2015), https://www.theatlantic.com/science/archive/2015/12/ paris-climate-deal-military-carbon-emissions-exemption/420399/.

Neville, Kate, and Peter Dauvergne, 'The Problematic of Biofuels for Development', in *The Palgrave Handbook of International Development*, ed. by Jean Grugel and Daniel Hammett (London: Palgrave Macmillan, 2016), pp. 649–68.

Sullivan, Michael, 'F-35 Joint Strike Fighter: Preliminary Observations on Program Progress' (Gao.gov, 2016), https://www.gao.gov/products/ GAO-16-489T.

Surprise, Keven, 'Geopolitical Ecology of Solar Geoengineering: From a 'Logic of Multilateralism' to Logics of Militarization', *Journal of Political Ecology*, 27 (2020), 213–35, https://doi.org/10.2458/v27i1.23583.

Turse, Nick, 'U.S. Military Says It Has a "Light Footprint" in Africa. These Documents Show a Vast Network of Bases' (Theintercept.com 2018), https:// theintercept.com/2018/12/01/u-s-military-says-it-has-a-light-footprint-in- africa-these-documents-show-a-vast-network-of-bases/.

The White House *National Security Strategy of the United States* (White House, 1991), https://history.defense.gov/Portals/70/Documents/nss/nss1991. pdf?ver=3sIpLiQwmknO-RplyPeAHw%3d%3d.

Vine, David, *Base Nation: How US Military Bases Abroad Harm America and the World* (New York: Metropolitan Books, 2016).

Westing, Arthur, 'The Impact of War on the Environment', in *War and Public Health*, ed. by Barry Levy and Victor Sidel (Oxford: Oxford University Press, 2008), pp. 69–86.

Unruh, Gregory, 'Understanding Carbon Lock-in', *Energy Policy*, 28(12) (2000), 817–30.

Urry, John, *Global Complexity* (Cambridge: Polity, 2003).

6. Climate Migration Is about People, Not Numbers

D. Durand-Delacre, G. Bettini, S. L. Nash, H. Sterly, G. Gioli, E. Hut, I. Boas, C. Farbotko, P. Sakdapolrak, M. de Bruijn, B. Tripathy Furlong, K. van der Geest, S. Lietaer and M. Hulme

It has become increasingly common to argue that climate change will lead to mass migrations. In this chapter, we examine the large numbers often invoked to underline alarming climate migration narratives. We outline the methodological limitations to their production. We argue for a greater diversity of knowledges about climate migration, rooted in qualitative and mixed methods. We also question the usefulness of numbers to progressive agendas for climate action. Large numbers are used for rhetorical effect to create fear of climate migration, but this approach backfires when they are used to justify security-oriented, anti-migrant agendas. In addition, quantification helps present migration as a management problem with decisions based on meeting quantitative targets, instead of prioritising peoples' needs, rights, and freedoms.

Introduction

Perhaps counterintuitively—in a volume calling for actions to tackle the climate crisis—this contribution cautions against the casual use of one of the primary narratives through which the climate crisis is signified and urgent action invoked. That is, the dramatic estimates and projections of a looming migration crisis caused by climate change. We problematise the numbers through which the spectre of such a crisis is supported and

https://doi.org/10.11647/OBP.0265.06

communicated. Our critique of these numbers takes place on several levels. We begin by pointing to the many methodological challenges in producing robust numbers. Estimates remain imprecise and highly uncertain, despite some significant developments in methods and datasets. We also diagnose more fundamental epistemological issues about the *kinds* of knowledges required to understand the climate-migration nexus. Numbers and quantitative estimates fail to capture crucial dimensions of human mobility. Migrants' decisions to move can be forced but also voluntary, are highly subjective, and need to be understood as situated, political, and non-deterministic.

Ultimately, however, our concern has less to do with what numbers can or cannot tell us about climate migration than with the ways in which numbers are (mis)used. On the one hand, a focus on mass migration numbers is intended to construct climate migration as a *crisis*. However, framing this crisis as a humanitarian issue has done little to protect migrants and more to stoke the fires of anti-immigrant populism, providing arguments for more stringent border controls and increasingly restrictive migration policies across the Global North. At the same time, the promise of quantification creates the impression that this crisis can be clearly defined, and *managed*, as long as better numbers are made available (also see Hannis, this volume). Attempts to use numbers to address issues of climate justice and responsibility are undercut by the focus on quantification itself, which tends to limit debates to technical questions about how many will move and how this movement can be organised.

This critique of headline estimates should not be misinterpreted as a denial of the impacts that climate change is having and will continue to have on peoples' mobilities. Climate change impacts related to sea-level rise, drought, increased frequency of wildfires and storms—and the associated declines in livelihoods—pose serious and differentiated challenges with which we must contend (as also highlighted by Lendelvo et al., this volume). Rather, our aim is to point to how a focus on numbers reduces political imaginaries of our response to climate migration to a narrow range of possibilities. We argue that a different approach is needed.

A Brief Overview of Climate Migration Numbers and their Methodological Limitations

The environmentalist Norman Myers (1934–2019) initiated efforts to estimate the impact of climate change on migration when he predicted in the early 1990s that there would be 150 million "environmental refugees" by 2050 (Myers 1993). He later updated his estimate to 200 million by 2050 (Myers and Kent 1995; Myers 1997, 2002). The latter figure remains one of the most widely cited climate migration numbers to date. Myers' estimations were based on linear extrapolations of demographic and displacement figures in what he considered "environmental hotspots". These methods were rapidly challenged as too simplistic, notably because they assumed a linear relationship between environmental impacts (such as sea-level rise or desertification) and out-migration from affected areas. They were also not based on any actual inquiry into the causal mechanisms involved and ignored potential *in-situ* adaptation strategies. Myers' approach relied on aggregate global forecasts, rather than specific case studies that could bring empirical grounding to these assumptions (Black 1994, 2001; Suhrke 1994; Castles 2002). Myers' numbers have been reproduced in many prominent reports since their publication (as critiqued by Saunders 2000). More recently, numbers larger than a billion people have also been disseminated in academic articles, NGO or think tank reports, and the press (see Table 1). Myers himself later admitted that coming up with the final estimates required "heroic extrapolations" (Brown 2008).

Despite this situation, many subsequent reports—mostly published by NGOs and (inter)governmental organisations—either reproduced Myers' numbers or provided other estimates based on analogous methods (see methodological notes in Table 1). These numbers are rounded to the nearest ten or hundred million, an indication of the crude methods employed to derive them. Most problematic is the prevalence of simple additions of annual figures and an extrapolation of such trends into the future, which can produce nothing other than a continuously rising graph (for a recent example, see Institute for Economics and Peace 2020).

Estimated number of migrants due to climate change (upper range)	Time period	Source	Notes on methodology
10 million	-	Jacobson 1988	Terminology: "Environmental refugees"
150 million	2050	Myers 1993	Terminology: "Environmental refugees" Geographical focus: both internal and international (specifically: Bangladesh, Egypt, China, India, other deltas and coastal zones, island states, and "agriculturally dislocated areas")
200 million	2050	Myers 1997, 2002; Myers and Kent 1995	Updated estimate based on Myers (1993).
250 million – "at least 1 billion"	2050	Christian Aid 2007	250 million number based on an interview with Norman Myers. Terminology: "people forced from their homes", "forced migration" Geographical focus: both internal and international
200 million	2050	Friends of the Earth 2007	Direct reference to Myers (1993) and updated figures.
150–200 million	2050	Stern 2007	Direct reference to Myers and Kent (1995).
200–250 million	2050	Biermann and Boas 2010	Direct references to Myers and Kent (1995) and interview with Myers in Christian Aid (2007).
250 million	2050	CARE Danmark 2016	Refers to Christian Aid (2007) interview with Myers as latest estimate.

1.4 billion	2060	Geisler and Currens 2017	Spatial vulnerability model exploring the impact of global mean sea-level rise (SLR) on coastal settlements. Mentions Neumann et al. (2015) as the source for 1.4bn people exposed to SLR. Terminology: "climate migrants" Geographical scope: global, in low elevation coastal zones
"millions or even billions"	-	Environmental Justice Foundation 2017	Cites Geisler and Currens (2017) as a source. Terminology: "climate refugees" Geographical scope: international and internal
143 million	2050	Rigaud et al. 2018 (Groundswell Report)	State-of-the-art gravity model of climate migration. Terminology: "internal climate migrants" Geographical focus: Sub-Saharan Africa, South Asia, Latin America
1.2 billion	2050	Institute for Economics and Peace 2020	Terminology: "people at risk of displacement" Geographical focus: forty-three countries

Table 1. Commonly cited climate migration numbers. This table is far from exhaustive but captures some of the more prominent and commonly cited examples. Sources: Authors' review (drawing on Brown 2008; Gemenne 2011; Ionesco et al. 2016; Luetz 2019).

Recognising the deep flaws in the methods employed to derive these numbers, quantitative social scientists and modellers have sought to develop more sophisticated datasets and methods to improve the credibility of numerical estimates. In doing this, they contend with the unavailability of data, particularly at small scales or in disaggregated form, its poor quality, and its limited comparability (Brown 2008; Tejero et al. 2020). On the condition that the necessary data could be gathered, reviewers of the field have repeatedly emphasised the need for more longitudinal studies, a multi-scalar research, analysis disaggregated along gender, age, ethnic, caste, and class lines, and consideration of a wider range of environmental drivers beyond precipitation changes or sea-level rise (Brown 2008; Piguet 2010; Obokata et al. 2014; Vinke and Hoffmann 2020).

As a result of these responses, some refinements to datasets and methods have been made. Most researchers also now present overall climate migration numbers with much greater care than was the case for earlier estimates. They largely acknowledge limitations and warn readers not to overinterpret results by pointing to numerical ranges and associated uncertainties. Nonetheless, reviews and meta-analyses of quantitative studies still conclude that many analytical problems persist. They find that models allow few confident causal claims about the environment's influence on migration except "in broad terms and at fairly large spatial scales" (Obokata et al. 2014: 127); and that results remain heavily influenced by the methods used. Ultimately, they find it remains exceedingly difficult for modellers to defend any single factor as the primary driver of migration (Piguet et al. 2011; McLeman 2013; Obokata et al. 2014; Beine and Jeusette 2018; Cattaneo et al. 2019). In other words, it remains extremely difficult to definitely link migration to climate change.

A Greater Diversity of Knowledges of Climate Migration Is Needed

In addition to these methodological shortcomings, a focus on climate migration numbers obscures the need for other forms of knowledge about the climate-migration nexus. Producing these knowledges requires more use of mixed and qualitative methods. These are better suited than quantification alone to study the multi-causal, highly situated, and subjective dimensions of climate migration. Attempts to isolate migration drivers related to climate change, and to identify and count 'climate migrants', are at odds with most social scientists' understanding of migration. From a social science perspective, human mobility is well-accepted to be a multi-causal and complex phenomenon. This implies that migration drivers—be they social, political, economic, environmental, or demographic—interact with and mutually influence each other (Black et al. 2011).

Our trouble with numbers is also motivated by fundamental questions on how migration itself should be understood. Attempting to reduce migration to a number is akin to the "migration map trap" whereby individual experiences of migration become "faceless pixel[s] in a big threatening arrow" (Van Houtum and Bueno Lacy 2020: 210). This approach not only overlooks the situated complexities already highlighted, but also obliterates the important subjective dimensions of migration.

To be clear, accounting for those subjective dimensions is not merely about looking at the faces behind the numbers. The problem with this methodological 'reversal' would be to individualise processes that are in fact emergent and collective, thereby interpreting peoples' experiences and extrapolating singular stories as representative of millions. Rather, we suggest that it is important to fully acknowledge that mobility itself is a cultural construct, and that distinct ontologies and epistemologies of mobility are embodied in and inform migrants' choices and experiences (also see Sullivan Chapter 3, and Dieckmann's chapter, this volume on the relevance of ontological concerns for situating choices and understanding). This subjective dimension of migration stresses its non-deterministic character and, in line with a general shift in migration studies, calls into question the dominant focus on 'root causes' and migration 'management'. Viewing migration and mobilities as autonomous (which is not a synonym for voluntary or individual) practices suggests that they must be investigated well beyond institutional constraints and categories (De Genova and Peutz 2010; Mezzadra and Neilson 2013; Scheel 2013).

In practical terms, we are calling for approaches that engage with the subjective diversity of migrant mobilities and situate people involved and their perspectives more prominently in the research process (Casas-Cortes et al. 2015). This orientation requires deep qualitative work—usually based on interviews and ethnographic fieldwork—and a general move away from an emphasis on legal and governmental frameworks and the economic determinism of the labour market.

A Climate Mobilities Approach to Diverse and Situated Mobilities

A climate mobilities approach is a promising way to understand the mechanisms behind the decision to migrate (or to stay) (Boas et al. 2019). This approach, based largely on qualitative research, does not aim to cut through causality and isolate 'the environment' from other 'contextual' factors, or to identify the dominant factor in migration decisions. Instead, it considers causation as always multi-faceted, situated, and nonlinear. The changing climate remains a relevant factor, but climate only exerts its influence on the world through the matrix of social, economic, environmental, cultural, historical, and political processes that comprise the social world (Hulme 2011). In this way, the climate is not privileged as an influence on mobilities but is also recognised as a pluralistic phenomenon worthy of multi-pronged empirical investigations.

These investigations need to be pursued using a rich vocabulary capturing the many nuances and forms that (im)mobilities take. Indeed, social science research has shown that climate mobilities can be short-term or long-term, but also circular or seasonal (Zickgraf 2018). What may start as a short-distance, temporary move can turn into a long-distance, permanent one (Van der Geest 2010). Some people choose to remain in their homes in full cognisance of the risks involved (McNamara and Gibson 2009; Adams 2016; Ayeb-Karlsson et al. 2016; Farbotko 2018), while others are trapped and experience involuntary immobility (Black et al. 2013), or embark on long-distance movements because everyday short-distance mobilities to markets or healthcare are disrupted (Blondin 2020). In addition, it is crucial to understand the situated dynamics and local contexts in which migration (or immobility) occurs. Socio-cultural, political, and environmental dynamics change from place to place, and can differ significantly even within a single community.

To better understand the complexity of migratory movements, we can also turn for example to "trajectory ethnographies" (Schapendonk and Steel 2014; Schapendonk et al. 2018), "geographies in and of movement" (Brachet 2012), or life history approaches (Singh et al. 2019). These in-depth approaches study how mobility unfolds, what shapes it, and how mobility decisions are made. The first two involve interviews in multiple locations, at different moments and stages relevant to a journey, while the latter explores personal narratives of mobility. Such methods provide a detailed picture of the circuitous routes people take; of the obstacles, meetings, and separations that punctuate them; and of peoples' perceptions, aspirations, and memories.

Of course, such methods also have their limitations. Questions can be raised about the power dynamics between researchers and research participants, and the latter's degree of representation (Cabot 2016; Khosravi 2018; Boas et al. 2020). Can we—as often privileged academics—really put ourselves in the shoes of affected individuals? Trajectory approaches allow researchers (to some degree) "to practice mobility and to reveal mobility-immobility relations that otherwise would remain hidden", but it is important to stay reflexive (Boas et al. 2020: 144).

This is not to say that better numbers cannot be produced in the course of estimating climate change and migration interconnections. Besides improving quantitative methods and data, however, we argue that progress will only be achieved through greater collaboration with qualitative social sciences of the kind just described. We see as promising the mixing of methods in work that 'grounds' big data with site-based fieldwork, so as to challenge assumptions made from afar and detect important dynamics that big-data research would otherwise miss (Boas et al. 2019). Collaborative work integrating behavioural migration theories and concepts of "place attachment" into agent-based models is also helping to increase the sensitivity of model results to variations in individual and community-level responses to environmental hazards (Adams and Kay 2019). Lastly, the recent uptake in mobilities research of tools such as Q methodology[1] also reveals a range of shared subjective understandings, attitudes and perceptions of climate change and human mobility (Van der Geest et al. 2019; Oakes 2019).

The Misuses of Climate Migration Numbers

If we insist on epistemological and methodological diversity beyond quantification alone, it is because social science studies on numbers— not just of climate migration but in other spheres too—have repeatedly shown their potential for misuse (Porter 1995). The point is not to do away with numbers, but to exercise caution at all stages of their production, communication, and use. Below we detail two specific areas of concern.

Climate Mobility Is Not a Crisis

Our first concern with headline numbers is how they are used to construct climate migration as a *crisis*. The intention behind such rhetoric may be laudable: to stimulate action on climate change and to assist its victims. But there are many problems with using fear of mass displacement as a rallying cry. There is absolutely no guarantee that crisis narratives underpinned by large numbers are an effective way to achieve these aims.

On the contrary, press releases and the news media tend to highlight a single number, usually drawn from the upper range of estimates presented in the original source. In some cases, the numbers lose all specificity, with headlines pointing only to 'millions' or even 'billions' of people on the move (also see discussion of the constructed and historical dimensions of 'environmental refugees' in Saunders 2000). Sometimes, they do not even refer to a specific source. In such a discursive context, the specifics of the number and the underlying

---

1   The aim of Q methodology is to identify the shared views of study participants on a given issue. Participants are asked, as a group, to rank a set of statements on a scale from "strongly disagree" to "strongly agree". The statements are selected to be representative of known existing opinions on the issue. In addition, participants are asked to explain their decisions, providing qualitative commentary to the quantitative sorting exercise.

methodology matter little, as long as it is high. Numbers are used to rhetorical effect.

The dystopian futures thus created are often associated with a tendency to discuss migratory flows in hydrological metaphors, such as in assertions of 'waves', 'floods' or 'rising tides' of migrants. Such invocations connect with a wider racist motif that contributes to an alarming imaginary wherein the 'Global North' will be overwhelmed by migrants from 'Global South' contexts (Bettini 2013; Methmann 2014; Pallister-Wilkins 2019). Worse still, the 'hotspots' approach on which these narratives are based tends to erase colonial histories and naturalise structural violence, neutralising local contexts and conditions. By reducing large parts of the Global South to a "hotspot", this perpetuates a dangerous othering exercise, denigrating large parts of the world as merely disaster-ridden, dangerous, overpopulated places from which people can only aspire to flee (Giuliani 2017). This narrative actively reproduces the figure of the climate migrant as a security threat, and ultimately, as a highly racialised entity (Baldwin 2013).

Researchers who seek to publicise their work on the climate-migration nexus must grasp these racist and simplifying dangers, as too should policymakers, practitioners, and journalists who promote these narratives. We argue that the stories of affected people ought to be central in such reporting, even if—perhaps especially if—their stories run counter to our intuitions and estimated numbers. It is people who matter, not numbers.

Climate Mobility Is Not a Management Problem

The promise of better numbers also reinforces the impression that the climate migration crisis can be managed *because* it can be quantified. Numbers hold a privileged place in contemporary political discourses and policymaking, as they are associated with rigour and objectivity in the public and scientific imagination (Porter 1995; Espeland and Stevens 2008; Hansen and Porter 2012). Migration policy has not escaped this trend. To a large extent migration expertise is understood to be associated with researchers' ability to quantify their findings, so they can be used in managerial practices that place statistical methods at the centre of decision-making (Takle 2017). While numbers have their place in policymaking, we argue that the qualitative methods and subjective perspectives highlighted above should be centred in policymaking on climate mobilities. This will only be possible if the rhetorical power of numbers is acknowledged and challenged.

Adopting a managerial approach to climate migration, guided by flawed numbers, risks disregarding many of the dangers associated with climate migration numbers discussed above. This practice can be understood in terms of "strategic ignorance". While policymakers are widely aware of migration data quality issues, and associated uncertainties, they still maintain a picture of migration as an "easily measurable, intelligible reality" that can therefore be managed by numbers (Scheel and Ustek-Spilda 2019). In this way, they avoid the difficult political questions—notably around responsibility—that a more head-on engagement with migration's complex realities would require (Betts and Pilath 2017; Kelman 2019).

In a political context where critical migration scholars struggle to make themselves heard, this situation is particularly concerning. Mobilities scholars often find that while many policymakers are willing to engage in discussion, these exchanges do not have any meaningful bearing on policy design (Baldwin-Edwards et al. 2019; Héran 2020). Researchers have even expressed concerns that findings of the migration research funded by the EU are disregarded by the EU's own policy processes (Kalir and Cantat 2020). In this light, the constant drive under migration management processes for "more and better data" can have the counterproductive effect of ignoring already well-established knowledge in migration research—whether quantitative or based on other methods.

Conclusion

Estimates of climate migration numbers present a facade of objective, authoritative and unemotional facts. But these numbers are highly contentious. Such estimates need to be recognised as being grounded in normative, epistemological and methodological assumptions which are often hidden and rarely challenged. At the very least, migration scholars should expose these assumptions so that audiences better understand how migration numbers are constructed (for a recent example, see McMichael et al. 2020). In many cases, the estimates of people moving in the context of climate change are methodologically questionable or else nebulous.

Migration numbers need to be understood against this background. But they also do not stand alone. They are attached to various narratives and are presented in multiple ways. Even rigorous, cautiously communicated estimates can become decoupled from the complexities of human mobilities. The significance of a discourse is not simply about how it is constructed, and by whom, but also lies in how the discourse is received, and how the narratives which package migration numbers are filtered and interpreted. Complexities that may initially be presented alongside the numerical estimates become erased. Numbers alone become the headline, thus distracting from important political questions about humanity, justice, and responsibility.

A common narrative with which climate migration estimates are coupled (either by their authors or during their reproduction and dissemination) is one that conceptualises human mobility in the context of climate change as a crisis. The large numbers of people imagined to be on the move are employed to signify the looming crisis and used to invoke urgent action. There comes a point where these numbers are instrumentalised by political movements whose values— expressed in xenophobic narratives and anti-immigrant agendas—are at odds with those who champion these numbers to call for bolder climate action.

The relationship between climate change and human mobility should not be seen as a security problem, as a managerial issue, or as a number to be controlled. The policy focus should be on people—their vulnerabilities, rights and freedoms—so as to help prise open political spaces for policy interventions beyond building walls, real or rhetorical, designed to control rather than to care.

References

Adams, H., 'Why Populations Persist: Mobility, Place Attachment and Climate Change', *Population and Environment*, 37(4) (2016), 429–48, https://doi.org/10/f8mp84.

Adams, Helen, and Susan Kay, 'Migration as a Human Affair: Integrating Individual Stress Thresholds into Quantitative Models of Climate Migration', *Environmental Science & Policy*, 93 (2019), 129–38, https://doi.org/10/gftmsq.

Ayeb-Karlsson, Sonja, Kees Van der Geest, Istiakh Ahmed, Saleemul Huq, and Koko Warner, 'A People-Centred Perspective on Climate Change, Environmental Stress, and Livelihood Resilience in Bangladesh', *Sustainability Science*, 11(4) (2016), 679–94, https://doi.org/10.1007/s11625-016-0379-z.

Baldwin, Andrew, 'Racialisation and the Figure of the Climate-Change Migrant', *Environment and Planning A: Economy and Space*, 45(6) (2013), 1474–90, https://doi.org/10.1068/a45388.

Baldwin-Edwards, Martin, Brad K. Blitz, and Heaven Crawley, 'The Politics of Evidence-Based Policy in Europe's "Migration Crisis"', *Journal of Ethnic and Migration Studies*, 45(12) (2019), 2139–55, https://doi.org/10.1080/1369183X.2018.1468307.

Beine, Michel, and Lionel Jeusette, *A Meta-Analysis of the Literature on Climate Change and Migration*, CREA Discussion Paper Series (Ideas.repec.org, 2018), https://ideas.repec.org/p/luc/wpaper/18-05.html.

Bettini, Giovanni, 'Climate Barbarians at the Gate? A Critique of Apocalyptic Narratives on "Climate Refugees"', *Geoforum*, Risky natures, natures of risk, 45 (2013), 63–72, https://doi.org/10/f234jv.

Betts, Alexander, and Angela Pilath, 'The Politics of Causal Claims: The Case of Environmental Migration', *Journal of International Relations and Development*, 20(4) (2017), 782–804, https://doi.org/10.1057/s41268-016-0003-y.

Biermann, Frank, and Ingrid Boas, 'Preparing for a Warmer World: Towards a Global Governance System to Protect Climate Refugees', *Global Environmental Politics*, 10(1) (2010), 60–88, https://doi.org/10.1162/glep.2010.10.1.60.

Black, Richard, 'Forced Migration and Environmental Change: The Impact of Refugees on Host Environments', *Journal of Environmental Management*, 42(3) (1994), 261–77, https://doi.org/10/c8spfh.

Black, Richard, 'Environmental Refugees: Myth or Reality?', *New Issues in Refugee Research (UNHCR Research Paper)*, 34 (Unhcr.org, 2001), http://www.unhcr.org/uk/research/working/3ae6a0d00/environmental-refugees-myth-reality-richard-black.html.

Black, Richard, Nigel W. Arnell, Neil Adger, David Thomas, and Andrew Geddes, 'Migration, Immobility and Displacement Outcomes Following

Extreme Events', *Environmental Science and Policy*, 27(Supplement 1) (2013), 32–43, https://doi.org/10/f4t8zs.

Black, Richard, Dominic Kniveton, and Kerstin Schmidt-Verkerk, 'Migration and Climate Change: Towards an Integrated Assessment of Sensitivity', *Environment and Planning A*, 43(2) (2011), 431–50, https://doi.org/10/fmpfdb.

Blondin, Suzy, 'Understanding Involuntary Immobility in the Bartang Valley of Tajikistan through the Prism of Motility', *Mobilities*, 15(4) (2020), 543–58, https://doi.org/10.1080/17450101.2020.1746146.

Boas, Ingrid, Ruben Dahm, and David Wrathall, 'Grounding Big Data on Climate-Induced Human Mobility', *Geographical Review*, 110(1–2) (2019), 195–209, https://doi.org/10.1111/gere.12355.

Boas, Ingrid, Carol Farbotko, Helen Adams, Harald Sterly, Simon Bush, Kees Van der Geest, and others, 'Climate Migration Myths', *Nature Climate Change*, 9(12) (2019), 901–03, https://doi.org/10.1038/s41558-019-0633-3.

Boas, Ingrid, Joris Schapendonk, Suzy Blondin, and A. Pas, 'Methods as Moving Ground: Reflections on the "Doings" of Mobile Methodologies', *Social Inclusion*, 8(4) (2020), 136–46, https://doi.org/10.17645/si.v8i4.3326.

Brachet, Julien, 'Géographie du mouvement, géographie en mouvement. La mobilité comme dimension du terrain dans l'étude des migrations', *Annales de géographie*, 687–88(5–6) (2012), 543–60, https://doi.org/10/ggqpbk.

Brown, Oli, *Climate Change and Forced Migration: Observations, Projections and Implications* (Hdr.undp.org, 2008), http://hdr.undp.org/en/content/climate-change-and-forced-migration.

Cabot, Heath, '"Refugee Voices": Tragedy, Ghosts, and the Anthropology of Not Knowing', *Journal of Contemporary Ethnography*, 45(6) (2016), 645–72, https://doi.org/10.1177/0891241615625567.

CARE Danmark, *Fleeing Climate Change: Impacts on Migration and Displacement* (Careclimatechange.org, 2016), https://careclimatechange.org/fleeing-climate-change-impacts-migration-displacement/.

Casas-Cortes, Maribel, Sebastian Cobarrubias, and John Pickles, 'Riding Routes and Itinerant Borders: Autonomy of Migration and Border Externalization', *Antipode*, 47(4) (2015), 894–914, https://doi.org/10.1111/anti.12148.

Castles, Stephen, 'Environmental Change and Forced Migration: Making Sense of the Debate', *New Issues in Refugee Research* (*UNHCR Research Paper*), 70 (2002), http://www.unhcr.org/3de344fd9.html.

Cattaneo, Cristina, Michel Beine, Christiane J. Fröhlich, Dominic Kniveton, Inmaculada Martinez-Zarzoso, Marina Mastrorillo, et al., 'Human Migration in the Era of Climate Change', *Review of Environmental Economics and Policy*, 13(2) (2019), 189–206, https://doi.org/10/ggsxn6.

Christian Aid, *Human Tide: The Real Migration Crisis* (Christianaid.org. uk, 2007), https://www.christianaid.org.uk/resources/about-us/human-tide-real-migration-crisis-2007.

De Genova, Nicholas, and Nathalie Peutz (eds), *The Deportation Regime: Sovereignty, Space, and the Freedom of Movement* (Durham, NC: Duke University Press, 2010).

De Haas, Hein, 'Climate Refugees: The Fabrication of a Migration Threat' (Heindehaas.blogspot.com, 2020), https://heindehaas.blogspot.com/2020/01/climate-refugees-fabrication-of.html.

Environmental Justice Foundation, *Beyond Borders: Our Changing Climate—Its Role in Conflict and Displacement* (London: Environmental Justice Foundation, 2017), https://ejfoundation.org/reports/beyond-borders.

Espeland, Wendy Nelson, and Mitchell L. Stevens, 'A Sociology of Quantification', *European Journal of Sociology / Archives Européennes de Sociologie*, 49(3) (2008), 401–36, https://doi.org/10.1017/S0003975609000150.

Farbotko, Carol, 'Voluntary Immobility: Indigenous Voices in the Pacific', *Forced Migration Review*, 57, 2018, 81–83.

Friends of the Earth, 'A Citizen's Guide to Climate Refugees' (Friends of the Earth, 2007), https://www.safecom.org.au/pdfs/foe-climate-citizens-guide.pdf.

Geisler, Charles, and Ben Currens, 'Impediments to Inland Resettlement under Conditions of Accelerated Sea Level Rise', *Land Use Policy*, 66(Supplement C) (2017), 322–30, https://doi.org/10.1016/j.landusepol.2017.03.029.

Gemenne, François, 'Why the Numbers Don't Add Up: A Review of Estimates and Predictions of People Displaced by Environmental Changes', *Global Environmental Change*, Migration and Global Environmental Change— Review of Drivers of Migration, 21 (2011), 41–49, https://doi.org/10.1016/j.gloenvcha.2011.09.005.

Giuliani, Gaia, 'Afterword', in *Life Adrift: Climate Change, Migration, Critique*, ed. by Andrew Baldwin and Giovanni Bettini (London: Rowman & Littlefield International, 2017), pp. 227–41.

Hansen, Hans Krause, and Tony Porter, 'What Do Numbers Do in Transnational Governance?', *International Political Sociology*, 6(4) (2012), 409–26, https://doi.org/10/gf5cxw.

Héran, François, 'La Recherche Produit des Données de Qualité sur les Migrations et les Migrants: Utilisons-les pour un Débat Réellement Informé!' (Icmigrations.fr, 2020), http://icmigrations.fr/2020/02/03/defacto-015-03/.

Hulme, Mike, 'Reducing the Future to Climate: A Story of Climate Determinism and Reductionism', *Osiris*, 26(1) (2011), 245–66, https://doi.org/10.1086/661274.

Institute for Economics and Peace, *Ecological Threat Register 2020: Understanding Ecological Threats, Resilience and Peace* (Sydney: Institute for Economics and Peace, 2020), http://visionofhumanity.org/reports/.

Ionesco, Dina, Daria Mokhnacheva, and François Gemenne, *Atlas des migrations environnementales* (Paris: Presses de la fondation nationale des sciences politiques, 2016), https://publications.iom.int/books/atlas-des-migrations-environnementales.

Jacobson, Jodi L., 'Environmental Refugees: A Yardstick of Habitability', *Bulletin of Science, Technology & Society*, 8 (1988), 257–58, https://doi.org/10/bwk3qs.

Kalir, Barak, and Céline Cantat, 'Fund but Disregard: The EU's Relationship to Academic Research on Mobility' (Crisismag.net, 2020), https://crisismag.net/2020/05/09/fund-but-disregard-the-eus-relationship-to-academic-research-on-mobility/.

Kelman, Ilan, 'Imaginary Numbers of Climate Change Migrants?', *Social Sciences*, 8(5) (2019), 131, https://doi.org/10/gf86dc.

Khosravi, Shahram, 'Afterword. Experiences and Stories along the Way', *Geoforum*, 116 (2018), 292–95, https://doi.org/10.1016/j.geoforum.2018.05.021.

Luetz, Johannes M., 'Climate Refugees: Why Measuring the Immeasurable Makes Sense Beyond Measure', in *Climate Action*, ed. by Walter Leal Filho, Anabela Marisa Azul, Luciana Brandli, Pinar Gökcin Özuyar, and Tony Wall (Cham: Springer International Publishing, 2019), pp. 1–14, https://doi.org/10.1007/978-3-319-71063-1_81-1.

McLeman, Robert, 'Developments in Modelling of Climate Change-Related Migration', *Climatic Change*, 117(3) (2013), 599–611, https://doi.org/10.1007/s10584-012-0578-2.

McMichael, Celia, Shouro Dasgupta, Sonja Ayeb-Karlsson, and Ilan Kelman, 'A Review of Estimating Population Exposure to Sea-Level Rise and the Relevance for Migration', *Environmental Research Letters*, 15(120) (2020), 123005, https://doi.org/10.1088/1748-9326/abb398.

McNamara, Karen Elizabeth, and Chris Gibson, '"We Do Not Want to Leave Our Land": Pacific Ambassadors at the United Nations Resist the Category of "Climate Refugees"', *Geoforum*, Gramscian Political Ecologies Themed Issue: Understanding Networks at the Science-Policy Interface, 40(3) (2009), 475–83, https://doi.org/10.1016/j.geoforum.2009.03.006.

Methmann, Chris, 'Visualizing Climate-Refugees: Race, Vulnerability, and Resilience in Global Liberal Politics', *International Political Sociology*, 8(4) (2014), 416–35, https://doi.org/10.1111/ips.12071.

Mezzadra, Sandro, and Brett Neilson, *Border as Method, or, the Multiplication of Labor* (Durham: Duke University Press, 2013).

Myers, Norman, 'Environmental Refugees in a Globally Warmed World', *BioScience*, 43(11) (1993), 752–61, https://doi.org/10.2307/1312319.

Myers, Norman, 'Environmental Refugees', *Population and Environment*, 19(2) (1997), 167–82, https://doi.org/10.1023/A:1024623431924.

Myers, Norman, 'Environmental Refugees: A Growing Phenomenon of the 21st Century', *Philosophical Transactions of the Royal Society of London B: Biological Sciences*, 357(1420) (2002), 609–13, https://doi.org/10.1098/rstb.2001.0953.

Myers, Norman, and Jennifer Kent, *Environmental Exodus: An Emergent Crisis in the Global Arena* (Washington, DC: The Climate Institute, 1995).

Neumann, Barbara, Athanasios T. Vafeidis, Juliane Zimmermann, and Robert J. Nicholls, 'Future Coastal Population Growth and Exposure to Sea-Level Rise and Coastal Flooding—A Global Assessment', *PLoS ONE*, 10(3) (2015), e0131375, https://doi.org/10.1371/journal.pone.0118571.

Oakes, Robert, 'Culture, Climate Change and Mobility Decisions in Pacific Small Island Developing States', *Population and Environment*, 40(4) (2019), 480–503, https://doi.org/10.1007/s11111-019-00321-w.

Obokata, Reiko, Luisa Veronis, and Robert McLeman, 'Empirical Research on International Environmental Migration: A Systematic Review', *Population and Environment*, 36(1) (2014), 111–35, https://doi.org/10.1007/s11111-014-0210-7.

Pallister-Wilkins, Polly, 'Walking, Not Flowing: The Migrant Caravan And The Geoinfrastructuring Of Unequal Mobility' (Societyandspace.org, 2019), https://www.societyandspace.org/articles/walking-not-flowing-the-migrant-caravan-and-the-geoinfrastructuring-of-unequal-mobility.

Pearce, Fred, 'Climate Refugees Exist but We Don't Know How Many There Are', *New Scientist*, 210(2810) (2011), 6–7, https://doi.org/10/csz8ss.

Piguet, Etienne, 'Linking Climate Change, Environmental Degradation, and Migration: A Methodological Overview', *Wiley Interdisciplinary Reviews: Climate Change*, 1(4) (2010), 517–24, https://doi.org/10.1002/wcc.54.

Piguet, Etienne, Pécoud, and Paul de Guchteneire, 'Migration and Climate Change: An Overview', *Refugee Survey Quarterly*, 30(3) (2011), 1–23, https://doi.org/10.1093/rsq/hdr006.

Porter, Theodore M., *Trust in Numbers: The Pursuit of Objectivity in Science and Public Life* (Princeton, NJ: Princeton University Press, 1995).

Rigaud, Kanta Kumari, Alex de Sherbinin, Bryan Jones, Jonas Bergmann, Viviane Clement, Kayly Ober, et al., *Groundswell: Preparing for Internal Climate Migration* (World Bank, 2018), https://doi.org/10.1596/29461.

Saunders, Patricia, L., 'Environmental Refugees: Origins of a Construct', in *Political Ecology: Science, Myth, and Power*, ed. by P. Stott and Sian Sullivan (London: Edward Arnold, 2000), pp. 218–46.

Schapendonk, Joris, Ilse van Liempt, Inga Schwarz, and Griet Steel, 'Re-Routing Migration Geographies: Migrants, Trajectories and Mobility

Regimes', *Geoforum*, 116 (2018), 211–16, https://doi.org/10.1016/j. geoforum.2018.06.007.

Schapendonk, Joris, and Griet Steel, 'Following Migrant Trajectories: The Im/ Mobility of Sub-Saharan Africans En Route to the European Union', *Annals of the Association of American Geographers*, 104(2) (2014), 262–70, https://doi. org/10.1080/00045608.2013.862135.

Scheel, Stephan, 'Studying Embodied Encounters: Autonomy of Migration beyond Its Romanticization', *Postcolonial Studies*, 16(3) (2013), 279–88, https://doi.org/10.1080/13688790.2013.850046.

Scheel, Stephan, and Funda Ustek-Spilda, 'The Politics of Expertise and Ignorance in the Field of Migration Management', *Environment and Planning D: Society and Space*, 37(4) (2019), 663–81, https://doi.org/10.1177/0263775819843677.

Singh, Chandni, Mark Tebboth, Dian Spear, Prince Ansah, and Adelina Mensah, 'Exploring Methodological Approaches to Assess Climate Change Vulnerability and Adaptation: Reflections from Using Life History Approaches', *Regional Environmental Change*, 19(8) (2019), 2667–82, https:// doi.org/10.1007/s10113-019-01562-z.

Stern, N. H. (ed.), *The Economics of Climate Change: The Stern Review* (Cambridge: Cambridge University Press, 2007).

Suhrke, Astri, 'Environmental Degradation and Population Flows', *Journal of International Affairs*, 47(2) (1994), 473–96.

Takle, Marianne, 'Migration and Asylum Statistics as a Basis for European Border Control', *Migration Studies*, 5(2) (2017), 267–85, https://doi.org/10.1093/ migration/mnx028.

Tejero, Debora Gonzalez, Lorenzo Guadagno, and Alessandro Nicoletti, 'Human Mobility and the Environment: Challenges for Data Collection and Policymaking', *Migration Policy Practice*, X(1) (2020), 30, 3–10.

Van der Geest, Kees, 'Local Perceptions Of Migration From North-West Ghana', *Africa*, 80(4) (2010), 595–619, https://doi.org/10.3366/afr.2010.0404.

Van der Geest, Kees, Maxine Burkett, Juno Fitzpatrick, Mark Stege, and Brittany Wheeler, *Marshallese Perspectives on Migration in the Context of Climate Change* (Geneva: International Organization for Migration, 2019).

Van Houtum, Henk, and Rodrigo Bueno Lacy, 'The Migration Map Trap. On the Invasion Arrows in the Cartography of Migration', *Mobilities*, 15(2) (2020), 196–219, https://doi.org/10.1080/17450101.2019.1676031.

Vinke, Kira, and Roman Hoffmann, 'Data for a Difficult Subject: Climate Change and Human Migration', *Migration Policy Practice*, X.1 (2020), 30, 16–22.

Zickgraf, Caroline, 'The Fish Migrate and so Must We': The Relationship between International and Internal Environmental Mobility in a Senegalese Fishing Community', *Journal of International Relations*, 16(1) (2018), 5–21.

7. We'll Always Have Paris

Mike Hannis

Many were rightly sceptical of the Paris Agreement's choreographed performance of success, given its reliance on theoretical carbon trading, fantastical Negative Emission Technologies (NETs), and voluntary national 'contributions'. But was COP21 the high-water mark of climate co-operation? Can COP26 rekindle the internationalist spirit required to keep even the idea of a globally co-ordinated effort alive, in the face of resurgent nationalism and the proliferation of apparently more immediate crises? This article explores the chances of COP26 reinvigorating international co-operation, and with it the flagging credibility of the whole Paris process. It focuses in particular on the Paris Agreement's controversial Article 6 rules on voluntary carbon trading, and the urgent need to prevent emissions traded across international borders from counting towards Nationally Determined Contributions (NDCs).

All Eyes on Paris[1]

For a few weeks in late 2015, all eyes were on Paris. High-level delegates from almost every country on Earth attended the 21st Conference of the Parties (COP) to the UN Framework Convention on Climate Change (UNFCCC)—more snappily known as COP21, or the Paris Climate Conference. Civil society and media swarmed in too. There are subsidiary climate COPs every year, but major ones follow a five-year cycle, making COP26 in Glasgow the next 'last chance to save the world'.

---

1   Parts of this chapter were first published in *The Land* magazine, 27 (2020), 18–20; 28 (2021), 4.

https://doi.org/10.11647/OBP.0265.07

Before Paris, 2009's COP15 in Denmark had failed to live up to its ill-advised branding as Hopenhagen. A tentative goal of limiting temperature increases to two degrees above pre-industrial levels was agreed, subject to review in 2015, but no progress was made on any practical steps towards actually achieving this, or towards any kind of legally binding agreement.

2015's COP21 in Paris did see a genuine breakthrough. This was achieved, however, by abandoning attempts to create a legally binding system, and instead adopting the voluntary Paris Agreement (UNFCCC 2015). The Agreement upheld the below 2°C warming target, and even added an aspiration to keep warming within 1.5 degrees. Each Party (UNFCCC member states, plus the EU) agreed to set out its planned reductions in emissions, now termed Nationally Determined Contributions (NDCs). Reviewed every five years, NDCs can be amended to be 'more ambitious', but are never supposed to be revised downwards. This review process is intended to 'ratchet up' commitment to emissions reduction, but detailed discussion of how the new system would actually work in practice was deferred.

Before diaries were ripped up by the COVID-19 pandemic, COP26 was scheduled for November 2020. In time for this Glasgow meeting, Parties had been asked to set long-term decarbonisation goals, as well as to undertake the first five-yearly review of their shorter-term NDCs (Gabbatiss 2021). This process was to be governed by a 'Paris rulebook', details of which were intended to have been agreed and finalised in advance of COP26. These rules are supposed to make NDCs transparent, fair and robust by ensuring that all countries calculate them using agreed common methodologies, rather than doing the sums in whatever way works to their advantage. Standardisation would also allow the Intergovernmental Panel on Climate Change (IPCC) and others to plausibly translate the aggregated NDCs into global temperature change forecasts.

Carbon Trading Rules

Intermediate COPs since Paris largely saw fudges and grandstanding rather than real progress, but nonetheless the Paris rulebook was mostly agreed by the end of COP24 (held in Katowice, Poland in 2018), albeit with the significant exception of rules on voluntary carbon trading. These are known as Article 6 rules, after the somewhat obscure but critically important part of the Paris Agreement they relate to (UNFCCC 2015: 4–5).

This trading issue is significant. The Paris Agreement explicitly allows countries calculating their NDCs to include emissions reductions elsewhere over which they have somehow gained 'ownership', as well as those actually achieved within their own territory. In so doing, it arguably makes voluntary carbon trading a more prominent mechanism for delivering emissions reductions than it ought to be. For clarity, the relevant UNFCCC COP21 Agreement Article 6 rules are as follows:

Article 6

1. Parties recognize that some Parties choose to pursue voluntary cooperation in the implementation of their nationally determined contributions to allow for higher ambition in their mitigation and adaptation actions and to promote sustainable development and environmental integrity.

2. Parties shall, where engaging on a voluntary basis in cooperative approaches that involve the use of internationally transferred mitigation outcomes towards nationally determined contributions, promote sustainable development and ensure environmental integrity and transparency, including in governance, and shall apply robust accounting to ensure, inter alia, the avoidance of double counting, consistent with guidance adopted by the Conference of the Parties serving as the meeting of the Parties to this Agreement.

3. The use of internationally transferred mitigation outcomes to achieve nationally determined contributions under this Agreement shall be voluntary and authorized by participating Parties.

4. A mechanism to contribute to the mitigation of greenhouse gas emissions and support sustainable development is hereby established under the authority and guidance of the Conference of the Parties serving as the meeting of the Parties to this Agreement for use by Parties on a voluntary basis. It shall be supervised by a body designated by the Conference of the Parties serving as the meeting of the Parties to this Agreement, and shall aim:

(a) To promote the mitigation of greenhouse gas emissions while fostering sustainable development;

(b) To incentivize and facilitate participation in the mitigation of greenhouse gas emissions by public and private entities authorized by a Party;

(c) To contribute to the reduction of emission levels in the host Party, which will benefit from mitigation activities resulting in emission reductions that can also be used by another Party to fulfil its nationally determined contribution;

and (d) To deliver an overall mitigation in global emissions.

5. Emission reductions resulting from the mechanism referred to in paragraph 4 of this Article shall not be used to demonstrate achievement of the host Party's nationally determined contribution if used by another Party to demonstrate achievement of its nationally determined contribution (UNFCCC 2015: 4–5).

The voluntary carbon trading framework established by Article 6 risks legitimising a wide range of questionable practices whereby richer countries offset their polluting activities by paying for allegedly emission-reducing or carbon-capturing activities in poorer ones. For instance, if the Norwegian government pays for some reforestation in Indonesia, or a British company pays to install a scrubber to remove potential carbon emissions from the chimney of a chemical plant in India, the resulting greenhouse gas reductions will be reported as part of the Norwegian or British NDCs (as provided for in Article 6.5). One problem immediately arising here is that the 'host Party', in this example Indonesia or India, might also want to report the resulting reductions in their own NDC.

Article 6 is clear that such double counting would not be allowed (see 6.2 and 6.5 above), which at first sight seems fair. But is it? Should the host really have to also identify and enact a *second* set of reductions which will count towards its own target? Would this not mean it was taking on a disproportionate share of the overall work? This kind of argument has been forcefully put, for example by Brazilian delegates concerned with retaining sovereignty over the Amazon Forest, but also by consultants (e.g. Streck 2020) whose creative interpretation of what constitutes double counting relies on the idea that private sector voluntary trading schemes (as opposed to government actions) should *not* be counted as part of NDCs. This apparently technical dispute masks a bitter standoff between countries hoping to offset their own activities by, for example, funding preservation of the Amazon through carbon offset purchases, and a Brazilian government insisting on full sovereignty over the area's resources. Neither is on the side of the angels.[2]

OMG(E)!

There are also tricky practical questions about standards and verification. Who is supposed to make sure that any given scheme is not counted twice? How exactly might this be done? The answer may well be different for bilateral deals between two countries (covered by Article 6.2), and for trades undertaken within the regulated global carbon trading market mechanism envisaged by Article 6.4, but neither is clear.

The latter mechanism is intended to supersede the earlier Clean Development Mechanism (CDM), a dysfunctional voluntary trading scheme established by the 1997 Kyoto Protocol of the UNFCCC. Some countries hold vast numbers of old CDM credits, and have argued that they should be able to count these against future NDCs. In the case of Australia, cashing in these CDM credits would at a stroke have achieved (on paper) more than half the emissions cuts required to meet its NDC target of reducing emissions to 26–28% below 2005 levels by 2030. One Australian economist has described this situation as "tantamount to a drunk guy waving an expired Starbucks coffee voucher around in a McDonald's and acting surprised that nobody wanted to give him a coffee" (Denniss 2020). In the face of such ridicule this strategy was eventually dropped (Doherty 2020).

The overarching issue here is whether voluntary trading actually results in 'Overall Mitigation in Global Emissions' (OMGE), meaning a genuine net reduction, rather than just serving as a way for emissions in one place to be offset elsewhere, thereby allowing business-as-usual to continue (as also foregrounded in Lankford's discussion in this

2   This summary of post-Paris COP negotiations is largely derived from comprehensive coverage at www.carbonbrief.org. Article 6 of the Paris Agreement (UNFCCC 2015: 4–5) addresses "[voluntary] use of internationally transferred mitigation outcomes [ITMOs] to achieve nationally determined contributions": for background, see https://www.carbon-mechanisms.de/en/introduction/the-paris-agreement-and-article-6. For an insider account of the recent bilateral agreement claimed to be "the first instrument that provides access to the voluntary carbon market to ITMOs under the provisions of the Paris Agreement", see Elgart and Secada 2020.

volume regarding the necessity and difficulty of securing permanent CO_2 'salvage', and in Dyke et al.'s chapter on the difficulties of achieving net zero emissions overall).

This issue was discussed at the Paris COP, and while detail is lacking, the principle of OMGE is acknowledged in Article 6.4 as an objective of the proposed global trading scheme. Worryingly, however, no such objective is included in Article 6.2 on bilateral trading, although this Article does explicitly mention the need to avoid "double accounting" of emissions reductions. As things stand there appears to be no watertight obligation to ensure that any given offsetting transaction between two countries actually results in a net emissions reduction. It has been largely left to the Association of Small Island States, whose territories are already literally disappearing under the waves, to point out how disastrous this could be (Dizzanne 2019).

COP25 (held in Madrid in 2019) was supposed to see all these arguments settled. To the dismay of many but the surprise of few, once again this did not happen. Newly minted climate celebrity Greta Thunberg captured the mood when she told a restive plenary hall that the COP seemed to have "turned into some kind of opportunity for countries to negotiate loopholes" (Evans and Gabbatis 2019: online). The lamentable failure to agree Article 6 rules before COP26 means that the five-yearly review of NDCs is happening without agreement on crucial elements of what these can or cannot contain. This raises a real danger that Parties' emissions reductions may be inflated beyond what has actually been achieved, meaning that the world is even further from achieving 'net zero carbon' than reported figures suggest.

Better Late than on Time?

Quite apart from the sorry outcome of preparatory negotiations, there was widespread relief that COP26 did not have to take place against the backdrop of the US's formal withdrawal from the Paris Agreement, which came into effect on 4 November 2020, just before the original Glasgow dates. Trump's contrarian refusal to co-operate on climate had of course been an elephant in every COP negotiating room since 2017. The incoming Biden administration wasted no time in rejoining the Paris Agreement, and has submitted a relatively ambitious NDC promising a 50–52% reduction in greenhouse gas emissions by 2030 from a 2005 baseline, alongside goals to create a "carbon pollution-free power sector by 2035" and a "net zero greenhouse gas emissions economy by no later than 2050" (White House 2021: online). Nonetheless while US re-engagement provides a much-needed boost to the flagging credibility of the whole Paris process, there are several lost years to make up before optimistic Democrat claims about 're-establishing climate leadership' will appear credible. Veterans of former US President Obama's Paris negotiating team have been recruited to assist, and will be working hard in Glasgow.

Climate economist Nicholas Stern claimed that the delay gave time to prepare for a big push at the Glasgow COP towards ensuring that rather than propping up business as usual, the massive funds being poured into pandemic recovery fund a transition to a sustainable and resilient economy (Harvey 2020). Was his optimism realistic? How will the new world of post-COVID international relations handle the need to co-operate? It is still too soon to say whether Stern's vision of pandemic recovery funds kick-starting a new global economy "in closer harmony with the natural world" (Harvey 2020: online) will come true, but early signs are not good. While there have been encouraging noises from the EU about making sure its COVID recovery plans are at least congruent with its NDC targets, there are (as ever) questions around whether this rhetoric will be matched by action. Elsewhere, many countries have given massive bailouts and loans to airlines and fossil fuel companies, without even attaching conditions on improved environmental performance (Bailoutwatch 2021; Transport and Environment 2020). This financial stimulus risks locking in business-as-usual for decades to come. Meanwhile, researchers identify a growing trend towards 'cutting green tape' as politicians accept arguments from business that climate-related regulation is hindering economic recovery (Bond et al. 2020). This tendency is happening in many countries but again the most egregious example has been the US, where the former Trump-led Republican administration seized on recovery rhetoric to justify its existing plans to rescind or weaken a truly alarming number of environmental regulations (Popovich et al. 2020).

The context here was not only climate denial and cronyism (as also flagged by Bigger et al. in this volume), but a wholesale repudiation of international agreements of all kinds. Trump withdrew the US not only from Paris, but also from numerous other international treaties on issues from nuclear arms control, to human rights, to the militarisation of space. His administration even quit the World Health Organisation in the middle of a pandemic. Clearly, internationalism of any kind was firmly off the table and it will not be politically easy to turn this supertanker around. It would take a very brave Democrat to stand up and say that America is no longer First.

Meanwhile, varieties of Trumpism live on in countries such as Brazil, Australia and India—all major players in climate negotiations. China is setting stronger domestic targets, but has never been noted either for its multilateralism or for its altruistic stance on global affairs. Despite the fall of Trump, this does not appear to be an auspicious time for Paris-style voluntary co-operation (Sachs 2019).

UK, EU, CO$_2$

Squirming in the COP26 host's spotlight will be a UK government incongruously obsessed with the idea that the country should make its own buccaneering way in the world, beholden to no-one. A desperate scramble to sign trade deals with anyone other than the EU led UK negotiators to accept that any US/UK deal must not even mention climate change (Hannis 2020). Whether this will change under Biden remains to be seen.

As the host of COP26, the UK is expected to set an example, and virtue-signalling on climate is also seen as an easy way for a newly isolated blond populist to build bridges with the Biden administration. Unfortunately, talking up the UK's climate commitment now takes the distinctly Trumpian form of claiming that a clean, high-tech Britain is forging ahead of the dirty old EU, not to mention the rest of the world.

In the proud new era of unchallenged sovereignty, no opportunity is missed to make clear that Britain is Best. Even when the UK drugs regulator licensed a COVID vaccine created in Germany by a Turkish couple, for a US drug company to manufacture in Belgium, puppyish cabinet minister Gavin Williamson explained that this proved British scientists were "the best in the world", and that "we're a much better country than every single one of them" (Euronews 2020: online). If still in post, Williamson may perhaps get deployed to Glasgow. His breathless, fact-free enthusiasm would be perfect for press releases like this:

> The UK's new target to reduce greenhouse gas emissions—our Nationally Determined Contribution (NDC) under the Paris Climate Agreement—is among the highest in the world and commits the UK to cutting emissions at the fastest rate of any major economy so far. Today's target is the first set by the UK following its departure from the EU, demonstrating the UK's leadership in tackling climate change (BEIS 2020a: online).

The UK NDC promises to reduce emissions "by at least 68% by 2030, compared to 1990 levels", meaning that "UK emissions per person will fall from around 14 tCO2e [tonnes of CO2 equivalent] in 1990 to fewer than 4 tCO2e in 2030" (BEIS 2020c: 1, 28). This sounds impressive, but most of this reduction has already happened, due largely to the historic move from coal to gas power stations between 1990 and 2015 (Thomas et al. 2019). The per person figure for 2019 was 5.3 tonnes, so the new target in fact proposes a less impressive cut of only around a quarter of this amount over the next decade (Evans 2020). The equivalent EU target is a cut of 55% by 2030 (Climate Action Tracker 2020). This is certainly a lower headline figure than the UK's proposed 68%, but it is also an average across twenty-seven countries facing many different challenges. Current EU average per capita annual emissions are around 6.7 tonnes, and if the 2030 target were met this would reduce to around five, meaning that a similar drop of around a quarter from today's levels is envisaged by 2030 (Eurostat 2021).

Even ten years ahead seems a very long time under present conditions, but these 2030 targets are intended as stepping stones towards reaching the current holy grail of 'zero carbon by 2050', to which both the UK and the EU have committed, along with a growing list of other countries. A timely measure of how hard this will be is provided by work estimating that the dramatic drop in global economic activity caused by the pandemic will impact global temperature by no more than 0.01 degrees (Forster et al. 2020).

Oddly, the UK's NDC announcement carried endorsements from banks, energy companies, Tesco and Coca Cola Europe (BEIS 2020a). Their enthusiasm may in part be explained by the fact that the NDC target does not include any emissions elsewhere in the world, such as those arising from the production of goods for UK consumption, or from overseas activities of UK-registered companies. It also excludes international aviation and shipping.

The Return of Article 6

On the face of it, the UK's NDC does at least commit to achieving reductions by actually emitting fewer greenhouse gases, rather than by international offsetting or voluntary carbon trading. But perhaps inevitably, voodoo carbon economics reappear in the small print:

> While the UK intends to meet its NDC target through reducing emissions domestically, it reserves the right to use voluntary cooperation under Article 6 of the Paris Agreement. Such use could occur through the linking of a potential UK emissions trading system to another emissions trading system or through the use of emissions reductions or removals units (BEIS 2020b: 27).

Meanwhile the recent Energy White Paper proclaims in now-familiar triumphal tones:

> Having left the EU, we are ready to lead the world again.

> We will establish a UK Emissions Trading Scheme (ETS) to replace the UK's participation in the EU ETS. [...] the UK is open to linking the UK ETS internationally [...] we are considering a range of options, but no decision on our preferred linking partners has yet been made (BEIS 2020c: online).

An unlinked UK ETS would be implausible, and the only realistic 'linking partner' will be the much bigger EU scheme, so it seems inevitable that any UK ETS will effectively become an offshoot of the EU ETS, sharing the many flaws of that scheme while having lost the ability to influence it (Gabbatiss 2020). Post-Brexit threats to genuine decarbonisation will of course come not only from the disingenuous carbon trading facilitated by Article 6, but also from old-fashioned physical trade. Importing and exporting goods across the world rather than across the English Channel is not exactly going to help with reducing carbon emissions. More broadly, EU law and oversight have been the key upward drivers of UK climate and environmental standards for decades. No-one seriously believes that the sacred 'divergence' will result in UK environmental standards being higher than those in the EU. But they will be *British* standards, so of course they will be better.

Beware the Bubble

If the retreat from global co-operation continues in Glasgow, one outcome might be that Article 6.4's projected global carbon trading mechanism never gets off the ground. As discussed above this might well encourage further growth in poorly regulated bilateral offset deals under Article 6.2. There are worrying signs that the UK may seek to become a hub for brokering such deals, given its stated aspiration to "position the UK, and the City of London, as a leader in the global voluntary carbon markets" (BEIS 2020d; for background see also Mikolajczyk and 't Gilde 2020).

There may, however, be a more constructive option, and indeed one which should appeal to those sceptical about international co-operation. Could failure to achieve a global trading regime encourage the Parties to actually take responsibility for their own emissions? Nothing in the Paris Agreement stops a country producing an honest NDC based on genuine reductions in emissions, with no reliance either on carbon trading or on fanciful 'negative emissions technologies' (Hannis 2017; Herzog 2018). Such honesty would also mean including emissions associated with everything the country consumes, no matter where it was produced. The Johnson Government has so far shown little interest in work mapping out what this could mean for the UK (see, for example, Allwood et al. 2019; Allen et al. 2019). Genuine leadership at COP26 could start here, rather than with attempts to reflate the carbon trading bubble.

References

Allen, Paul, Laura Blake, Peter Harper, Alice Hooker-Stroud, Philip James, Tobi Kellner, et al., *Zero Carbon Britain: Rising to the Climate Emergency* (Machynlleth: Centre for Alternative Technology, 2019), https://www.cat.org.uk/info-resources/zero-carbon-britain/research-reports/zero-carbon-britain-rising-to-the-climate-emergency/.

Allwood, Julian, M., C. F. Dunant, R. C. Lupton, C. J. Cleaver, A. C. H. Serrenho, J. M. C. Azevedo, et al., *Absolute Zero: Delivering the UK's Climate Change Commitment with Incremental Changes to Today's Technologies* (Cambridge:

University of Cambridge, 2019), http://www.eng.cam.ac.uk/news/absolute-zero, https://doi.org/10.17863/CAM.46075.

Bailoutwatch, *What Fossil Fuels Really Did with Their Bailouts* (Bailoutwatch.org, 2021), https://bailoutwatch.org/analysis/what-fossil-fuels-really-did-with-bailouts.

BEIS (Department for Business, Energy and Industrial Strategy), *UK Sets Ambitious New Climate Target Ahead of UN Summit* (Gov.uk, 2020a), https://www.gov.uk/government/news/uk-sets-ambitious-new-climate-target-ahead-of-un-summit.

BEIS (Department for Business, Energy and Industrial Strategy), *United Kingdom of Great Britain and Northern Ireland's Nationally Determined Contribution* (Gov.uk, 2020b), https://assets.publishing.service.gov.uk/government/uploads/system/uploads/attachment_data/file/943618/uk-2030-ndc.pdf.

BEIS (Department for Business, Energy and Industrial Strategy), *Energy White Paper: Powering Our Net Zero Future* (Gov.uk, 2020c), https://www.gov.uk/government/publications/energy-white-paper-powering-our-net-zero-future.

BEIS (Department for Business, Energy and Industrial Strategy), *The Ten Point Plan for a Green Industrial Revolution* (Gov.uk, 2020d), https://assets.publishing.service.gov.uk/government/uploads/system/uploads/attachment_data/file/936567/10_POINT_PLAN_BOOKLET.pdf.

Billy, Dizzanne, 'Controversial Carbon Markets and Small Island States' (Earthjournalism.net, 2019), https://earthjournalism.net/stories/cop25-focus-controversial-carbon-markets-and-small-island-states.

Bond, Alan J., Angus Morrison-Saunders, Francois Pieter Retief, and Meinhard Doelle, 'Environmental Regulations Likely to be First Casualties in Post-pandemic Recovery' (Theconversation.com, 2020), https://theconversation.com/environmental-regulations-likely-to-be-first-casualties-in-post-pandemic-recovery-137941.

Climate Action Tracker, *CAT Climate Target Update Tracker: EU* (Climateactiontracker.org, 2020), https://climateactiontracker.org/climate-target-update-tracker/.

Denniss, R., 'Until Recently, Pressure on Australia to Drop Carryover Credits Had Little Impact. But Times Change' (Theguardian.com, 2020), https://www.theguardian.com/commentisfree/2020/dec/09/until-recently-pressure-on-australia-to-drop-carryover-credits-had-little-impact-but-times-change.

Doherty, Ben, 'Australia Won't Use Kyoto Carryover Credits to Meet Paris Climate Targets, Scott Morrison Confirms' (Theguardian.com, 2020), https://www.theguardian.com/world/2020/dec/11/australia-wont-use-kyoto-carryover-credits-to-meet-paris-climate-targets-scott-morrison-confirms.

Elgart, Veronika, and Secada, Laura, 'The Peruvian-Swiss Article 6 Agreement: How it Came About and How it Works', *Carbon Mechanisms Review* (4)

(2020), 10–18, https://www.carbon-mechanisms.de/en/publications/details/carbon-mechanisms-review-04-2020-managing-the-interim.

Evans, Simon, 'UK's CO2 Emissions Have Fallen 29% Over the Past Decade' (Carbonbrief.org, 2020), https://www.carbonbrief.org/analysis-uks-co2-emissions-have-fallen-29-per-cent-over-the-past-decade.

Evans, Simon, and Gabbatis, Josh, 'Key Outcomes Agreed at the UN Climate Talks in Madrid' (Carbonbrief.org, 2019), http://www.carbonbrief.org/cop25-key-outcomes-agreed-at-the-un-climate-talks-in-madrid.

Euronews, 'Gavin Williamson: UK Better Than France, US and Belgium, Says Minister on Vaccine Race Win' (Euronews, 2020), https://www.euronews.com/2020/12/03/gavin-williamson-uk-better-than-france-us-and-belgium-says-minister-on-vaccine-race-win.

Eurostat, *Greenhouse Gas Emission Statistics—Carbon Footprints* (Ec.europa.eu, 2021), https://ec.europa.eu/eurostat/statistics-explained/index.php?title=Greenhouse_gas_emission_statistics_-_carbon_footprints.

Forster, Piers M., Harriet I. Forster, Mat J. Evans, Matthew J. Gidden, Chris D. Jones and Christoph A. Keller, et al., 'Current and Future Global Climate Impacts Resulting From COVID-19', *Nature Climate Change*, 10 (2020), 10, 913–19, https://doi.org/10.1038/s41558-020-0883-0.

Gabbatiss, Josh, 'How Does the UK's 'Energy White Paper' Aim to Tackle Climate Change?' (Carbonbrief.org, 2020), https://www.carbonbrief.org/in-depth-qa-how-does-the-uks-energy-white-paper-aim-to-tackle-climate-change#brexit.

Gabbatiss, Josh, 'Which Countries Met the UN's 2020 Deadline to Raise 'Climate Ambition'?' (Carbonbrief.org, 2021), https://www.carbonbrief.org/analysis-which-countries-met-the-uns-2020-deadline-to-raise-climate-ambition.

Hannis, Mike, 'After Development? In Defence of Sustainability' *Global Discourse*, 7(1) (2017), 28–38, https://doi.org/10.1080/23269995.2017.1300404.

Hannis, Mike, 'Night of the Long Spoons', *The Land*, 26 (2020), 4, https://www.thelandmagazine.org.uk/articles/night-long-spoons.

Harvey, Fiona, ''We Must Use This Time Well': Climate Experts Hopeful After Cop26 Delay' (Theguardian.com, 2020), http://www.theguardian.com/environment/2020/apr/02/we-must-use-this-time-well-climate-experts-hopeful-after-cop26-delay-coronavirus.

Herzog, Howard, 'Why We Can't Reverse Climate Change with 'Negative Emissions' Technologies' (Theconversation.com, 2018), https://theconversation.com/why-we-cant-reverse-climate-change-with-negative-emissions-technologies-103504.

Mikolajczyk, Szymon, and 't Gilde, Lieke, *Leveraging Ambition Through Carbon Markets: Effectiveness of Abatement Action Through International Carbon Markets* (Ebrd.com, 2020), https://www.ebrd.com/carbon-markets.pdf.

Popovich, Nadja, Livia Albeck-Ripka, and Kendra Pierre-Louis, 'The Trump Administration Is Reversing Nearly 100 Environmental Rules. Here's the Full List' (Nytimes.com, 2020), https://www.nytimes.com/interactive/2020/climate/trump-environment-rollbacks.html.

Sachs, Noah, 'The Paris Agreement in the 2020s: Breakdown or Breakup?', *Ecology Law Quarterly*, 46 (2019), 865909, https://doi.org/10.15779/Z38H708140.

Streck, Charlotte, 'Corresponding Adjustments for Voluntary Markets—Seriously?' (Ecosystemmarketplace.com, 2020), https://www.ecosystemmarketplace.com/articles/shades-of-redd-corresponding-adjustments-for-voluntary-markets-seriously/.

Transport and Environment, *Bailout Tracker* (Transportenvironment.org, 2020), https://www.transportenvironment.org/what-we-do/flying-and-climate-change/bailout-tracker.

Thomas, Nathalie, Leslie Hooke, and Chris Tighe, 'How Britain Ended its Coal Addiction' (Ft.com, 2019), https://www.ft.com/content/a05d1dd4-dddd-11e9-9743-db5a370481bc.

United Nations Framework Convention on Climate Change (UNFCCC), *Paris Agreement* 2015, https://unfccc.int/sites/default/files/english_paris_agreement.pdf.

White House Press Release, *President Biden Sets 2030 Greenhouse Gas Pollution Reduction Target Aimed at Creating Good Paying Union Jobs and Securing US Leadership on Clean Energy Technologies* (Whitehouse.gov, 2021), https://www.whitehouse.gov/briefing-room/statements-releases/2021/04/22/fact-sheet-president-biden-sets-2030-greenhouse-gas-pollution-reduction-target-aimed-at-creating-good-paying-union-jobs-and-securing-u-s-leadership-on-clean-energy-technologies/.

8. The Atmospheric Carbon Commons in Transition

Bruce Lankford

Originally conceived to discuss water in irrigation systems, this chapter adapts the concept of 'paracommons' to CO_2 governance. The paracommons is 'a commons of material salvages', occurring within the context of multiple pathways for resources salvaged from wastage/waste and via reduced consumption. The carbon/atmospheric commons can be framed in three consecutive stages, with implications for how carbon dioxide is conceived, counted and managed to achieve reductions in global emissions and levels: a 'sink-type atmospheric commons' occurring prior to the 1980/90s, a 'husbandry-type carbon commons' lasting from the 1980/90s to the 2030s, and an emergency 'carbon paracommons' post-2030s. The first stage sees the atmosphere treated as a dump or sink for carbon dioxide (CO_2) 'waste' resulting in rising CO_2 levels. The second stage sees climate change mitigation (e.g. carbon sequestration in forests) as Ostromian-commons husbandry that attempts to reduce CO_2 emission rates but continues to result in levels remaining above 400 ppm. In the third stage, the paracommons treats CO_2 and its 'salvaging' as a matter of urgency leading to permanent sequestration, non-use and transformation.

A 'Commons' Framing of Atmospheric Carbon Dioxide

This article frames carbon in the Earth's atmosphere as three sequential stages of commons,[1] as illustrated in Figure 4: a sink-type commons

---

1   An area or collection of resources for use by individuals and groups often held 'in common' but subject to varying pressures and ownership modalities.

Negotiating Climate Change in Crisis

for carbon dioxide[2] waste;[3] a 'husbandry-type' Ostromian commons for governing CO_2 emissions; and a 'paracommons' where salvaged CO_2 products (such as liquid or frozen carbon dioxide) are created, permanently sequestered and un-used. The first stage, where wastes of combusting fossil fuel were dumped with little regard for their impact on climate change, occurred prior to the 1990s (but has continued), causing increases in atmospheric CO_2 levels. The second stage, running from the 1990s to the near future (2030), sees increasing management or 'husbandry' of terrestrial and atmospheric carbon and carbon emissions. It is suggested that the third stage will consolidate over the next twenty to thirty years as a scarcity or emergency-driven 'paracommons' concerned with controlling the means, amounts, pathways, and ownership of CO_2 'salvages' in order to drive down atmospheric levels. These commons are described in more detail below.

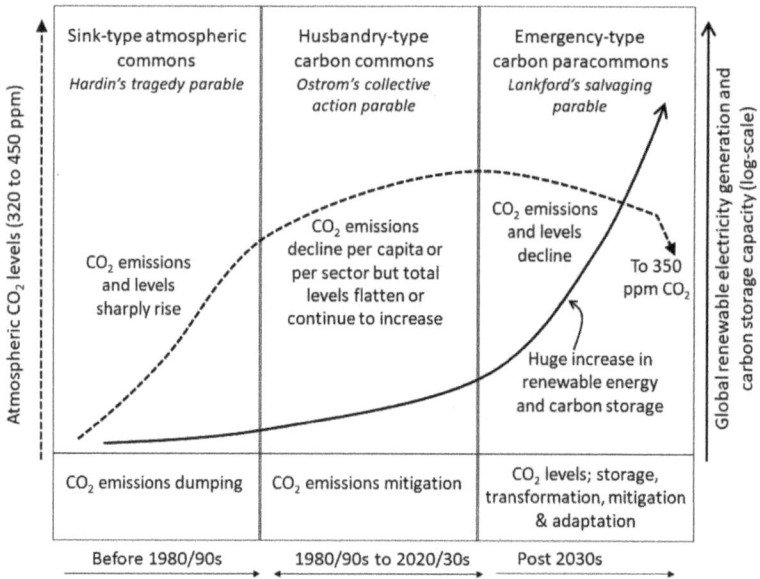

Fig. 4. Three frames and stages of the global atmospheric/carbon commons. Image by chapter author.

2   With limited space only carbon dioxide (CO_2) amongst greenhouse gases is discussed.

3   CO_2 is a gas wasted during fossil fuel conversion that is difficult to capture and recycle (usually termed a wastage), but can be captured with technological innovation.

The Atmospheric Commons: A Sink-Type Commons

The term 'atmospheric commons' observes carbon dioxide as a wastage/ waste dumped into the atmosphere:

> One may argue that the atmosphere can also be regarded as a commons, exploited by all yet owned by none. Most significantly the atmosphere has been abused as a 'common sink'. Until relatively recently it provided a completely free waste disposal system for a whole range of anthropogenic pollutants. It also constitutes the ultimate 'public good', that is to say if resources are expended on improving air quality, it is impossible to exclude people from enjoying the benefits (Vogler 2001: 2427).

The word 'sink' has been used by others to cast the atmosphere as a commons (Edenhofer et al. 2012) consistent with Vogler's "exploitation by all" and also invoking Hardin's controversial Tragedy of the Commons hypothesis (Hardin 1968). As Brown et al. (2019: 61) argue, however, "this pervasive framing of climate change as a commons tragedy limits how we confront the climate challenge".

The Carbon Commons: A Husbandry-Type Commons

Brown et al. (2019) thus critique this sink-type view of the commons by reminding us of Ostrom's 'parable' of collective action regarding resources held in common and their management (Ostrom 1990) and joint governance (Schrijver 2016). Observing three characteristics, this 'carbon commons' invokes an Ostromian 'husbandry' of carbon and CO_2 emissions and levels, including: 1) recognition of the limited atmospheric headroom for further increases in CO_2 because of its causal contribution to climate change; 2) implementation of CO_2 mitigation tools to reduce carbon dioxide emissions (e.g. by carbon offsetting and raising fossil fuel efficiency), and; 3) questions of distribution regarding who is using and has used the carbon commons the most by emitting the most atmospheric carbon (Meyerson 1998; Pierrehumbert 2012).

Given accelerated climate change, however, the concern here is that 'business-as-usual' husbandry of the carbon commons is increasingly insufficient. Current governance will be ineffective, or not effective enough, in bringing down CO_2 levels within a rapid time-frame. Although carbon emissions will flatten out under present approaches to mitigation (Lovell et al. 2009), this 'emissions-focused' husbandry will not deliver stringent decreases in CO_2 concentration levels to less than 350 ppm, as required in order to avert climate volatility (cf. Rockstrom et al. 2009). Extending Brown et al.'s (2019) critique of a 'Tragedy of the Commons' framing of the 'sink' of atmospheric carbon, then, we should consider how a husbandry-type Ostromian commons also limits how we conceive of and confront escalating climate change (Rabinowitz 2010).

A Carbon Paracommons: An Emergency-Driven Resource-Salvaging Commons

This is where an adaptation of the 'paracommons' concept comes in. Drawing from analyses of efficiency gains and their variable uptake in irrigation systems (Lankford 2013, 2018), the term 'paracommons' describes a commons of 'conserved' or 'salvaged' resources arising from efficiency gains and managed non-consumption of natural resources. The Greek prefix 'para', meaning 'alongside', is used to signal that the commons here is of salvaged wastages that emerge from, and sit alongside, the primary commons resource under consideration (i.e. water in irrigation systems and atmospheric carbon in fossil-fuel-dependent industrial production).

Inspiration for the paracommons idea came from identifying why and how water resources believed to be 'lost' in inefficient irrigation systems became the focus of competition if these 'losses' could be 'salvaged'. Conceptually, four parties may compete for these salvages: a 'proprietor' making the efficiency gain (usually the irrigators managing an irrigation system); their 'immediate neighbour' (communities often placed near the periphery of the irrigation system gaining from or losing water 'losses' emanating from the irrigation system); 'society' more broadly; and 'nature', when water is freed up to benefit environmental processes beyond the irrigation system.

In adapting the paracommons concept to global carbon management, the following eight features can be identified:

Scarcity and emergency. The irrigation paracommons sees that 'salvages of irrigation losses' become valuable when water scarcity boosts competition for 'losses freed up' through efficiency gains. In a post-2030 climate future, however, circumstances for carbon management will presumably be different, although analogous in certain respects. Atmospheric CO_2 will not become valuable or sought after under demand-driven conditions of scarcity, but by viewing climate change as an "emergency" (Gills and Morgan 2020), and as a form of scarcity (Asayama et al. 2019), three features will reshape how we view CO_2 'salvage' from the atmosphere. First, a regulated scarcity of CO_2 headroom emission possibilities will mean that CO_2 levels can no longer grow. Second, greater financial and societal values afforded to CO_2 salvages permanently removed from the atmosphere will potentially create much greater interest in taking such salvage action. Connected to this valorisation, broad spectrum ownership of effective carbon salvaging technologies at all levels of society will give an appearance of competition for salvaging CO_2 amongst many players, sitting within a broader cooperative endeavour.

Salvage. A definition for 'salvage' as a verb is to retrieve, utilise or preserve something from potential loss, with 'salvage' as a noun being short-hand here for any means by which CO_2 within, or destined for, the atmosphere is removed from, or stopped from passing to, the atmosphere, thus signalling the production of negative emissions. Examples of CO_2 salvage include its transformation and sequestration into organic liquids and solids (e.g. trees and algae-based fuels), or into liquid or frozen and stored CO_2, or fossil and man-made organic solids and liquids whose oxidation or burning is avoided or minimised (on the complexities posed by such 'salvage' technologies, also see Dyke et al. this volume).

Transience and impermanence. In complex systems represented by water and carbon, the amounts, boundaries and pathways of the resources and their salvaging are leaky and transient (Murray et al. 2007). This means that without strict controls, most husbandry attempts to sequester carbon dioxide into, for example, soil organic matter or trees, are impermanent beyond a time scale of twenty to fifty years. A related problem is the difficulty of accurately accounting for carbon in ways that value and record permanent sequestration (Gifford 2020), as also signalled in the chapters in this volume by Bigger et al. and Hannis.

Consumption rebound. An effect of transience and leakiness is that unused or temporarily salvaged products in one part of the economy may be prone to a consumption rebound elsewhere. This is akin to the Jevons paradox (Stoknes and Rockström 2018; Ruzzenenti et al. 2019), arising, for example, when an increase in vehicle fuel efficiency is undermined by increases in the number of vehicles in use.

Exteriorising and making visible the wastes. In the paracommons, potential wastes/wastages need to be made visible and exteriorised (Lankford 2018), meaning their presence needs to be seen as an integral part of the unused and untransformed resource. Progress in the public understanding of climate change, for example, means that many people now see that oil and coal reserves are not only fuels for energy but also constitute the future atmospheric loading of CO_2 'wastes'. The CO_2 in the fossil fuel has been made visible, and society's changing relationship with the properties of fossil fuels has become a discussion about what becomes exteriorised as they are used, and with what socio-environmental effects.

Exteriorising (making visible) the salvage. The second 'making visible' that exists in the CO_2 paracommons involves the transformation of wastes/wastages into salvage. In irrigation the waste that previously is 'lost' water becomes a gain, because through efficiency innovations losses are recovered or water withdrawals (and their internal losses) are foregone, making more water available for reuse and repurposing (Lankford 2018). With carbon dioxide, various visible salvages exist: CO_2 is permanently evacuated in the form of timber or frozen CO_2; CO_2 is not produced because fossil fuel extraction and burning is foregone; or CO_2 is turned into carbon-salvaged fuels (more or less emission-neutral) that replace fossil fuels (generating additional CO_2 emissions).

A distributive and destination puzzle. Being concerned with controlling CO_2 salvages in a notoriously leaky environment where many possible carbon pathways, distributions and destinations exist, a paracommons framing asks 'who gets the final salvage'. As introduced above, this question identifies four parties acting as destinations for the salvages. Determining these CO_2 pathways and destinations is about bearing down on leakage and the rebound effect in order to ensure salvages permanently end up where they need to be so as to prevent further emissions. As an illustrative example, in the 'husbandry-commons' (as per the description above) vehicle, fuel efficiency results in reduced emissions per driver-kilometre in the short term, but may lead to an uptick in fuel consumption elsewhere either with the same driver (the 'proprietor') or their partner (the 'immediate neighbour') making more journeys, or with more people driving overall ('society'), because it has become more efficient and less costly to do so. Here, then, the salvage (fuel not burnt) passes to the proprietor, their immediate neighbour, or to society, but is not passed to, or withheld permanently, 'in nature'. This example can be extrapolated analogously to the sorts of thorny discussions regarding whether or not voluntary carbon trading in actuality supports effective CO_2 'salvage', or if it instead mostly passes the emitted CO_2 elsewhere (as also highlighted in Hannis's chapter, this volume).

What or who is nature here? The paracommons concept proposes schematically that 'nature' is one of four parties that may benefit from conserved resource salvages, either by recovering the salvages to 'the environment' or by not consuming resources in the first place. In the case of irrigated river basins, irrigation 'wastes' recovered to nature should see ecological/environmental water flows restored. In a CO_2 paracommons, nature is defined schematically as a benefitting party when levels of CO_2 in the atmosphere are decisively reduced. Nature in this simplified CO_2 budget is thus not forest, biochar or organic sediment, if these carbon stores are set to re-release their CO_2 back into the atmosphere within twenty-five to thirty years, such that the trajectory of carbon dioxide levels will remain upwards, undulating or flat (Figure 4). In these urgent terms, mechanisms for permanently sequestering CO_2, such as, for example, in warehouses containing frozen CO_2, would proffer a clearer salvage pathway 'to nature'. That said, such 'fixes' pose their own CO_2 and other complexities, since to industrially process and store CO_2 in ways that do not increase CO_2 levels requires a considerable growth in renewable energy (these concerns are also highlighted in the chapters by Dyke et al., Bigger et al., and Dunlap in this volume). However, in this unreserved 'crisis' definition of 'nature' in a CO_2 paracommons is complicated by the many overlapping 'ecosystem services' also harmed or benefited by a shifting carbon cycle (O'Connor 2008).

Illustrating the Three Commons for Carbon/CO_2 Management

These three types of commons can be illustrated by imagining an industrial mining company that owns one gigatonne of carbon dioxide in coal reserves. In the sink-type commons, the coal is entirely burnt within ten years, dumping waste CO_2 into the Earth's atmosphere.

In the husbandry-type commons, attempts to reduce the company's emissions of CO_2 are made. The gigatonne of coal CO_2 is emitted over a longer twenty-year period because the company reduces annual combustion responding to pricing charges for emissions. Lower emission rates are also offset with a programme of afforestation leading to a forest with a lifespan of thirty years. But after thirty years practically all of the coal's CO_2 ends up in the atmosphere.

A paracommons view of the husbandry-type commons asks where CO_2 'salvaged wastes' (including coal not burnt) end up during attempts to manage, offset or be more efficient with this coal and its yet-to-be-released CO_2. The paracommons argues that four parties compete over the salvaged gain but 'predicts' that with the leakiness of the husbandry-type commons none of these options is easy to trace or constitutes the 'salvage' needed to meet the 350ppm target.

For example: the industrial company (the proprietor) may sell or burn any non-consumed coal after the period under focus; an immediate neighbour (e.g. a community connected to land afforested through offsetting mechanisms) may use the forest resulting in this carbon released back into the atmosphere; 'society' may use fossil carbon from other sources, thereby failing to make the necessary reductions in net consumption; and unused coal retained in the ground may produce 'atmosphere-nature' gains from non-released CO_2. The paracommons model envisages that gains are most likely to pass to the proprietor, its immediate neighbour, and society. Without strong social, political and economic regulation, a permanent salvage is least likely to protect, pass to, or be retained within 'nature'. Unsurprisingly, then, these observations suggest that a nature-safeguarding paracommons needs to be actively designed and regulated so as to genuinely lower atmospheric CO_2 levels.

Designing an Effective Paracommons to Serve 'Nature'?

How, then, might we treat the Earth's atmosphere as a purposively governed paracommons wherein future carbon/CO_2 salvages assuredly protect nature? The points below sketch some principles in moving forward, demonstrating the very real challenges faced in creating meaningful societal structures that combine both CO_2 and decarbonisation:

- Recognising the leakiness and impermanence of the carbon cycle, the paracommons emphasises carbon dioxide *levels* over emissions. A vision for averting dangerous climate change is that salvaged CO_2 must be permanently locked away, as defined by a lowering of CO_2 concentration below 350 ppm within a defined time period (e.g. one hundred years).

- The carbon paracommons asks for a switch to an economy and society that highly values salvaged carbon dioxide products, thus calling for a substantial enablement, reward and valorisation (Luque and Clark 2013) of CO_2 salvage. At the same time, however, such valorisation needs to be considered against the sorts of financialising dynamics, complexities and inequities considered in the chapters by Bracking and Kaplan and Levy, this volume.

- Carbon storage could be enabled by volume-dense cold-storage carbon warehousing, created and managed by a mix of public and private entities and companies. Carbon storage is a provocative 'techno-fix', but consider the following figures. Trees and tree-planting to lock up CO_2 work well in the right conditions: they have a low unit price, are scalable, can be planted by many actors, and of course already exist. But they are slow growing, relatively impermanent, and not 'CO_2 dense enough' to constitute carbon salvage at the rates needed. This is of course not to argue 'against trees', but to draw a storage comparison with large-scale warehousing of CO_2. Assuming an effective annual tree-based CO_2 sequester of 30 t/ha/year (from absorption per tree of 20 kg/year and a tree density of 1500/hectare), a target of CO_2 removal at 2.2 ppm/year (17.2 Gt/yr) would require 5,727,333 sq km.[4] (In other words, an area three-quarters the size of Australia would effectively need to be afforested, kept forested on a rolling basis, plus the timber products would need to be locked away after trees had been harvested). The same hundred-year total of 1718,2 Gt of CO_2 in warehousing at 60% effective storage would require a total of 2864 cubic kilometres of volume or a warehouse footprint of 28,637 km^2 to be built (in one-hundred-metre-high buildings at an approximate density of 1 tonne CO_2 to 1 cubic metre CO_2) which is 0.376% of the size of Australia.[5] Put another way, CO_2 warehousing outclasses trees on an area basis by 200 to one. The permanence of warehousing of CO_2 would, however, need to be powered by a considerable increase of renewable energy generation with its own associated environmental impacts.[6]

- Household storage of permanently evacuated carbon could become a normalised everyday activity, with the storing of several tonnes of liquid or frozen carbon dioxide on a private property (and provision of energy to do so) becoming a rational response to the urgent need for carbon salvage.

- As already noted, carbon salvage would require an immense expansion of renewable energy to power carbon transformation, and direct air capture and carbon storage. There is a serious paradox here in that energy, and the cost of energy, cannot be the limiting factor in creating a carbon-salvaging paracommons. This paradox,

---

4   Drawing on Smith et al. (2006). Furthermore, the figure of 2.2 ppm/year for a span of 100 years is set as an example of a rate that would both counter on-going emissions plus bring net reductions in atmospheric CO_2 levels to 350 ppm within 100 years. This is equivalent to a target sequestration of 1718.2 Gt over 100 years.

5   A volume of 2864 cubic kilometres in 100 years is equivalent to building approximately 2600 Tesla Nevada gigafactories each year at 60% effective storage.

6   Alternatives to warehousing include deep-sea storage (Hume 2018) and evacuation to space.

and its accompanying emissions-linked complexities, is scrutinised in Dunlap's chapter, this volume.

- Carbon dioxide transformation is also exemplified by all carbon fuels being sourced from man-made biological sources (powered by renewable energy).

- Highly accurate carbon accounting to track and trace the products, size, pathways and final destinations of salvaged CO_2 is required with tangible monitoring and targets vital to an exteriorised 'making visible' of salvaged/stored carbon 'gains' (Allen 2009). As Hannis, this volume, clarifies, however, it is fiendishly difficult to secure accounting practices that provide certainty in this regard.

- Onerous standards and specifications on paracommoners and new institutional rules to salvage carbon will be required (Bosselmann 2019), in a context where environmental 'red tape' is elsewhere being contested as a constraint to economic growth and post COVID-19 economic recovery.

Concluding Remarks

Governing the global atmosphere as a sink- or husbandry-type carbon commons brings attendant concerns over whether and how we will reduce carbon dioxide levels sufficiently and quickly enough. Clearly, we should see the permanent and rapid reduction of CO_2 levels in the atmosphere as a matter of urgency. In a paracommons framing, carbon dioxide, its conversion products, storage and non-generation are seen through the lens of emergency and scarcity that results in new economic, financial, legislative, technological and behavioural solutions to bring down its concentration in the atmosphere. By being aware of the different leaky/impermanent or permanent pathways that CO_2 takes, the paracommons then asks how we solve this leaky pathway uncertainty to ensure that we put carbon dioxide permanently away when attempts to salvage CO_2 are made.

The framing of an atmospheric, climate and carbon commons needs to be expanded, but also better defined (Schrijver 2016; Edenhofer et al. 2012). Debating this conceptual challenge will bring forward alternative framings fit for the next hundred years. New commons metaphors and parables, of which the paracommons is an example, should aim to stretch our conceptual space in which the target of <350 ppm CO_2 is to be achieved.

References

Allen, Myles, 'Planetary Boundaries: Tangible Targets Are Critical', *Nature Climate Change*, 1 (2009), 114–15, https://doi.org/10.1038/climate.2009.95.

Asayama, Shinichiro, Rob Bellamy, Oliver Geden, Warren Pearce, and Mike Hulme, 'Why Setting a Climate Deadline Is Dangerous', *Nature Climate Change*, 9 (2019), 570–72, https://doi.org/10.1038/s41558-019-0543-4.

Bosselmann, Klaus, 'The Atmosphere as a Global Commons', in *Research Handbook on Global Climate Constitutionalism*, ed. by Jordi Jaria-Manzano and Susana Borrás (Cheltenham: Edward Elgar Publishing, 2019), pp. 75–87, https://doi.org/10.4337/9781788115810.

Brown, Katrina, W. Neil Adger, and Joshua E. Cinner, 'Moving Climate Change Beyond the Tragedy of the Commons', *Global Environmental Change*, 54 (2019), 61–63, https://doi.org/10.1016/j.gloenvcha.2018.11.009.

Edenhofer, Ottmar, Christian Flachsland, and Bernhard Lorentz, 'The Atmosphere as a Global Commons', in *The Wealth of the Commons. A World Beyond Market & State*, ed. by David Bollier and Silke Helfrich (Amherst, MA: Levellers Press, 2012).

Gifford, Lauren, '"You Can't Value What You Can't Measure": A Critical Look at Forest Carbon Accounting', *Climatic Change*, 161 (2020), 291–306, https://doi.org/10.1007/s10584-020-02653-1.

Gills, Barry, and Jamie Morgan, 'Global Climate Emergency: After Cop24, Climate Science, Urgency, and the Threat to Humanity', *Globalizations*, 17 (2020), 885–902, https://doi.org/10.1080/14747731.2019.1669915.

Hardin, Garrett, 'The Tragedy of the Commons', *Science*, 162 (1968), 1243–48.

Hume, David, 'Ocean Storage of CO2' (Theliquidgrid.com, 2018), https://theliquidgrid.com/2018/07/22/ocean-storage-co2/.

Lankford, Bruce, 'The Liminal Paracommons of Future Natural Resource Efficiency Gains', in *Releasing the Commons. Rethinking the Futures of the Commons.*, ed. by Ash Amin and Philip Howell (London: Routledge, 2018), pp. 66–68.

Lankford, Bruce, *Resource Efficiency Complexity and the Commons: The Paracommons and Paradoxes of Natural Resource Losses, Wastes and Wastages* (Abingdon: Routledge, 2013).

Lovell, Heather, Harriet Bulkeley, and Diana Liverman, 'Carbon Offsetting: Sustaining Consumption?', *Environment and Planning A: Economy and Space*, 41 (2009), 2357–79, https://doi.org/10.1068/a40345.

Luque, Rafael, and James H. Clark, 'Valorisation of Food Residues: Waste to Wealth Using Green Chemical Technologies', *Sustainable Chemical Processes*, 1 (2013), 10, https://doi.org/10.1186/2043-7129-1-10.

Meyerson, Frederick A. B., 'Population, Development and Global Warming: Averting the Tragedy of the Climate Commons', *Population and Environment*, 19 (1998), 443–63, https://doi.org/10.1023/A:1024622220962.

Murray, Brian C., Brent Sohngen, and Martin T. Ross, 'Economic Consequences of Consideration of Permanence, Leakage and Additionality for Soil Carbon Sequestration Projects', *Climatic Change*, 80 (2007), 127–43, https://doi.org/10.1007/s10584-006-9169-4.

O'Connor, David, 'Governing the Global Commons: Linking Carbon Sequestration and Biodiversity Conservation in Tropical Forests', *Global Environmental Change*, 18 (2008), 368–74, https://doi.org/10.1016/j.gloenvcha.2008.07.012.

Ostrom, Elinor, *Governing the Commons: The Evolution of Institutions for Collective Action* (Cambridge: Cambridge University Press, 1990).

Pierrehumbert, Raymond T., 'Cumulative Carbon and Just Allocation of the Global Carbon Commons', *Chicago Journal of International Law*, 13 (2012), 527–48.

Rabinowitz, Dan, 'Ostrom, the Commons, and the Anthropology of "Earthlings" and Their Atmosphere', *Focaal*, 57 (2010), 104–08, https://doi.org/10.3167/fcl.2010.570108.

Rockström, Johan, Will Steffen, Kevin Noone, Åsa Persson, F. Stuart Chapin, Eric Lambin, et al., 'Planetary Boundaries Exploring the Safe Operating Space for Humanity', *Ecology and Society*, 14(2) (2009), 32, http://www.ecologyandsociety.org/vol14/iss2/art32/.

Ruzzenenti, Franco, David Font Vivanco, Ray Galvin, Steve Sorrell, Aleksandra Wagner, and Hans Jakob Walnum, 'Editorial: The Rebound Effect and the Jevons' Paradox: Beyond the Conventional Wisdom', *Frontiers in Energy Research*, 10 (2019), https://doi.org/10.3389/fenrg.2019.00090.

Schrijver, Nico, 'Managing the Global Commons: Common Good or Common Sink?', *Third World Quarterly*, 37 (2016), 1252–67, https://doi.org/10.1080/01436597.2016.1154441.

Smith, James, Linda Heath, Kenneth Skog, and Richard Birdsey, 'Methods for Calculating Forest Ecosystem and Harvested Carbon with Standard Estimates for Forest Types of the United States', Gen. Tech. Rep. NE-343 (Newtown Square, PA: US Department of Agriculture, Forest Service, Northeastern Research Station, 2006).

Stoknes, Per Espen, and Johan Rockström, 'Redefining Green Growth within Planetary Boundaries', *Energy Research & Social Science*, 44 (2018), 41–49, https://doi.org/10.1016/j.erss.2018.04.030.

Vogler, John, 'Future Directions: The Atmosphere as a Global Commons', *Atmospheric Environment*, 35 (2001), 2427–28, https://doi.org/10.1016/S1352-2310(01)00127-3.

III

EXTRACTION

9. The Mobilisation of Extractivism: The Social and Political Influence of the Fossil Fuel Industry

Christopher Wright and Daniel Nyberg

The worsening climate crisis has led to growing social and political demands for meaningful climate action and the decarbonisation of economies. And yet, the modern global economy is defined by fossil fuel energy which has shaped the last two centuries of economic growth and development. In this chapter, we outline how the fossil fuel industry has defined the global economy and defended its position as the most powerful industry in the world. We examine how assumptions of corporate self-regulation as the logical response to the climate crisis allow for the continuation of a 'business as usual' approach in which fossil fuel energy is maintained. We argue that this approach deliberately ignores the urgent need for government regulation of carbon emissions, and that current corporate responses to the climate crisis rely on the politics of 'predatory delay'.

Introduction

A new wave of activism has emerged in response to the worsening climate crisis. Following popular environmental protest movements such as Extinction Rebellion and the School Strike for Climate, a growing range of lawsuits are now targeting governments and fossil fuel corporations for their contributions to the climate crisis. Moreover, there is now active discussion amongst governments and global organisations about the need for urgent reductions in greenhouse gas emissions, with even fossil fuel corporations committing to carbon neutrality by 2050. However, there have been similar commitments before (e.g. BP's Beyond Petroleum rebranding) and since the 1970s the fossil fuel industry has actively misled societies about the impact of its activities, using its innovative capacities to open up new carbon frontiers such as deep-water and Arctic oil drilling, tar sands processing and shale gas fracking (Wright and Nyberg 2015). The domination of the fossil fuel industry is based on its political tactics and this needs to be laid bare in order to be effectively resisted.

In this chapter, we outline how the fossil fuel industry has defined the global economy and defended its position as the most powerful industry in the world. We examine how assumptions of corporate self-regulation as the logical response to the climate crisis allow for the continuation of a 'business as usual' approach in which fossil fuel energy is maintained. We argue that this deliberately ignores the urgent need for government regulation of carbon emissions and that current corporate responses to the climate crisis rely on the politics of 'predatory delay'.

Fossil Energy and the Climate Crisis

The origins of the global fossil-fuelled economy date back to the beginnings of the industrial revolution in Britain in the late-eighteenth century and the development of the coal-fired steam engine. Coal power provided the basis for rapid industrialisation across manufacturing and expanded global markets through the transformation of transport (Malm 2016). With the growth of the railway, steel and chemical industries during the late-nineteenth and early-twentieth centuries, oil emerged as a further fossil fuel underpinning economic expansion. In the post-World War II decades, the power of the fossil fuel industry grew dramatically, driving economic expansion and the emergence of Western consumer lifestyles requiring a growing energy demand (Mitchell 2013). In recent decades, the globalisation of the economy and continued economic growth have relied upon the ever-increasing consumption of the world's fossil fuel reserves. Global distribution networks of pipelines, tankers, refineries, ports and rail systems have further reinforced fossil fuel investments and path dependency (as also highlighted in Bigger et al.'s chapter, this volume). The emergence of Asian economic powerhouses, such as Japan, South Korea and, most recently, China and India, has broadened the scale of fossil fuel consumption.

The pervasive impact of fossil fuels across energy, resource extraction, manufacturing, transport, agriculture, and food production make it hard to imagine how our society could be organised differently. National governments are key supporters of the expansion of fossil fuel energy through public financing of infrastructure, financial subsidies, discounted royalties and favourable tax regimes; a system critics have labelled "fossil fuel welfare" (Lenferna 2019). Fossil fuels provide over 80% of the world's total primary energy supply and underpin the global financial system not only as the most heavily capitalised sector but also a dominant source of finance and investment for the world's banks, insurance companies and pension funds (RAN 2020). The global market economy is thus fundamentally defined by fossil fuels; we live within what some have termed a "petro-market civilization" (DiMuzio 2012).

Fossil fuel energy provided the basis for the expansion of global capitalism, but has also incurred an existential environmental cost. The extraction and combustion of coal, oil and gas has over the last two centuries resulted in the release of huge quantities of greenhouse gases (GHGs) (principally carbon dioxide CO_2 and methane CH_4), resulting in an unprecedented human perturbation of the Earth's carbon cycle and energy balance (Mann and Kump 2015). From a pre-industrial level of around 280 parts per million (ppm), the combustion of fossil fuels and the diminution of forests and other carbon sinks has led to a rapid increase in the atmospheric concentration of carbon dioxide. In 2018, atmospheric CO_2 concentrations exceeded 410ppm, a level not seen on this planet for several million years (Mooney 2018). Moreover, research has found that close to two-thirds of cumulative worldwide emissions of industrial GHGs between 1751 and 2010 are the result of just ninety 'carbon major entities' (large fossil fuel corporations and state-owned entities), with half of these emissions released since 1986 (Heede 2014).

Organising a Climate Change Denial Industry

In 1966, the US coal industry publication *Mining Congress Journal* published an article which identified with surprising candour the link between coal as an energy source and the disruptive effects of the resulting carbon emissions upon the Earth's climate (Young 2019). This article, along with similar documents produced within the oil industry during the 1970s, highlighted the fact that major fossil fuel companies have long known of the terrible impact that their products were having on the planet's climate system (Supran and Oreskes 2017). Rather than developing adaptive strategies to transition to a low-carbon economy, however, the fossil fuel industry created a politically organised climate denial movement, which has proven remarkably successful in preventing any meaningful form of emissions mitigation (Oreskes and Conway 2010).

In the United States (US), corporations from the fossil fuel, energy and manufacturing sectors came together during the early 1990s to form the Global Climate Coalition (GCC) in order to push back against proposals for the regulation of carbon emissions (Levy and Egan 1998). Wider corporate resistance included financial contributions to political parties, funding for major advertising campaigns and appeals to broader conservative ideological values. Fossil fuel interests played a key role in swaying conservative politicians against carbon regulation, stressing 'uncertainty' and 'doubts' over climate science, highlighting the economic costs of cutting emissions, and promoting the views of climate 'sceptics' in government representations, media and publications (Dunlap and McCright 2011; Oreskes and Conway 2010). This vehement opposition to emissions reductions by the global fossil fuel industry not only hobbled national governments' attempts to respond to climate change, but also undermined international collaboration.

From Denial to Delay

While the fossil fuel industry has proven remarkably successful over the last thirty years in limiting carbon regulation through a strategy of organised climate denial, the growing social and political discourse around climate change now appears to fundamentally threaten the industry. Following the release of catastrophic scientific projections of the world's climate future (IPCC 2018), a new wave of climate activism has erupted around the world through groups such as Extinction Rebellion (Blackall 2019), and the school climate strikes initiated by Swedish teenage climate activist Greta Thunberg (Watts 2019) (see chapters by Gardham, North and Paterson, this volume). Combined with social movements for fossil fuel divestment (Mangat et al. 2018), legal actions against governments and fossil fuel corporations (Powers 2018), and growing concerns amongst regulators and institutional investors over the financial implications of climate change (Carney 2015), a tipping point may well have been reached. Indicative of recent shifts in the political and legal context have been: a growing procession of nations publicly proclaiming a commitment to achieve 'net zero' carbon emissions (Black et al. 2021; on the complexities of achieving net zero, however, see Dyke et al., this volume); a recent report by the International Energy Agency which declared no new coal, oil or gas extraction can occur if the world is to reach a net zero emission goal by 2050 (IEA, 2021); and the recent decision by a Dutch court that oil giant Shell is liable for its contributions to climate change which undermine basic human rights and require dramatic reductions in its global carbon dioxide emissions by the end of the decade (Juhasz 2021). This accelerating social and political critique of the fossil fuel industry has forced the industry to develop new justifications to defend its position.

The first step in this changed industry response has been public statements accepting the reality of climate change and a desire for 'climate action' broadly defined. This progressive stance was first highlighted by the European oil majors BP and Shell which developed a more engaged public stance on climate change than their conservative US counterparts (Levy and Egan 2003). Most famously, in 2000 BP rebranded itself as 'beyond petroleum' which promoted the oil giant as an environmentally aware energy company. While this marketing pivot soon faltered under a change of CEO and the infamous 2010 Deepwater Horizon oil disaster, more recently BP has returned to its focus on climate change, emphasising reductions in operational emissions and commitments to climate science (Ferns and Amaeshi 2019; Ferns et al. 2019).

Indeed in 2020, with an unprecedented downturn in oil demand following the outbreak of the global coronavirus pandemic, oil companies unleashed a bevy of public announcements about their climate ambitions. First out of the gate was new BP CEO Bernard Looney who, in February 2020, announced to stunned media and analysts that he intended to make the company a net zero carbon emitter by 2050 through reductions in direct and embedded emissions. This goal involved a projected ten-fold increase in green energy investment and development, an approximately 40% reduction of oil and gas production, an end to new oil exploration, and the selling-off of its petrochemicals business. Within months, Shell and Total also announced 2050 net zero emissions goals and even American oil giants ExxonMobil and Chevron made announcements signalling an intention to reduce their contribution to carbon emissions (Kusnetz 2020).

Fossil fuel corporations have also sought to promote the moral worthiness of their activities through marketing and public relations activities, which stress the benefits of fossil energy. For instance, US coal giant Peabody Energy's 'Advanced Energy for Life' campaign (developed by New York public relations firm Burson-Marsteller) has proclaimed the benefits of coal-based electricity for citizens in developing countries as a way in which the industry is contributing to solving global energy poverty (Sheppard 2014). In a similar manner, oil companies promoting the exploitation of the Canadian Alberta tar sands, developed an extensive public relations campaign promoting what they have termed "ethical oil", which promotes Canada's liberal democratic political system as a more morally worthy context for fossil fuel extraction (Hickman 2011).

In acknowledging the reality of climate change, coal, oil and gas companies have also marketed possible technological 'solutions' while maintaining fossil fuel extraction and use. Key amongst these has been the discourse of 'clean coal' which argues that more efficient coal-fired electricity production can dramatically reduce the industry's climate impacts. Examples include the promotion of 'high efficiency low emissions' (HELE) coal-fired power plants and carbon capture and storage (CCS) technologies (Hudson 2017). The declining competitiveness of coal-fired electricity has also led to the promotion of rival fossil fuels such as methane (promoted by the industry as 'natural gas') as a 'transition fuel' in the move towards future decarbonisation.

However, rather than reducing the world's carbon emissions, these industry responses can be seen more as a process of 'predatory delay' in which the fossil fuel industry seeks to slow the process of decarbonisation to maximise their financial returns in the short term while appearing as concerned corporate citizens (Nyberg et al. 2013; Steffen 2016). For instance, the rush by different companies to declare a goal of net zero emissions by mid-century is also a way to placate the growing social criticism of corporate climate denial. More specifically, the details of these commitments and future responses are reliant on technologies that have yet to be developed (e.g. bioenergy with carbon capture and storage (BECCS) and direct air capture) (Anderson and Peters 2016) (for more detail regarding BECCS, see Dyke et al.'s chapter, this volume). Despite the discourses of 'natural gas' and 'clean coal', researchers note that when fugitive emissions are accounted for, methane has a similar climate impact to coal, and, despite billions of dollars of government funding worldwide, only two large-scale Carbon Capture and Storage (CCS) plants have ever been completed with limited emissions capture, and these have turned out to be far more expensive than renewable solar and wind energy. Indeed, a recent analysis of the practice of the major oil corporations over the last fifteen years found no evidence of operational decarbonisation and that, while the public discourse had shifted towards a more climate-focused stance, the most progressive European-based oil companies were simply hedging their bets through limited diversification and risk mitigation (Green et al. 2020). Viewed from this perspective, the oil majors' recent apparent conversion on the issue of climate change appears more part of a longer-term pattern of skilful marketing and defensive justification in the face of growing social and political critique (Brulle et al. 2020).

A Turning Point for Fossil Energy?

Since February 2020, the world has been plunged into the greatest energy shock since World War II as a result of the coronavirus pandemic. As the International Energy Agency has outlined, the economic contraction resulting from the pandemic led to the biggest fall in global energy investment in history, with plunging demand for coal, oil and gas resulting in dramatic reductions in the value of fossil fuel stocks, and the likelihood of fossil fuel reserves becoming 'stranded assets' while renewable energy costs continue to fall (IEA 2020). This rapid decline in the value of fossil fuel corporations was vividly demonstrated in August 2020 when one of the biggest of the oil majors, ExxonMobil, ended its ninety-two-year run on the Dow Jones industrial average as its market value collapsed to about a third of its 2008 high-point of US$500 billion. Oil and gas companies now make up only 2.3% of the Standard and Poor's (S&P) 500, compared with 15% in 2008 (Grandoni 2020). At the time of writing, the COVID-19 pandemic shows no sign of abating and the severity of the global economic impacts have yet to fully play out. Together with the market, technological and political challenges to hydrocarbon energy, the fossil fuel sector is now facing its greatest challenge since the beginning of industrialisation.

Despite growing awareness of a worsening climate crisis, however, tangible action in terms of mitigating carbon emissions, let alone reining in fossil fuel production and use, has been limited. Proposals for carbon emissions reductions have continued to rely upon market-based measures that have failed to dent the steady increase in global emissions. Up until the 2020 pandemic, the global fossil fuel burn continued to increase year by year, hitting an all-time high in 2019 of 11.7 Gtoe (billion tonnes of oil equivalent), up from 7.1 Gtoe in 1990 (Saxifrage, 2020). While the pandemic has resulted in a short-term contraction in global carbon emissions during 2020 of 6% on 2019 levels (Tollefson 2021), atmospheric concentrations of greenhouse gases continue to increase (World Meteorological Organization 2020).

In carbon-intensive economies, such as the US, China, Canada, Australia and Saudi Arabia, the fossil fuel industry continues to expand, assisted by government subsidies and financial incentives (Lenferna 2019). In the US, the so-called 'fracking revolution' has led to the country becoming the world's largest producer of oil and gas (Downie 2019). China is now the world's largest producer and consumer of coal (constituting over half of the world's total consumption), with significant foreign investments in new fossil fuel developments in developing economies through its 'Belt and Road' initiative (Umbach and Yu 2016). In Canada and Australia, expansion of fossil fuel extraction has led to dramatic growth in energy exports, with Canada's Alberta tar sands delivering oil to the US and China (Bloomberg 2019), and Australia is now the world's largest exporter of coal and gas (Kilvert 2019). Moreover, the election of Donald Trump in 2016 as President of the US provided a huge boost for the fossil fuel industry, as evidenced by the US withdrawal from the Paris Climate Agreement, the removal of environmental regulations dating back to the 1970s, and the promotion of prominent fossil fuel executives and climate deniers to key government positions (De Pryck and Gemenne 2017)—and discussed further in the chapter by Hannis, this volume. While the recent election of Democrat President Joe Biden in late 2020 has led to far more progressive announcements from the US on climate change and growing international momentum for reductions in carbon emissions, it is unclear whether this will be sufficient to overcome the profound political divisions that still exist in many countries over climate change, let alone lead to the decarbonisation and reinvention of a global economy defined by and reliant upon fossil energy.

Conclusion

The unprecedented decline in the demand for fossil fuel energy resulting from the current worldwide pandemic offers a unique opportunity for the global economy to break free of its carbon addiction and commit to a genuine and far-reaching energy transition. As we have pointed out, however, global capitalism and the assumptions of compound economic growth have to date been constructed upon the extracted energy of fossil fuels such as coal, oil and gas. This structural dynamic has in turn made the fossil fuel industry amongst the most powerful actors in the world, able to draw on political capital in determining the policy decisions which shape the future of human civilisation. While the recent social demands for meaningful climate action and growing political commitments to avoid dangerous climate change are heartening, it remains to be seen whether we are at a watershed moment in confronting the climate crisis, or whether the fossil fuel industry will succeed in its strategy of 'predatory delay', such that this decade becomes another missed opportunity to reduce the harm of a rapidly worsening climate crisis.

References

Anderson, Kevin, and Glenn Peters, 'The Trouble with Negative Emissions', *Science*, 354(6309) (2016), 182–18, https://doi.org/10.1126/science.aah4567.

Black, Richard, Kate Cullen, Byron Fay, Thomas Hale, John Lang, Saba Mahmood, and Steve Smith, *Taking Stock: A Global Assessment of Net Zero Targets* (Energy & Climate Intelligence Unit, 2021), https://ca1-eci.edcdn.com/reports/ECIU-Oxford_Taking_Stock.pdf.

Blackall, Molly, 'Extinction Rebellion Protests Block Traffic in Five UK Cities' (Theguardian.com, 2019), https://www.theguardian.com/environment/2019/jul/15/extinction-rebellion-protests-block-traffic-in-five-uk-cities.

Bloomberg, Robert, 'Pipelines Add Room on "unrelenting" Demand for Canada's Oil' (Thestar.com, 2019), https://www.thestar.com/business/2019/08/02/pipelines-add-room-on-unrelenting-demand-for-canadas-oil.html.

Brulle, Robert, Melissa Aronczyk, and Jason Carmichael, 'Corporate Promotion and Climate Change: An Analysis of Key Variables Affecting Advertising Spending by Major Oil Corporations, 1986–2015', *Climatic Change*, 159(1) (2020), 87–101, https://doi.org/10.1007/s10584-019-02582-8.

Carney, Mark, *Breaking the Tragedy of the Horizon—Climate Change and Financial Stability* (Bankofengland.co.uk, 2015), https://www.bankofengland.co.uk/-/media/boe/files/speech/2015/breaking-the-tragedy-of-the-horizon-climate-change-and-financial-stability.pdf.

De Pryck, Kari, and François Gemenne, 'The Denier-in-chief: Climate Change, Science and the Election of Donald J. Trump', *Law and Critique*, 28(2) (2017), 119–26, https://doi.org/10.1007/s10978-017-9207-6.

DiMuzio, Tim, 'Capitalizing a Future Unsustainable: Finance, Energy and the Fate of Market Civilization', *Review of International Political Economy*, 19(3) (2012), 363–88, https://doi.org/10.1080/09692290.2011.570604.

Downie, Christian, *Business Battles in the US Energy Sector: Lessons for a Clean Energy Transition* (Abingdon: Routledge, 2019).

Dunlap, Riley E., and Aaron M. McCright, 'Organized Climate Change Denial', in *The Oxford Handbook of Climate Change and Society*, ed. by John S. Dryzek, Richard B. Norgaard, and David Schlosberg (Oxford: Oxford University Press, 2011), pp. 144–60.

Ferns, George, and Kenneth Amaeshi, 'Fueling Climate (In)action: How Organizations Engage in Hegemonization to Avoid Transformational Action on Climate Change', *Organization Studies*, 42(7) (2019), 1005–29, https://doi.org/10.1177/0170840619855744.

Ferns, George, Kenneth Amaeshi, and Aliette Lambert, 'Drilling Their Own Graves: How the European Oil and Gas Supermajors Avoid Sustainability

Tensions Through Mythmaking', *Journal of Business Ethics*, 158(1) (2019), 201–31, https://doi.org/10.1007/s10551-017-3733-x.

Grandoni, Dino, 'Big Oil Just Isn't as Big as it Once Was' (Washingtonpost. com, 2020), https://www.washingtonpost.com/business/2020/09/04/exxon-dow-jones/.

Green, Jessica F., Jennifer Hadden, Thomas Hale, and Paasha Mahdavi, 'Transition, Hedge, or Resist? Understanding Political and Economic Behavior Toward Decarbonization in the Oil and Gas Industry', *SSRN* (2020), https://papers.ssrn.com/sol3/papers.cfm?abstract_id=3694447.

Heede, Richard, 'Tracing Anthropogenic Carbon Dioxide and Methane Emissions to Fossil Fuel and Cement Producers, 1854–2010', *Climatic Change*, 122(1–2) (2014), 229–41, https://doi.org/10.1007/s10584-013-0986-y.

Hickman, Leo, 'Canadian Campaign Puts the Spin on ethical oil' (Theguardian. com, 2011), http://www.theguardian.com/environment/blog/2011/jul/28/oil-tar-sands-canada-ethical.

Hudson, Marc, 'Ultra, Super, Clean Coal Power? We've Heard it Before' (Theconversation.com, 2017), https://theconversation.com/ultra-super-clean-coal-power-weve-heard-it-before-71468.

IEA, 'The Covid-19 Crisis is Causing the Biggest Fall in Global Energy Investment in History' (Iea.org, 2020), https://www.iea.org/news/the-covid-19-crisis-is-causing-the-biggest-fall-in-global-energy-investment-in-history.

IEA, *Net Zero by 2050: A Roadmap for the Global Energy Sector* (Geneva: International Energy Agency, 2021), https://www.iea.org/reports/net-zero-by-2050.

IPCC, *Global Warming of 1.5°C: An IPCC Special Report on the Impacts of Global Warming of 1.5 °C Above Pre-Industrial Levels and Related Global Greenhouse Gas Emission Pathways, in the Context of Strengthening the Global Response to the Threat of Climate Change, Sustainable Development, and Efforts to Eradicate Poverty* (Geneva: Intergovernmental Panel on Climate Change, 2018).

Juhasz, Antonia, 'A Court Ruled Shell is Liable for its Contributions to Climate Change. What Happens Now?' (Rollingstone.com, 2021), https://www.rollingstone.com/politics/politics-features/shell-climate-change-oil-dutch-court-1175404/.

Kilvert, Nick, 'Australia is the World's Third-largest Exporter of CO_2 in Fossil Fuels, Report Finds' (Abc.net.au, 2019), https://www.abc.net.au/news/science/2019-08-19/australia-co2-exports-third-highest-worldwide/11420654.

Kusnetz, Nicholas, 'What Does Net Zero Emissions Mean for Big Oil? Not What You'd Think' (InsideClimateNews.org, 2020), https://insideclimatenews.org/news/15072020/oil-gas-climate-pledges-bp-shell-exxon.

Lenferna, Alex, 'Fossil Fuel Welfare Versus the Climate', in *Palgrave Handbook on Managing Fossil Fuels and Energy Transitions*, ed. by Geoffrey Wood and Geoffrey Baker (London: Palgrave Macmillan, 2019), pp. 551–67.

Levy, David L., and Daniel Egan, 'Capital Contests: National and Transnational Channels of Corporate Influence on the Climate Change Negotiations', *Politics and Society*, 26(3) (1998), 337–61, https://doi.org/10.1177/00323292 98026003003.

Levy, David L., and Daniel Egan, 'A Neo-Gramscian Approach to Corporate Political Strategy: Conflict and Accommodation in the Climate Change Negotiations', *Journal of Management Studies*, 40(4) (2003), 803–29, https://doi.org/10.1111/1467-6486.00361.

Malm, Andreas, *Fossil Capital: The Rise of Steam Power and the Roots of Global Warming* (London: Verso, 2016).

Mangat, Rupinder, Simon Dalby, and Matthew Paterson, 'Divestment Discourse: War, Justice, Morality and Money', *Environmental Politics*, 27(2) (2018), 187–08, https://doi.org/10.1080/09644016.2017.1413725.

Mann, Michael E., and Lee R. Kump, *Dire Predictions: Understanding Climate Change*, 2nd ed. (New York: DK Publishing, 2015).

Mitchell, Tim, *Carbon Democracy: Political Power in the Age of Oil* (London: Verso, 2013).

Mooney, Chris, 'Earth's Atmosphere Just Crossed Another Troubling Climate Change Threshold' (Washingtonpost.com, 2018), https://www.washingtonpost.com/news/energy-environment/wp/2018/05/03/earths-atmosphere-just-crossed-another-troubling-climate-change-threshold/.

Nyberg, Daniel, André Spicer, and Christopher Wright, 'Incorporating Citizens: Corporate Political Engagement with Climate Change in Australia', *Organization*, 20(3) (2013), 433–53, https://doi.org/10.1177/1350508413478585.

Oreskes, Naomi, and Eric M. Conway, *Merchants of Doubt: How a Handful of Scientists Obscured the Truth on Issues from Tobacco Smoke to Global Warming* (New York: Bloomsbury Press, 2010).

Powers, Melissa, 'Juliana v United States: The Next Frontier in US Climate Mitigation?', *Review of European, Comparative & International Environmental Law*, 27(2) (2018), 199–204, https://doi.org/10.1111/reel.12248.

RAN, *Banking on Climate Change: Fossil Fuel Finance Report 2020* (San Francisco, CA: Rainforest Action Network, 2020).

Saxifrage, Barry, 'Global Fossil Fuel Burning Breaks Record in 2019' (Nationalobserver.com, 2020), https://www.nationalobserver.com/2020/07/16/opinion/global-fossil-burning-breaks-record-2019-canadians-top-1.

Sheppard, Kate, 'World's Biggest Coal Company, World's Biggest PR Firm Pair up to Promote Coal for Poor People' (Huffingtonpost.com.au, 2014), https://www.huffingtonpost.com.au/2014/03/27/peabody-burson-marstellar-coal_n_5044962.html.

Steffen, Alex, 'Predatory Delay and the Rights of Future Generations' (Medium.com, 2016), https://medium.com/@AlexSteffen/predatory-delay-and-the-rights-of-future-generations-69b06094a16.

Supran, Geoffrey, and Naomi Oreskes, 'Assessing ExxonMobil's Climate Change Communications (1977–2014)', *Environmental Research Letters*, 12(8) (2017), 084019, https://doi.org/10.1088/1748-9326/ab89d5.

Tollefson, Jeff, 'COVID Curbed Carbon Emissions in 2020—but not by Much' (Nature.com, 2021), https://www.nature.com/articles/d41586-021-00090-3.

Umbach, Frank, and Ka-ho Yu, *China's Expanding Overseas Coal Power Industry: New Strategic Opportunities, Commercial Risks, Climate Challenges and Geopolitical Implications* (London: European Centre for Energy and Resource Security, 2016).

Watts, Jonathan, '"The Beginning of Great Change": Greta Thunberg Hails School Climate Strikes' (Theguardian.com, 2019), https://www.theguardian.com/environment/2019/feb/15/the-beginning-of-great-change-greta-thunberg-hails-school-climate-strikes.

World Meteorological Organization, *United in Science: A Multi-organization High-level Compilation of the Latest Climate Science Information* (Geneva: WMO, 2020).

Wright, Christopher, and Daniel Nyberg, *Climate Change, Capitalism and Corporations: Processes of Creative Self-destruction* (Cambridge: Cambridge University Press, 2015).

Young, Élan, 'Coal Knew, Too' (Huffingtonpost.com.au, 2019), https://www.huffingtonpost.com.au/entry/coal-industry-climate-change_n_5dd6bbebe4b0e29d7280984f.

10. End the 'Green' Delusions: Industrial-Scale Renewable Energy is Fossil Fuel+

Alexander Dunlap

Industrial-scale renewable energy does nothing to remake exploitative relationships with the Earth, and instead represents the renewal and expansion of the present capitalist order. This chapter argues that industrial-scale renewable energy is more accurately understood as 'fossil fuel+'. The purpose is to re-think the socio-ecological reality of so-called renewable energy to create space for the step-change of strategies needed to mitigate and avoid climate and ecological catastrophe.

Industrial-Scale 'Renewable Energy' Is a False Solution

Renewable energy is not the solution we think it is. We have inherited the bad/good energy dichotomy of fossil fuels versus renewable energy, a holdover from the environmental movement of the 1970s that is misleading, if not false. Fossil fuels are correctly understood to be at the heart of capitalism, industrialism, and state formation, the results of which have been ecologically catastrophic (Malm 2016). Meanwhile, industrial-scale renewable energy has emerged as the protagonist of our times, positioned as the solution to our ever-increasing energy consumption. Along with market-based conservation and 'natural capital' policy making, it is taken to be among the central mitigating forces against climate change and ecological degradation (as critiqued by Sullivan 2009; Huff and Brock 2017).

  https://doi.org/10.11647/OBP.0265.10

With the rise of the green economy and climate change legislation, renewable energy includes harnessing wind, solar, and other apparently infinite 'natural resources' to meet energy consumption on an unprecedented, ever expanding scale. Contrary to the claims of its proponents, however, it by no means adequately addresses the materially real problems posed by current levels of energy consumption, which are driven by capitalist growth imperatives that ultimately cause the ecological degradation and climate change we see today. A focus on the technocratic issue of energy consumption often leaves unchallenged the political-economic violence intrinsic to the production system that such energy powers (as also highlighted in Sullivan Chapter 11, this volume).

Industrial-scale renewable energy does nothing to remake the exploitative relationships with the earth and ecosystems created and reproduced by 'industrialised humans'—people acclimated to, and dependent upon, an industrial, capitalist way of life. The excessive concern with possible energy solutions within capitalism, as opposed to more fundamental social transformations, expresses our inability to imagine any other way of living, blinding us to the deeper socio-ecological insurrection that climate change makes necessary.

Industrial-scale renewable energy and the grid-centric systems it powers represent the renewal and expansion of the present political and capitalist order. Not only are existing social discontents such as inequality, discrimination, and exploitation reinforced by renewable energy, but the amount of infrastructure it presently requires clearly indicates the ecological costs involved in its full implementation. The wind and solar parks that span across fields and hillsides as far as the eye can see are harbingers of what this new energy system looks like. Questions need asking: where does all this metal come from? How much energy can this new energy system produce? What ecological impact does it have? And what kind of society does it propel and enable?

On Energy Extractivism

In 1980, American Indian Movement (AIM) activist Russell Means explained the uncomfortable reality of energy extractivism relating especially to uranium mining in Native territory (and echoed in the similar concerns expressed by AIM activist the late John Trudell, in Sullivan Chapter 3, this volume). Confronting a room of revolutionary Communists about their desire for industrialism, Means (1985: 25) said:

> Right now, today, we who live on the Pine Ridge Reservation are living in what Euro society has designated a "national sacrifice area." What this means is that we have a lot of uranium deposits here and Euro culture (not us) needs this uranium as energy production material. The cheapest, most efficient way for industry to extract and deal with the processing of this uranium is to dump the waste byproducts right here at the digging sites. Right here where we live. This waste is radioactive and will make the entire region uninhabitable forever. This is considered by industry, and the white society which created this industry, to be an "acceptable" price to pay for energy resource development. Along the way they also plan to drain the water-table under this area of South Dakota as part of the industrial process, so the region becomes doubly uninhabitable. The same sort of thing is happening down in the land of the Navajo and Hopi, up in the land of the Northern Cheyenne and Crow, and elsewhere. Over 60 percent of all U.S. energy resources have been found to lie under reservation land, so there's no way this can be called a minor issue. For American Indians it's a question of survival in the purest sense of the term. For white society and its industry it's a question of being able to continue to exist in their present form.
>
> We are resisting being turned into a national sacrifice area. We're resisting being turned into a national sacrifice people. The costs of this industrial process are not acceptable to us. It is genocide to dig the uranium here and to drain the water-table, no more, no less. So the reasons for our resistance are obvious enough and shouldn't have to be explained further. To anyone.

As with the mining of fossil fuels and uranium, the siting and implementation of renewable energy systems entails the creation of such sacrifice zones, often on Indigenous land. These projects have thus confronted considerable pushback from rural and Indigenous populations, and the struggles around extraction outlined by Means have continued to intensify (Avila 2018; Dunlap 2017, 2019; Franquesa 2018; Lawrence 2014; Lucio 2016; Siamanta 2019). By clinging to ideas like 'sustainable development' and the 'green economy', progressives and other conscientious citizens are staking the future of the planet on dubious mechanisms of oversight, rife with conflicts of interest. The proliferation of voluntary UN standards, corporate social responsibility initiatives, private auditing firms (Brock and Dunlap 2018), and free, prior, and informed consent (FPIC) are but "band aids of good intentions" (Dunlap 2018). They ultimately conceal the true costs of extractivism, especially for the Indigenous people most affected by it.

The distinctions drawn between fossil fuels and renewable energy involve a sleight of hand that masks the continued ecological degradations necessary for the continuation of consumer society and its ecological modernisation (see Bond and Downey 2012). Renewable energy requires immense amounts of mineral and fossil fuel resources, both in the construction of machinery necessary for extraction and for the manufacturing, transportation, construction and operation of wind turbines and other industrial-scale renewable energy systems.

For all these reasons, instead of conceiving renewable energy as a 'green' environmental solution, industrial or utility-scale renewable energy is more accurately referred to as 'Fossil Fuel+'.

Wind Energy as Fossil Fuel+

Let us focus the discussion on a single source of renewable energy: wind. Wind energy is something of a poster child for renewable energy in general, and is increasingly becoming a preeminent approach to climate change mitigation. Through fieldwork in the Isthmus of Tehuantepec region of Oaxaca, Mexico, where I was embedded for six months in a *polícia comunitaria*,[1] I witnessed firsthand the struggles and negative outcomes involved in the implementation of this form of renewable energy, even as it continues to be encouraged and incentivised by national and international climate change mitigation programmes.

Consider, for example, the resources required to construct a single two-megawatt wind turbine. One of these turbines uses roughly 150 metric tons of steel for reinforced concrete foundations, 250 metric tons for the rotor hubs and nacelles, and 500 metric tons for the tower (Smil 2016a), as well as 3.6 tons of copper per megawatt (Smith 2014).

---

1   The communitarian police were initiated after local Zapotec and Ikoots took over a town hall and expropriated the municipality's police truck (in 2013). This incident spawned the self-organisation of an unpaid community police force (*polícia comunitaria*) by Zapotec farmers and fishers to stop wind companies, politicians and others from entering the region to exploit their habitat/ecosystem.

Industrial steel production is currently impossible without burning coal, as metallurgical or coking coal is a vital ingredient in the process (Diez and Barriocanal 2002; Smil 2016b). Now, imagine regions like the Isthmus of Tehuantpec, where roughly 1,700 wind turbines operate to provide energy to Walmart, Grupo Bimbo, industrial construction, mining, and other companies and industries (Dunlap 2019). These turbines require significant amounts of mining, and every stage of the mining process—from extraction, to processing, manufacturing, transport, construction and, to some degree, operation—requires a large expenditure of fossil fuels, a fact often neglected in the ecological accounting of wind energy (as similarly observed for 'clean energy' produced from uranium in Sullivan 2013). According to Guezuraga et al. (2012: 40) the main consumers of energy and producers of CO_2 for the turbines are "the production of stainless steel, followed by concrete and cast iron," while "plastic production represents the most energy intensive process of all materials".

From the perspective of carbon accounting, steel, concrete, and cast iron production are the main consumers of energy, with the ecological costs of mining and processing the rare earth minerals required to create permanent magnet generators in wind turbines being relatively disregarded. But where do these minerals come from, and what is the ecological cost of their extraction? Many of the rare earth minerals required for the operation of the turbines—such as dysprosium, praseodymium neodymium, terbium—come from places like Baotou, Inner Mongolia, and Ganzhou, South East China, which have produced some 85–98% of rare earth minerals used in wind turbines, electric cars, smart phones and other technologies between the late 1980s and 2015 (Hongiao 2016). The socio-ecological costs of this extraction are high.

The Costs?

A BBC report from 2015 called the Baotou mining and processing area "hell on Earth": a terrifying, dystopian industrial environment filled with pollution and cluttered with factories, pipelines, high-tension wires, and artificial lakes oozing "black, barely-liquid, toxic sludge" that "tested at around three times background radiation" (Maughan 2015; also see Klinger 2017).

Rare earth mining is also disastrously inefficient. Mined with open pit, underground, or leached in-situ methods, rare earth ore deposits contain "low concentrations [of desired minerals] ranging from 10 to a few hundred parts per million by weight" (Yang et al. 2013: 133). Most concerning, however, is that,

> [t]he mining and processing steps for refining of rare earths tend to be energy, water and chemical intensive with significant environment risks affecting water discharges (radionuclides, mainly thorium and uranium; heavy metals; acid; fluorides), tailing management and air emissions (Haque et al. 2014: 621).

Echoing the observations on uranium extraction by Russell Means quoted above, wind energy thus similarly involves socially and ecologically destructive mining processes that produce large amounts of mining tailings (or waste) containing heavy metals, thorium, and radioactive materials that enter the air, water, soil, animals and people. The quantity and intensity of this pollution is difficult to measure, for both political and scientific reasons, making accounting for all ecosystem impacts not only costly, but impossible.

While in theory wind turbines can be built without rare earth minerals (as in geared turbines), this is not currently the case for the majority of utility-scale wind parks—especially wind turbines located offshore or in areas with extremely strong winds. This is because rare-earth-based PMSG (permanent magnet synchronous generator) turbine technology allows for the construction of more compact turbines which require less maintenance, making them more profitable to operate. The bigger the turbine, the more there is to gain for the operator by installing PMSG models (Lovins 2017).

Like other industrial enchantments (such as computers or smart technologies), wind farms continue to require levels of extraction that generate toxic and radioactive waste excluded from carbon accounting and often exempt from outdated life-cycle assessments (Kiezebrink et al. 2017; Klinger 2017).

While further research on the exact levels of ecocide and political violence is necessary, the fact remains that the green economy is expanding demand for destructive mining of iron ore, copper, oil, and rare earth minerals. This in turn is part and parcel of the creation and expansion of sacrifice areas engulfing entire regions of China and the mountains, rivers, and forests across the world.

The political and environmental costs of implementing these renewable wind energy systems are also high. Scale, placement, mitigation practices, and energy-use are foundational for assessing the viability and long-term socio-ecological sustainability of wind turbines. This means taking cognisance of the quantity and location of large-scale turbines, as well as the various political and socio-geographical factors involved in their construction.

For example, while it is ill advised to place them on lands used for semi-subsistence production by Indigenous groups, within 1.5 kilometres of people's homes, or in areas with fresh groundwater, farming, and fishing areas, this is precisely what has happened on the Isthmus of Tehuantepec (Dunlap 2019), from which the following observations derive. The construction and placement of wind turbines requires the creation of roads that clear trees and animal habitats and compact soil. They also require the creation of wind turbine foundations that range, depending on the site, between 7–14 metres (32–45 ft.) deep and about 16–21 metres (52–68 ft.) in diameter. The foundations require the filling of groundwater with solidifying chemicals before filling them with steel reinforced concrete. During operation, leaking oil seeps into the ground where animals graze and into water wells where people drink. And this leaves aside the effects of concrete production, as well as the violence involved in building wind or other renewable energy systems on Indigenous territory. On top of all this, each wind turbine only has roughly a 30–40 thirty-to-forty-year lifespan before it needs to be decommissioned and, hopefully, recycled (Habib and Wenzel 2014).

Fossil Fuel+

These unpleasant facts are why renewable energy should really be called fossil fuel+. The plus sign indicates the added benefit of the renewable component or multiplier present in renewable energy systems, while simultaneously acknowledging their dependence on fossil fuel based-technologies and extractivism. The '+', or renewable, component is dependent on fossil fuels, and thus is not entirely positive in CO_2 emissions terms. A focus on the benefits of renewable energy systems additionally overlooks the simple but paramount question: what is all this energy used for?

Renewable energy is opening and widening new wind, solar, and other natural resource frontiers, and in doing so it is *renewing capitalism* as well. In addition to private industry, militaries are beginning to take an increased interest in renewable energy systems (as also observed in Bigger et al.'s chapter, this volume). The same techniques and technologies that are helping corporations expand in ostensibly 'green' directions will be applied to power military infrastructures and equipment. Whether the question is of solar in the Middle East, wind power in Mexico, or aircraft carriers that run on biofuels, these relations support the expansion of capitalism whilst obscuring its gut-wrenching crises and obstructing effective action (Al-Waeli et al. 2017; Bigger and Niemark 2017; Dunlap 2017).

Industry and security forces are beginning to acknowledge their ecologically destructive operations, and repressive forces are looking for ways to become ecologically 'sustainable'. Such "sustainable violence" is not just the result of "bad governance" (Dunlap 2017). It is inextricably bound with industrial extraction and efforts to economise on the destructive and repressive actions of governments, industry and security forces involved in the expansion of industrial-scale renewable energy systems.

Fossil fuel industries—whether coal, natural gas, or oil—are also beginning to invest and use renewable industry to legitimise their resource extraction operations and diversify their energy-related holdings (as outlined by Wright and Nyberg, this volume). Examples range from Gas Natural Fenosa, which is investing in wind parks in Mexico (Dunlap 2019), to RWE in Germany, operator of the largest coal mine in the country, which is setting up their own green daughter company—Innogy—to invest in wind energy and other 'renewables' after spending years subverting and lobbying against them (Brock and Dunlap 2018). Grupo Mexico is also buying wind in Mexico and solar parks in the US to cloak their company in a 'green' image. Meanwhile, they are powering the extraction of raw materials with renewable sources (Dunlap 2019; GrupoMexico 2016). With Andrea Brock, I have called this the "renewable energy-extraction nexus", which demonstrates the intimate relationship between forms of extraction—whether wind, natural gas, coal, or copper—necessary for renewable energy development and the continued subsumption of the Earth and its inhabitants to industrial society (Dunlap and Brock 2021).

The Renewable Energy-Extraction Nexus

The renewable energy-extraction nexus also renews the multiple and self-reinforcing extractivisms comprising the material structure of the state: becoming part of the intricate web of subsidies, collaboration and, at times, competition that feeds the techno-industrial machine, spreading its infrastructure and values across the planet. This expansion happens at a great disregard for the costs involved, whether for people (particularly Indigenous or rural communities in both the Global North and South), animals, plants or geophysical nature.

The preceding considerations allow us to recognise renewable energy as renewing destruction (Dunlap 2019). It entails revived and intensified relations of domination that have much in common with colonial and centre-periphery dynamics. When people embrace renewable energy systems, many do not realise that they require various forms of violence against people, environments, and animals, which must remain hidden for obvious reasons. These systems, which require concrete, steel, copper, rare earth minerals and, by extension, fossil fuel and mineral extraction, are made to appear acceptable through their placement out of sight and out of mind, in the materially poor, rural, and Indigenous territories of the Global South and North.

When liberals, progressives, 'the Left', and even environmental justice activists applaud the large-scale transition to renewable energy, they ignore the many hazards that would otherwise be unacceptable to them.

Displacing fossil fuel industries to the Global South, where there are fewer environmental regulations and political rights, also enables the use of excessive forms of state-private security violence against anyone who might protest them. The material necessary for renewable energy can only result in an increase in extractivism in the Global South and all the negative consequences this entails for people on the ground. If we do not confront these facts, then the solution of today—like previous energy systems and regime changes—will likely result in the complicated tyrannies of tomorrow. Recognising renewable energy as Fossil Fuel+ is a first step to combat the fairytale of renewable energy. By highlighting the myths surrounding renewable energy, we also create the groundwork for greater environmental considerations and the enactment of radical ecological alternatives that address the roots of consumer society and its marketed solutions.

References

Al-Waeli, Ali A. K., Kadhem A. Al-Asadi, and Mariyam M, Fazleena, 'The Impact of Iraq Climate Condition on the Use of Solar Energy Applications in Iraq: A Review', *International Journal of Science and Engineering Investigations*, 6 (2017), 64–73.

Avila, Sofia, 'Environmental Justice and the Expanding Geography of Wind Power Conflicts', *Sustainability Science*, 13 (2018), 599–616, https://doi.org/10.1007/s11625-018-0547-4.

Bigger, Patrick, and Benjamin D. Neimark, 'Weaponizing Nature: The Geopolitical Ecology of the US Navy's Biofuel Program', *Political Geography*, 60 (2017), 13–22, https://doi.org/10.1016/j.polgeo.2017.03.007.

Bonds, Eric, and Liam Downey, 'Green Technology and Ecologically Unequal Exchange: The Environmental and Social Consequences of Ecological Modernization in the World-System', *Journal of World-Systems Research*, 18 (2012), 167–86, https://doi.org/10.5195/jwsr.2012.482.

Brock, Andrea, and Alexander Dunlap, 'Normalising Corporate Counterinsurgency: Engineering Consent, Managing Resistance and Greening Destruction Around the Hambach Coal Mine and Beyond', *Political Geography*, 62 (2018), 33–47, https://doi.org/10.1016/j.polgeo.2017.09.018.

Diez, M. A., R. Alvarez, and C. Barriocanal, 'Coal for Metallurgical Coke Production: Predictions of Coke Quality and Future Requirements for Cokemaking', *International Journal of Coal Geology*, 50 (2002), 389–412, https://doi.org/10.1016/S0166-5162(02)00123-4.

Dunlap, Alexander, 'Wind Energy: Toward a "Sustainable Violence" in Oaxaca, Mexico', *NACLA*, 49 (2017), 483–88, https://doi.org/10.1080/10714839.2017.1409378.

Dunlap, Alexander, '"A Bureaucratic Trap:" Free, Prior and Informed Consent (FPIC) and Wind Energy Development in Juchitán, Mexico', *Capitalism Nature Socialism*, 29 (2018), 88–108, https://doi.org/10.1080/10455752.2017.1334219.

Dunlap, Alexander, and Andrea Brock, 'When the Wolf Guards the Sheep: Green Extractivism in Germany and Mexico', University of Sussex CGPE

Working Paper 21 (29 Jan 2021), https://www.sussex.ac.uk/webteam/gateway/file.php?name=when-the-wolf-guards-the-sheep-dunlap-and-brock.pdf&site=359.

Franquesa, Jaume, *Power Struggles: Dignity, Value, and the Renewable Energy Frontier in Spain* (Bloomington: Indiana University Press, 2018).

GrupoMexico, 'Ingenia 02' (Gmexico.com, 2016), http://www.gmexico.com/site/images/documentos/ingenia/INGENIA02.pdf.

Guezuraga, Begoña, Rudolf Zauner, and Werner Pölz, 'Life Cycle Assessment of Two Different 2 MW Class Wind Turbines', *Renewable Energy*, 37 (2012), 37–44, https://doi.org/10.1016/j.renene.2011.05.008.

Habib, Komal, and Henrik Wenzel, 'Exploring Rare Earth Supply Constraints for the Emerging Clean Energy Technologies and the Role of Recycling', *Journal of Cleaner Production*, 84 (2014), 348–59, https://doi.org/10.1016/j.jclepro.2014.04.035.

Haque, Nawshad, Anthony Hughes, Seng Lim, and Chris Vernon, 'Rare Earth Elements: Overview of Mining, Mineralogy, Uses, Sustainability and Environmental Impact', *Resources*, 3 (2014), 614–35, https://doi.org/10.3390/resources3040614.

Hongiao, Liu, 'The bottleneck of a low-carbon future' (Chinadialogue.net, 2016), https://chinadialogue.net/article/9209-The-bottleneck-of-a-low-carbon-future.

Huff, Amber, and Andrea Brock, 'Accumulation by Restoration: Degradation Neutrality and the Faustian Bargain of Conservation Finance' (Antipodefoundation.org, 2017), https://antipodefoundation.org/2017/11/06/accumulation-by-restoration/.

Kiezebrink, Vincent, Joseph Wilde-Ramsing, and Gisela ten Kate, 'Human rights in wind turbine supply chains: Towards a truly sustainable energy transition' (Somo.nl, 2018), https://www.somo.nl/wp-content/uploads/2018/01/Final-ActionAid_Report-Human-Rights-in-Wind-Turbine-Supply-Chains.pdf.

Klinger, Julie Michelle, *Rare Earth Frontiers: From Terrestrial Subsoils to Lunar Landscapes* (Cornell University Press, Ithaca, 2018).

Lawrence, Rebecca, 'Internal Colonisation and Indigenous Resource Sovereignty: Wind Power Developments on Traditional Saami Lands', *Environment and Planning D: Society and Space*, 32 (2014), 1036–53, https://doi.org/10.1068/d9012.

Lovins, Amory, 'Clean Energy and Rare Earths: Why Not to Worry', *Bulletin of the Atomic Scientists* (Thebulletin.org, 2017), https://thebulletin.org/clean-energy-and-rare-earths-why-not-worry10785.

Lucio, Carlos Federico, *Conflictos socioambientales, derechos humanos y movimiento indígena en el Istmo de Tehuantepec* (Zacatecas, Universidad Autónoma de Zacatecas, 2016).

Malm, Andreas, *Fossil Capital: The Rise of Steam Power and the Roots of Global Warming* (London: Verso Books, 2016).

Maughan, Tim, 'The Dystopian Lake Filled by the World's Tech Lust' (Bbc.com, 2015), http://www.bbc.com/future/story/20150402-the-worst-place-on-earth.

Means, Russell, 'The Same Old Song', in *Marxism and Native Americans*, ed. by Ward Churchill (Boston: South End Press, 1985), pp. 19–33.

Siamanta, Zoi Christina, 'Wind Parks in Post-crisis Greece: Neoliberalisation Vis-à-vis Green Grabbing', *Environment and Planning E: Nature and Space*, 2 (2019), 274–303, https://doi.org/10.1177/2514848619835156.

Smil, Vaclav, *Energy Transitions: Global and National Perspectives* (Santa Barbara: Praeger 2016).

Smil, Vaclav, 'To Get Wind Power You Need Oil' (Spectrum.ieee.org, 2016), http://spectrum.ieee.org/energy/renewables/to-get-wind-power-you-need-oil.

Smith, Patrick, 'Soaring Copper Prices Drive Wind Farm Crime' (Windpowermonthly, 2014), http://www.windpowermonthly.com/article/1281864/soaring-copper-prices-drive-wind-farm-crime.

Sullivan, Sian, 'Green Capitalism, and the Cultural Poverty of Constructing Nature as Service Provider', *Radical Anthropology*, 3 (2009), 18–27.

Sullivan, Sian, 'After the Green Rush? Biodiversity Offsets, Uranium Power and the 'Calculus of Casualties' in Greening Growth', *Human Geography*, 6(1) (2013), 80–101, https://doi.org/10.1177/194277861300600106.

11. I'm Sian, and I'm a Fossil Fuel Addict: On Paradox, Disavowal and (Im)Possibility in Changing Climate Change

Sian Sullivan

In recent years I have returned to west Namibia to work with elders of families I have known for more than two decades. Oral histories, recorded as we find and revisit places my companions knew as home, have increasingly struck a chord as a record of lives lived largely untouched by fossil fuels. As the complexity of these pasts has come further into focus, it has become impossible to avoid the gulf between this kind of attunement to environmental contexts and my own life, especially the reality that I am completely dependent on fossil fuels and the products they make possible. This essay is an attempt to fully face this paradox of maintaining hope for binding international climate agreements that have teeth, whilst being aware of my dependence on the fossil fuel extracting and emissions-spewing juggernaut that permeates all our lives. Drawing critically on twelve-step thinking and psychoanalysis literatures I reflect on fossil fuel addiction, and the destructive paradox of not being able to live up to internalised but unreachable values regarding environmental care in a fossil-fuelled world.

https://doi.org/10.11647/OBP.0265.11

Once Upon a Time in the Wild West

Sometimes life brings experiences that give pause for thought.

In recent years I have returned to west Namibia to work with elders of families I have known for almost thirty years—a legacy of a childhood split between Britain and southern Africa. We have been documenting histories of land connections prior to a series of clearances of people from large areas of the west Namibian landscape that occurred some decades ago (Sullivan and Ganuses 2020, 2021; Sullivan in press).[1] Often now perceived as an untouched and pristine wilderness, our work instead draws into focus a landscape intimately known, named and remembered by people who once lived there (as conveyed in Figure 5). Oral histories recorded as we find and revisit places my companions knew as home, have increasingly struck a chord as a record of lives lived largely untouched by fossil fuels.

In the contemporary terms defined by modernity, industrialisation and capital, theirs was an economically impoverished existence. But this is not how they define and describe their experience. Beyond the nostalgia that people tend to have for times past, their prior existence is valued in some of the following ways. For the freedom to move to locations where particular foods could be acquired, and for the pleasure of meeting and sharing food, songs and dances with friends associated with different places. For harvesting a series of highly appreciated foods enabling their subsistence in an extreme, dryland environment: the endemic cucurbit (melon-plant) !nara (*Acanthosicyos horridus*) processed from 'fields' managed far west in the dunes of the Northern Namib Desert (see Video 1); the seeds of sâui (*Stipagrostis* spp.) and bosûi (*Monsonia* spp.) collected from harvester ant nests found further inland (Sullivan 1999); and the fruits of xoris (*Salvadora persica*) found in ephemeral rivers traversing the landscape. For sweet honey (*danib*) pulled from hives harvested over decades, and diverse animals lived with, hunted and appreciated as sentient, intentional beings with whom people could communicate. For a life filled with nights of songs and healing dances when times of abundance were celebrated,[2] and when

---

1   I am grateful to Sesfontein residents Welhemina Suro Ganuses and Filemon |Nuab for their multi-year collaboration and leadership in this field research.
2   See *The Music Returns to Kai-as*, online at https://vimeo.com/486865709.

the skills for living in an environment considered one of the most hostile on earth were valued highly.

Fig. 5. ‖Oeb: Cousins Noag Mûgagara Ganaseb (L) and Franz |Haen ‖Hoëb (R) revisit places in the westward reaches of the Hoanib River where they used to live. Here they are close to ‖Oeb, now the site of a high-end eco-tourism lodge called Hoanib Camp, located on the south side of the bend in the Hoanib River just to the right of centre in this image. When Franz, Noag and their families lived in this area they would alternate between harvesting *!nara* (*Acanthosicyos horridus*) from their *!nara* plants near the springs of Auses / !Ui‖gams, and walking southwards to Kai-as and the !Uniab River where different foods as well as *!nara* could be found. In the 1950s the coastal dunes were opened for diamond mining. Then in 1971 the lower Hoanib River was gazetted as part of the Skeleton Coast National Park. As these areas became opened for extractive industry and conservation, they became closed to habitation by those who once lived there. Photo: Sian Sullivan, November 2015, composite made with Mike Hannis using three 10 x 10 km aerial images from Directorate of Survey and Mapping, Windhoek, July 2017, as part of a series of images for the exhibition *Future Pasts: Landscape, Memory and Music in West Namibia*: see https://www.futurepasts.net/memory. © Future Pasts, CC BY-NC-ND 4.0.

Video 1. Hildegaart |Nuas of Sesfontein / !Nani|aus, Kunene Region, north-west Namibia remembers harvesting *!nara* (*Acanthosicyos horridus*) in the dune fields of the Hoanib River. Video by Sian Sullivan (2019), https://vimeo.com/380044842, © Future Pasts, CC BY-NC-ND 4.0.

As the complexity of these pasts has come further into focus, it has become impossible to avoid the gulf between this kind of attunement to environmental contexts and my own life. Like many in 'the West', I would describe myself as concerned with environmental and social justice. All my work has been energised by such concerns, as well as by an animist sense of the natures with which we live, which seems resonant with aspects of the lifeworlds of those Indigenous people with whom I have interacted and worked (Sullivan 2019).

At the same time, the reality is that I am completely dependent on fossil fuels and the products they make possible. This dependence exists even as I simultaneously and publicly acknowledge the serious implications of pumping more climate-forcing gases into the atmosphere. There is almost nothing manufactured in the world around me, or in my life as it is currently structured, that exists independently of fossil fuels. The basic things with which I write and share these reflections—from the plastic refillable pencil I scribble notes with, to the laptop I write on and the Wi-Fi system I am now connected with—are shot through with fossil fuels.

Facing It

Under current structural circumstances, I am completely unable to unhook myself from fossil fuel production and consumption. Even consciously 'low-impact' and low-carbon lifestyles are bound with the fossil fuel industry and the apparent necessity of economic growth this supports (Böhm 2015). The solar panels on my roof at home installed to foster an 'off-grid' lifestyle are made and transported using fossil fuels—not to mention the host of other substances involve in their fabrication whose extraction and associated wastes are seriously environmentally-damaging (see Dunlap, this volume).

Documenting the fossil-fuel-free pasts of people like Franz and Noag above means I fly to Namibia and then drive a diesel-fuelled 4x4 so as to go with small groups of people to the far-flung locations where they lived. Maybe I should simply give up this research so as to be more congruent with a stringently decarbonised lifestyle? But apart from personal love for this research—for the places it takes me to, the people I work alongside, and the diversity I am exposed to and learn from—I do this work in a context of local desire for such pasts to be documented and made public,[3] institutional support for contributions along these lines, and professional pressures to continue with work that consolidates and internationalises earlier research effort. Like others working to engage with and bear witness to justice issues in various global contexts—issues frequently associated with fossil fuel extraction and emissions management—all my research and activist engagements are paradoxically fuelled by fossil fuels.

Of course, I can assuage my conscience by purchasing carbon offset credits—perhaps using a carbon credit card through which every dollar I spend will apparently reduce my carbon footprint[4]—or by planting trees somewhere else. But I do not really accept a model that sees the earth in terms of aggregates (an aggregate carbon budget, an aggregate level of 'natural capital', etc.[5]), the composition of which can be traded, exchanged and substituted between times and places so as somehow to cancel out emissions. I am generating carbon emissions through my life and work, full-stop. There is no 'elsewhere' for these emissions.

As someone who cares about planetary health and is also concerned about the perpetuation and deepening of grotesque economic inequity, I rationalise my activities by considering the documentation I am doing with others in Namibia to be worth it—in terms of making visible currently occluded pasts, experiences of displacement, and different possible ways of living with diverse natures-beyond-the-human. Such

---

3   Most recently this work has been drawn on by an Ancestral Land Commission appointed in 2019 by the Namibian Government, in its final report to the Office of the Prime Minister making recommendations for Parliament to enact "ancestral land rights claim and restitution legislation" (GRN 2020).

4   For example, 'Eco-friendly Credit Cards' (Thalesgroup.com, 2021), https://www. thalesgroup.com/en/markets/digital-identity-and-security/banking-payment/ cards/eco-friendly-credit-card.

5   As advocated, for example, in Helm (2015).

rationalisations, which are amongst those we engage in all the time so as to exist amidst unavoidable contradictions, go some way towards cognitively smoothing the dissonances described above.[6]

In facing these contradictions and the dependences they mask, it becomes harder to simultaneously maintain a stringently critical position towards fossil fuel extractors (I need the substances they produce, dammit!), states and negotiators in the worlds of climate change management. Indeed, motivating this essay is a sense that I am not alone in deploying psychic compartmentalisations so as to act affirmatively in the world, whilst simultaneously damning the fuels, technologies, organisations and structures that make these actions possible—thus ultimately also damning myself. I am wondering if it is increasingly important to recognise the prevalence of such internal divisions, particularly the destructive paradox of not being able to live up to internalised but unreachable values.

I am influenced here by a provocative meditation on the natures of authoritarianism: *The Guru Papers: Masks of Authoritarian Power*, by Joel Kramer and Diana Alstad.[7] In *The Guru Papers*, Kramer and Alstad (1993: 228) speak of "the hypocrisy masking so much of social interaction where people pretend to be far more virtuous than they are". They highlight the destructive authoritarianism that arises as internalised 'good' / 'bad' dualities pit aspects of the self against each other. And they connect this internalised conflict with social contexts in which assumptions and projections of superior morality maintain problematic authoritarianisms. They argue that such everyday authoritarianisms act to avert equality in social relationships, whilst also reducing possibilities for strengthening self-trust and for improving broader awareness of the structural dissonances preventing systemic change.

Acknowledging the implications of such internal and social conflicts and inconsistencies seems critical right now, when so many ecological, psychological and social indicators—not least a global pandemic that has erupted since I first drafted this essay—suggest we need systemic change in spades.

---

6   I discuss such dissonances in more detail in Sullivan 2018.
7   Thank you to Ya'acov Darling Kahn for drawing my attention to this text.

What on Earth Is Going on?

When re-emerging from periods of field research in west Namibia where Internet coverage is very sparse, into more-or-less constant Internet reality, I have often felt like I am viewing events unfolding in the world from a sort of bemused and horrified distance. An example comes from December 2015, in the moment when the 21st Conference of the Parties (COP) of the United Nations Framework Convention on Climate Change (UNFCCC) was taking place in Paris.

As I sat in a small hotel in Windhoek readying myself to return to the UK just after journeying between Sesfontein / !Nani-ǀaus and the Skeleton Coast with Franz and Noag pictured above, I wrote about the following constellation of events.

So, the world's government negotiators are meeting for the twenty-first time to attempt to agree to systemically adjust economic activities so as to decarbonise the global economy (UNFCCC 2015). For months now, climate justice activists have also been mobilising protests and actions (for example, Global Justice Now 2015). Of particular concern is the corporate agenda considered to be preventing the UNFCCC COP from reaching a binding international agreement that has real teeth in terms of emissions reductions. Corporate sponsors supporting the Paris COP include airlines, energy corporations and banks (Mcdonnnell 2015; Team Ecohustler 2015). Great effort by (h)activists has gone into designing possibilities for radical play to disrupt the 'mesh' of the formal summit and its associations with "austerity-dictating politicians, fossil fuel corporations, industry lobbyists, peddlers of false solutions and greenwashers" (McDonald 2015: online); as well as to deflect the policing "sidekicks" of the COP (referred to as "Team Blue"). Protests in Paris on the eve of the COP were met by "team blue" with tear gas and police baton charges (Fieldstadt and Grimson 2015).

Simultaneously, the chilling pre-summit attacks in Paris by ISIL ('Islamic State of Iraq and the Levant') in mid-November (BBC News 2015) have precipitated further fossil fuel-intensive military strikes by the West against sites in the Middle East (Marcus 2015), as well as a potentially indefinite state of emergency in France under which public demonstrations are banned (Osborne 2015). The right of public assembly so as to contest the climate change negotiations of COP21 is curbed under these emergency powers (Chan 2015).

Meanwhile, within hours of a yes vote in the UK parliament (Sparrow and Perraudin 2015), RAF bombers joined allies who have been bombing Syrian oil fields since 2014 (LoGiurato 2014). The speed by which bombers were deployed suggested it was farcical to think there could have been an alternative outcome to the parliamentary vote. Justified as striking at the source of oil finance for 'the terrorist group Daesh', it seems beyond irony that at this intense moment of global climate change negotiations in Paris, wells supplying the supposedly scarce and climate-forcing substance of oil are being bombed, entailing huge emissions into the atmosphere in a situation that would not look out of place in a post-apocalyptic *Mad Max* film.

And here I am nodding to another irony, in that the last *Mad Max* film—*Fury Road* (*The Future Belongs to the Mad*) of 2015—was filmed in the sensitive desert landscapes of west Namibia, not far from where I started this essay, its destructive impacts causing fury amongst environmentalists and scientists there. The film uses the stark beauty of the Namib Desert as the backdrop for a post-apocalyptic desert wasteland where the scarcest of resources are petrol, water and fertile women, and violence is the means whereby control of these precious items is maintained . . .

In any case, apart from the heart-breaking humanitarian disaster of military intervention in Middle East contexts over the last twenty-plus years of international climate negotiations (Sullivan 2003; UN Security Council 2015), such tactics surely contradict 'the West's' avowed allegiance to reduce climate change emissions (Graham 2015). The ferocity with which Western corporations carved up Iraq's oil fields as they worked to remove Saddam Hussein in 2002–2003 (Beaumont and Islam 2002) should remind us that aggressive access to fossil fuels infuses international policy too. Indeed, current military adventure by the West in Syria appears a bloodsoaked strategy to beat Russia and Iran to the significant 'hydrocarbon potential' of Syria's offshore resources (Ahmed 2015). These geopolitical issues are not even close to the public negotiating table in Paris.

Consider, as well, a couple of announcements made as government climate negotiators were meeting in Paris in 2015. The environment minister(!) of the Australian government justified recent approval of the $16bn Carmichael coal mine in Queensland, to be operated by Indian company Adani, on the grounds that Australia is not a neo-colonialist power that tells poor countries what to do (Taylor 2015). Construction of the mine is ongoing, amidst an array of challenges on environmental and Indigenous title grounds (Currell et al. 2020; Wangan and Jagalingou Family Council 2020). Botswana reportedly announced the sale of fracking rights to a UK company covering half the Kgalagadi Transfrontier Park, an area also associated with Indigenous San / Bushmen (Barbee 2015). Although refuted in 2016 by the Botswana government, alarm in the region has now shifted to exploratory drilling for oil in neighbouring northern Namibia, by a Canadian company—Recon Africa—that reportedly also has a licence to prospect for oil in northwestern Botswana (Tan 2020). Initial drilling has been touted in industry publications as potentially yielding "the biggest oil story of the decade" (Leigh 2021). Meanwhile, in the US in 2015, Oklahoma was experiencing an "earthquake boom" (Chow 2015: online), recognised as linked to some extent by oil and gas related processes, including the injection of water into basement rock in extracting natural gas from bedrock, i.e. fracking (US Geological Survey 2020).

Fast forwards to the present moment: the build up to the twenty-sixth COP of the UNFCCC, to be held in November 2021 in Glasgow, UK.[8] Postponed twice due to the COVID-19 pandemic, COP26 is taking place in a world that has tilted on its political axis towards right-wing populism and consolidated plutonomy (Greven 2016): *viz.* the elections of presidents Trump (US) (reprieved by the tight Democrat presidential win in December 2020), Bolsonaro (Brazil), Erdoğan (Turkey), and the ascent of Johnson to Prime Minister in the UK. The green hopes stimulated by a COVID-19-induced pause in especially flying, which is stranding fossil fuel assets everywhere (Kusnetz 2020), are being dashed as recovery packages for oil companies are announced (Harvey 2020), as also observed in Hannis, this volume. Clearly, the contradictions continue.

---

8   See https://ukcop26.org/.

Disavowal and Doublethink

In trying to generate a coherent picture from these fragments, I feel acutely sensitive to the difficulty of maintaining hope for binding international climate agreements that have teeth, whilst being aware of the fossil fuel extracting and emissions-spewing juggernaut that permeates all our lives.

Humans are adept at deploying the layers of our consciousness to simultaneously maintain sometimes diametrically opposed realities. The pioneer of psychoanalysis Sigmund Freud, in a succinct essay published in 1938, identified this human 'talent' as the *Splitting of the Ego in the Process of Defence*. He asserted that in order to accommodate traumatic and dangerous reality the ego may behave in remarkable ways. In short, a defensive splitting can be deployed such that the threat associated with particular behaviours is both acknowledged and systematically turned away from (Freud 2009[1938]). Attention is instead transferred to fetishised solutions that facilitate continuation of the dangerous but satisfying behavior, at the same time as constituting symptoms of the acknowledged reality of this danger (see discussion in Sullivan 2017). Freud used the term 'disavowal' to describe this simultaneous defence against, and displaced acknowledgement of, dangerous reality.

In carbon management, offsetting mechanisms designed to mitigate emissions production can be seen as paradigmatic of such a fetishised 'solution' (Fletcher 2013; Watt 2021). They signal simultaneous acknowledgement and sustenance of harms caused. As a fetishised solution to anthropogenic climate change, offsets are directing oceans of creative energy and resources towards the production of metrics to fabricate equivalence between carbon produced and stored at different sites, and away from achieving reductions in emissions production (Moreno and Speich Chassé 2015). This is why critics of such exchange and market-based approaches to emissions management cry "false solutions" (REDD Monitor 2014).

'Doublethink' was the term that George Orwell, in his dystopian novel *1984*, used for such practices of structural disavowal. He defined doublethink as "the power of holding two contradictory beliefs in one's mind simultaneously, and accepting both of them" (Orwell 2013[1949]: 244), identifying enforcement of this practice as at the heart of maintaining a systemically unequal totalitarian regime. He wrote further that the prevailing mental condition associated with doublethink was "controlled insanity"—a state necessary to forever avert human equality (Orwell 2013[1949]: 226).

Disavowal and doublethink enable hope for emissions-reducing agreements to succeed, even as oil fields are being bombed by the same powers making those promises. They enable acceptance of Western governments as liberal democracies, even as freedom of assembly is severely constrained under arrangements which in France in 2015 were precipitating rushed changes to the constitution. They perhaps also run through the internal psychological divisions enabling impassioned pleas for the cessation of fossil fuel emissions production to be made using technologies, gadgets and transports fuelled by fossil fuels.

Fossil-Fuel(led) Culture

All of us contesting climate change and railing against the activities of fossil fuel companies are doing this using technologies, infrastructures and materials that are fossil-fuelled. Every single one of our online posts working to organise social movements for climate justice are made possible by fossil fuels. They are embedded in our computers, in our mobile phones, in all our mechanised transport systems, in our bikes, in the transport of our foods, however organic and fairly traded they are.

Fossil fuel corporations might be blamed for their hunger to capture oil under the land of indigenous peoples in Ecuador and elsewhere (see Sullivan Chapter 3, and Fremeaux and Jordan, this volume), and cynicism may be justified regarding the will and/or ability of government negotiators to agree a summit text that is binding in terms of national emissions reductions and fossil fuel investments (see Hannis, this volume). But, given the systemic nature of our dependence on fossil fuel products and infrastructure it is starting to feel uncomfortable and inaccurate to engage in divisive communications around the issue of fossil fuel dependence. Just about everything around us and with which we are entangled—much of which we might appreciate and even love—is fuelled by and/or made with fossil fuel.

Addiction and Taking Steps

Recently, and quite frequently, our fossil-fuelled culture has been framed in terms of substance addiction (see also Wright and Nyberg, this volume). It seems clear that we need nothing short of a complete global energy, and thus societal, revolution to unhook us from fossil fuel addiction.

Fossil Fuel Addiction (FFA) and its associated denials, dismissals, disassociations and rationalisations has been identified as a key aspect of climate change negotiations, requiring intervention which climate justice activists, with their perhaps clearer grasp of the desperate reality approaching us, are considered well-placed to offer. Thus, "the climate justice movement must perform a planned intervention with a professional who helps the Addict to see the truth in their polluting" (Hornack 2015).

In the famous twelve-step process of Alcoholics Anonymous and associated programmes, the first step is to recognise that you are indeed addicted. That you are bound to a substance over which you do not have control, such that this substance has become your 'higher power', its material qualities and structures of access determining one's activities and choices in the world. Subsequent steps include acknowledging that external help is needed so as to disrupt patterns of habit and addiction.

Referring again to Kramer and Alstad in *The Guru Papers*, however, such solutions may also sustain perhaps destructive divisions between 'good' and 'bad' behaviours, promoting powerlessness and internal warring in ways that may prevent psychological integration and self-trust regarding choices. To take this back to fossil fuels, in integrating some realities about my own structural dependence on fossil fuels I am starting to feel something a little unfamiliar: something more akin to empathy for the challenge facing government negotiators in the various COPs, all of whom are as wrapped up in a system of intractable dependencies as myself. Even less familiar is something approaching appreciation for the power of fossil fuels and their provision by those organisations I tend to view with intense suspicion and dislike. As Gunster et al. (2018: 12) write, these kinds of perceptual shifts are relevant since frank recognition of the [climate] hypocrisy of those who possess environmental sympathies can open up space for understanding the structural forces that generate the gaps between intention and action and thus promote a more complex understanding of the relations between social and political change and individual practices. Embedded within reflexive, sympathetic and dialogic venues of communication, the (often uncomfortable) feelings that attend such recognition can become a spur to reflection, conversation and, most importantly, modes of agency and action that dismantle (rather than enforce) conventional liberal distinctions between public and private, political and economic, citizen and consumer.

In addition to demystifying and working to transform the structural forms driving deepening inequality and environmental damage, then, twelve-step thinking can add some ingredients of its own. In encouraging honest and open acknowledgement of the grip of fossil fuels, combined with inventory of the harms caused by this grip, it can further support whatever choices are possible to reduce consumption and decarbonise one's own life, without losing sight of the systemic and infrastructural nature of one's dependence. Importantly, whilst prompting honesty about one's own substance (ab)use, twelve-step thinking also foregrounds *compassion* for the (im)possibility of unhooking from an addiction that is societal and structural, as much as individual.

In doing so, the emphasis shifts to asking for, and providing support to, fellow addicts. This is not necessarily a competition. We are all in this together, although clearly there are gaping inequities in intentions, uses and impacts. We need to work with each other now in order to unhook ourselves, our groupings and our societies from fossil fuels. Only blaming and entering into conflict with those needed as allies in this transformation potentially hinders the possibility of systemic and supportive reimagining and reconstitution. This includes fossil fuel producers. I of course find this step difficult; but I cannot escape the fact that they have made both the delights and the difficulties of our current lives possible, and perhaps they too need non-violent communications and support in order to divest from and/or to fuel systemic change (although see Fremeaux and Jordan, this volume). At the same time, clearly they need to be radically reconfigured and regulated so as to be weighted more clearly towards people and planet, rather than profit

(Paterson, this volume, also grapples with these systemic connections and complexities).

A re-oriented perception of the nature of fossil fuels may also help. Indigenous peoples have known oil as the blood of a feminised nurturing earth and as saturated with spirit beings considered to nourish the spirits infusing all plant and animal life (see Sullivan Chapter 3, this volume). Such worldviews may limit the possibility of extraction when perceived as fundamentally damaging to the systemic and nourishing health of life, and thus as morally wrong. Whimsical perhaps? But knowing oil and other potent minerals to be precious energisers of the earth, as opposed to disembedded materials whose value exists only in their burning to fuel industrial processes and economic growth, may be part of a toolkit that foundationally shifts human relationships with these substances.

References

Ahmed, Nafeez, 'Western Firms Primed to Cash in on Syria's Oil and Gas "Frontier"' (Medium.com/insurge-intelligence, 2015), https://medium.com/insurge-intelligence/western-firms-plan-to-cash-in-on-syria-s-oil-and-gas-frontier-6c5fa4a72a92#.5vyvi1vf9.

Barbee, Jeff, 'Botswana Sells Fracking Rights in National Park' (Theguardian.com, 2015), https://www.theguardian.com/environment/2015/dec/02/botswana-sells-fracking-rights-in-national-park.

BBC News, 'Paris attacks: What happened on the night' (Bbc.co.uk, 2015), https://www.bbc.co.uk/news/world-europe-34818994.

Beaumont, Peter, and Faisal Islam, 'Carve-up of Oil Riches Begins' (Theguardian.com, 2002), https://www.theguardian.com/world/2002/nov/03/iraq.oil.

Böhm, Steffen, 'Why the Paris Climate Talks are Doomed to Failure, Like All the Others' (theconversation.com, 2015), https://theconversation.com/why-the-paris-climate-talks-are-doomed-to-failure-like-all-the-others-50815.

Chan, Sewell, 'France Uses Sweeping Powers to Curb Climate Protests, but Clashes Erupt' (Nytimes.com, 2015), https://www.nytimes.com/2015/11/30/world/europe/france-uses-sweeping-powers-to-curb-climate-protests-but-clashes-erupt.html.

Chow, Lorraine, 'Another Earthquake Hits Oklahoma: Officials Worry Stronger Quake Could Threaten National Security' (Ecowatch.com, 2015), https://www.ecowatch.com/another-earthquake-hits-oklahoma-officials-worry-stronger-quake-could--1882127597.html.

Currell, Matthew, Adrian Werner, Chris McGrath, and Dylan Irvine, 'Australia Listened to the Science on Coronavirus. Imagine if we Did the Same for Coal Mining (Theconversation.com, 2020), https://theconversation.com/australia-listened-to-the-science-on-coronavirus-imagine-if-we-did-the-same-for-coal-mining-138212.

Fieldstadt, Elisha, and Matthew Grimson, 'Global Climate March: Clashes in Paris as Protesters Rally Ahead of COP21' (Nbcnews.com, 2015), https://www.nbcnews.com/news/world/global-climate-march-record-numbers-turn-out-climate-protests-n470836.

Fletcher, Robert, 'How I Learned to Stop Worrying and Love the Market: Virtualism, Disavowal, and Public Secrecy in Neoliberal Environmental Conservation', *Environment and Planning D: Society and Space*, 31(5) (2013), 796–812, https://doi.org/10.1068/d11712.

Freud, Sigmund, 'Splitting of the Ego in the Process of Defence', in *On Freud's 'Splitting of the Ego in the Process of Defence'*, ed. by T. Bokanowski and S. Lewkovitz (London: Karnac Books, 2009[1938]), pp. 3–6.

Global Justice Now, *COP21 Paris Climate Mobilisation* (Globaljustice.org, 2015), https://www.globaljustice.org.uk/events/cop21-paris-climate-mobilisation.

Government of Namibia, *Report of the Commission of Inquiry into Claims of Ancestral Land Rights and Restitution* (Windhoek: GRN, 2020).

Graham, David, 'The Politics of Obama's Greenhouse-Gas Rule' (Theatlantic.com, 2015), https://www.theatlantic.com/politics/archive/2015/08/obama-greenhouse-gas-rule/400382/.

Greven, Thomas, *The Rise of Right-wing Populism in Europe and the United States: A Comparative Perspective* (Bonn: Friedrich Ebert Stiftung, 2016), http://dc.fes.de/fileadmin/user_upload/publications/RightwingPopulism.pdf.

Gunster, Shane, Darren Fleet, Matthew Paterson, and Paul Saurette, '"Why Don't You Act Like You Believe It?": Competing Visions of Climate Hypocrisy', *Frontiers in Communication*, 3 (2018), https://doi.org/10.3389/fcomm.2018.00049.

Harvey, Fiona, 'US Fossil Fuel Giants Set for a Coronavirus Bailout Bonanza' (Theguardian.com, 2020), https://www.theguardian.com/environment/2020/may/12/us-fossil-fuel-companies-coronavirus-bailout-oil-coal-fracking-giants-bond-scheme.

Helm, Dieter, *Natural Capital: Valuing the Planet* (London: Yale University Press, 2015).

Hornack, Brad, 'COP21 Fossil Fuel Addiction: A guide for intervention from the climate justice movement' (Socialistproject.ca, 2015), https://socialistproject.ca/2015/12/b1192/.

Kramer, Joel, and Diana Alstad, *The Guru Papers: Masks of Authoritarian Power* (Berkeley: North Atlantic Books, 1993).

Kusnetz, Nicholas, 'BP and Shell Write-Off Billions in Assets, Citing Covid-19 and Climate Change' (Insideclimatenews.org, 2020), https://insideclimatenews. org/news/02072020/bp-shell-coronavirus-climate-change/.

Leigh, Alex, 'Why Namibia Could Become The Biggest Oil Story of the Decade' (Oilprice.com, 2021), https://oilprice.com/Energy/Energy-General/Why-Namibia-Could-Become-The-Biggest-Oil-Story-of-the-Decade.amp.html.

LoGiurato, Brett, 'The US Has Begun Bombing ISIS Oil Refineries' (Businessinsider.com,   2014),   https://www.businessinsider.com/ us-bombs-isis-oil-fields-2014-9.

Marcus, Jonathan, 'Islamic State: Where Key Countries Stand' (Bbc.co.uk, 2015), https://www.bbc.co.uk/news/world-middle-east-29074514.

McDonald, Matt, 'The "Climate Games" aren't just activist stunts—they're politics beyond the UN' (Theconversation.com, 2015), https://theconversation. com/the-climate-games-arent-just-activist-stunts-theyre-politics-beyond-the-un-51872.

McDonnell,   Tim,   'The   Fossil   Fuel   Industry   Is   Bankrolling the   Paris   Climate   Talks'   (Motherjones.com,   2015), h t t p s : / / w w w . m o t h e r j o n e s . c o m / e n v i r o n m e n t / 2 0 1 5 / 1 2 / climate-change-summit-paris-cop21-fossil-fuels-sponsors/.

Moreno, Camila, and Daniel Speich Chassé, *Carbon Metrics: Global Abstractions and Ecological Epistemicide* (Berlin: Heinrich-Böll-Stiftung, 2015).

Orwell, George, *1984* (London: Penguin, 2013[1949]).

Osborne, Samuel, 'France's State of Emergency Could be Extended Indefinitely' (Independent.co.uk, 2015), https://www.independent.co.uk/news/world/ europe/france-s-state-emergency-could-be-extended-indefinitely-a6758686. html.

REDD Monitor 2014 'Nine Reasons Why REDD is a False Solution: Friends of the Earth International' (Redd-monitor.orgg, 2014), https://redd-monitor. org/2014/10/15/nine-reasons-why-redd-is-a-false-solution-friends-of-the-earth-international/.

Sparrow, Andrew, and Frances Perraudin, 'Cameron Wins Syria Airstrikes Vote by Majority of 174—As it Happened' (Theguardian.com, 2015), https://www.theguardian.com/politics/blog/live/2015/dec/02/ syria-airstrikes-mps-debate-vote-cameron-action-against-isis-live.

Sullivan, Sian, 'Folk and Formal, Local and National: Damara Cultural Knowledge and Community-based Conservation in Southern Kunene, Namibia', *Cimbebasia*, 15 (1999), 1–28.

Sullivan, Sian, 'Frontline(s)', *ephemera: critical dialogues on organization*, 3(1) (2003), 68–89, http://www.ephemerajournal.org/contribution/frontlines.

Sullivan, Sian, 'What's Ontology Got to Do with It? On Nature and Knowledge in a Political Ecology of "the Green Economy"', *Journal of Political Ecology*, 24 (2017), 217–42.

Sullivan, Sian, 'Dissonant sustainabilities? Politicising and psychologising antagonisms in the conservation-development nexus', *Future Pasts Working Paper Series*, 5 (2018), https://www.futurepasts.net/fpwp5-sullivan-2018.

Sullivan, Sian, 'Towards a Metaphysics of the Soul and a Participatory Aesthetics of Life: Mobilising Foucault, Affect and Animism for Caring Practices of Existence', *New Formations: A Journal of Culture, Theory & Politics*, 95(3) (2019), 5–21.

Sullivan, Sian, 'Maps and Memory, Rights and Relationships: Articulations of Global Modernity and Local Dwelling in Delineating Land for a Communal-area Conservancy in North-west Namibia', *Conserveries Mémorielles: Revue Transdisciplinaire*, 25 (in press).

Sullivan, Sian, and Welhemina Suro Ganuses, 'Understanding Damara / ǂNūkhoen and ǁUbun Indigeneity and Marginalisation in Namibia', in *'Neither Here nor There': Indigeneity, Marginalisation and Land Rights in Post-independence Namibia,* ed. by Willem Odendaal and Wolfgang Werner (Windhoek: Legal Assistance Centre, 2020), pp. 283–324.

Sullivan, Sian, and Welhemina Suro Ganuses, 'Densities of Meaning in West Namibian Landscapes: Genealogies, Ancestral Agencies, and Healing', in *Mapping the Unmappable? Cartographic Explorations with Indigenous Peoples in Africa*, ed. by Ute Dieckmann (Bielefeld: Transcript, 2021), pp. 139–90.

Tan, Jim, 'Alarm as Exploratory Drilling for Oil Begins in Northern Namibia' (News.mongabay.com, 2020), https://news.mongabay.com/2020/12/alarm-as-exploratory-drilling-for-oil-begins-in-northern-namibia/.

Taylor, Leanore, 'Australia approved coalmine because it isn't a "neo-colonialist" power, Greg Hunt claims' (Theguardian.com, 2015), https://www.theguardian.com/environment/2015/dec/02/australia-approved-coalmine-because-it-isnt-a-neo-colonialist-power-greg-hunt-claims.

Team Ecohustler, 'The world's largest Disobedient Action Adventure Game!' (Ecohustler.com, 2015), https://ecohustler.com/culture/climate-games/.

UNFCCC, *Paris Climate Change Conference—November 2015* (Unfccc. int, 2015), https://unfccc.int/process-and-meetings/conferences/past-conferences/paris-climate-change-conference-november-2015/paris-climate-change-conference-november-2015.

UN Security Council, 'All United Nations Tools Must Be Used to Reverse Downward Spiral of Instability in Middle East, North Africa, Secretary-General Tells Security Council' (Un.org, 2015), https://www.un.org/press/en/2015/sc12064.doc.htm.

US Geological Survey, 'Oklahoma has had a Surge of Earthquakes Since 2009. Are They Due to Fracking?' (Usgs.gov, 2020), https://www.usgs.gov/faqs/oklahoma-has-had-a-surge-earthquakes-2009-are-they-due-fracking.

Wangan and Jagalingou Family Council, 'If They Destroy our Country, They Will Destroy Us as a People (Wanganjagalingou.com, 2020), https://wanganjagalingou.com.au/if-they-destroy-our-country-they-will-destroy-us-as-a-people/.

Watt, Robert, 'The Fantasy of Carbon Offsetting', *Environmental Politics* (2021), 1–20, online first, https://doi.org/10.1080/09644016.2021.1877063.

IV

DISPATCHES FROM A CLIMATE CHANGE FRONTLINE COUNTRY— NAMIBIA, SOUTHERN AFRICA

12. Gendered Climate Change-Induced Human-Wildlife Conflicts amidst COVID-19 in Erongo Region, Namibia

Selma Lendelvo, Romie Nghitevelekwa and Mechtilde Pinto

The risks of climate change for drier countries have become more pronounced. Small increments in temperature changes are considered to pose serious consequences for dry countries such as Namibia and Botswana, both of which have also experienced significant drought in recent years. In this chapter, we discuss climate change-induced human-wildlife conflicts (HWC) as they relate to gender, for communities in Erongo Region, west Namibia. We draw attention to the experiences of women as a vulnerable social group that is bearing climate change-induced HWC, and foreground how they are adapting to these pressures.

Setting the Scene

Namibia is one of the driest countries in southern Africa with a semi-arid climate characterised by low and highly variable rainfall (Bann and Wood 2012). The average annual rainfall for Namibia is 350mm, ranging from 50mm in the west to 650mm in the north-eastern part of the country. In addition to variable rainfall, Namibia experiences increased unpredictability in weather patterns associated with extreme events such as frequent floods and droughts (for consideration of climate and environmental change and stability over the longer timeframe of the last 150 years, see Rohde et al.'s chapter, this volume).

https://doi.org/10.11647/OBP.0265.12

During the period from 2002 to 2013 drought spells were recorded (Kapolo 2014; Schnegg and Bollig 2016). In the successive years from 2013 through to 2019 severe droughts were experienced, necessitating the country's president to declare a state of emergency in 2013, 2016 and 2019. Farmers lost close to 90,000 livestock between October 2018 and June 2019, due to severe drought (Shikangalah 2020).

The impacts of these recent droughts were felt on many sectors of the economy including agriculture, livestock, water, and conservation. While these impacts can be measured quantitatively and effects on the country's Gross National Product (GDP) can be pronounced nationally, qualitative impacts at local communities' levels are equally important if somewhat less visible.

For the wildlife and biodiversity conservation sector, direct impacts of droughts have been documented. For instance, a report by Kaula Nhongo observed for Erongo Region that in drought conditions thirsty elephants (*Loxodonta africana*) raid villages to eat crops and drink water from the storage tanks, describing how "elephants destroy houses, water points, gardens and fences and this has caused an uproar with most communities who are now threatening to take matters into their own hands" (Nhongo 2019: online). The implications are not only for the material well-being of the communities and/or the country at large, but can also lead to the loss of human lives. The encounter related below lays bare the reality of such experiences 'on the ground':

> [o]ne of the farmers, who is still haunted by a vicious encounter with an elephant, is 37-year-old Tjitemiso (nickname) who was attacked in 2017 while accompanying his friend home in the night. Tjitemiso and his friend crossed paths with the elephant which chased them into the bush. They tried to hide behind a tree, but the elephant discovered them. He managed to escape but his friend was trampled to death. Although he lived to tell the tale, his life will never be the same again. "I watched my long-time friend get trampled to death by an elephant, a picture that will remain with me for the rest of my life... the government needs to act" (Nhongo 2019: online).

And again,

> in 2017, a man in Omatjete villages was killed by an elephant on his way from visiting friends. This incident combined with others triggered

this community to tender a petition to MET [Ministry of Environment and Tourism, now MEFT] to have elephants removed from the Daures Constituency (cited in Hartman 2017).

We interpret this incident as an example of climate change manifesting through drought-induced human-wildlife conflict (discussed further below). Namibia has been hailed for its progressive legislation and policies for wildlife protection, but events like this one may reverse the social acceptability of its conservation programme. Wildlife legislation in Namibia, through enactment of the Nature Conservation Amendment Act of 1996, allows for the involvement of local communities in Namibia's remaining communally managed areas if they organise themselves into formal organisations called conservancies (NACSO 2004; Weaver and Petersen 2008). Namibia's community-based conservation initiative has been further hailed for bringing economic benefits for communities and the country at large through tourism and conservation hunting alongside increased wildlife populations (Jacobsohn and Owen-Smith 2003; Bandyopadhyay et al. 2004; Lendelvo et al. 2012: Silva and Mosimane 2014).

Battling the Triple Threat: Climate Change + Human-Wildlife Conflict + COVID-19 Pandemic

While communities in the conservation sector are already battling the impacts of climate change-induced human-wildlife conflict (HWC), the outbreak of the COVID-19 pandemic has greatly exacerbated the situation. As we have indicated, Namibia's frequent droughts over the past few years have encouraged the movement of wildlife such as elephant closer to human settlements, searching especially for water and resulting in competition with people for already scarce resources. As in Kenya, where climate change has also been observed to contribute to HWC, particularly with elephants (Mukeka et al. 2019), these combined circumstances have escalated HWC in communities in Namibia's communal-area conservancies.[1] Instead of revenues from conservation

---

1   The location and tenure of communal-area conservancies are an outcome of Namibia's specific historical circumstances. This history gave rise to a division

hunting and tourism being spent on addressing developmental challenges facing communities, they are being rerouted to address climate change-induced human-wildlife conflict.

Recent funding acquired from the Green Climate Fund through the Environmental Investment Fund of Namibia has thus been prioritising water infrastructure maintenance in the conservancies to address the current water situation for conservation and reducing human-wildlife conflict.[2] Apart from supporting water infrastructure, different insurance schemes have also been put in place to cover the cost of HWC in communities in Namibia, as well as globally (Leslie et al. 2019).

During 2020, already hard-hit revenues from the conservation and tourism efforts of local communities, became further severely affected by the outbreak of the COVID-19 pandemic, which halted the inflow of tourism and conservation hunting revenue into conservancies, because of the closure of international borders which disrupted tourism arrivals. These events seriously limited revenue inflow and slowed down conservation and monitoring interventions including addressing human-wildlife conflicts, thereby additionally increasing communities' vulnerabilities in rural, wildlife-rich areas (Lendelvo et al. 2020).

We trace these interconnections in more detail for two specific communal-area conservancies below, drawing on a survey with which we were involved.[3] This survey formed part of the project *Integrated Approach to Proactive Management of Human-wildlife Conflict and Wildlife Crime in Hotspot Landscapes in Namibia* (UNDP 2020), carried out for Namibia's Ministry of Environment, Forestry and Tourism (MEFT), and supported by the Global Environment Facility (GEF) through the United Nations Development Programme (UNDP).

---

between surveyed freehold farms allocated to settlers by the country's colonial and apartheid governments, separated from areas forming so-called 'Native Reserves' and 'Homelands' where peoples autochthonous at the advent of colonial rule were constrained to live, and that have remained after independence under communal forms of tenure and management (Sullivan 2018).

2   Environmental Investment Fund of Namibia (EIF), CBNRM EDA Project (Eif.org. na, 2020), https://www.eif.org.na/project/eda1-project.

3   Lendelvo and Nghitevelekwa co-led the stakeholders' assessment and gender analysis for this project.

Local Experiences of Climate Change-Induced Human-Wildlife Conflict

Otjimboyo and Ohungu are communal-area conservancies located in the Erongo Region, west Namibia (see Figure 6). The two conservancies are registered and gazetted by the Namibian government and form part of the country's acclaimed Community-Based Natural Resources Management (CBNRM) Programme. The conservancies lie along major westward-flowing ephemeral rivers called the Ugab (!U‡gāb) and Huab (‖Huab), whose riverine vegetation constitutes important habitat for different wildlife species, including the desert-adapted elephant and lion (*Panthera leo*) (MET/NACSO 2018), as well as for livestock.

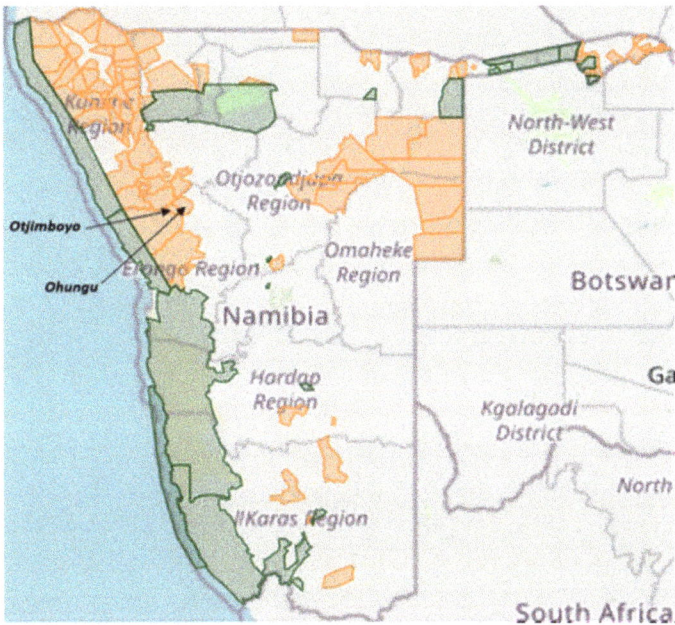

Fig. 6. Map showing the location of the Otjimboyo and Ohungu conservancies in Namibia, adapted by the authors from public domain image at http://www.nacso. org.na/conservation-and-conservancies. Areas shaded orange are communal-area conservancies; those in green are national parks and other state conservation areas. © NACSO, CC BY 4.0.

Because of their location in the driest part of the country, livestock farming rather than crop production is the primary source of local livelihood, with conservancies relying on the same landscapes for wildlife conservation to diversify and strengthen local livelihoods (MET/NACSO 2018; Kamupingene et al. 2016). The post-independence conservancy approach was embraced by these communities to generate income from tourism and conservation hunting, and particularly to engage in organised wildlife conservation so as to reduce the cost of livestock predation by indigenous fauna amongst these pastoralist communities (MET/NACSO 2018).

Gender-Differentiated Experiences of Climate Change-Induced Human-Wildlife Conflict

Women in so-called developing nations are especially vulnerable to climate-related events such as droughts, a situation that is further exacerbated by existing cultural norms, unequal distribution of roles and responsibilities at household levels, and inequalities in access to, and power over, resources (Lendelvo et al. 2018; Yavinsky 2012). In addition, economic disparities between men and women at household levels in many rural communities in Africa persist, and are further exacerbated by the triple threat of climate change, HWC and the COVID-19 pandemic. Women in the communal area conservancies of Otjimboyo and Ohungu associate climate change with the interruption of natural patterns and events upon which culture and traditions are reliant (as similarly recorded in Sullivan 2002). Norms and practices within these communities are strongly connected with seasons because seasonal rainfall patterns drive productivity and thus the availability of different resources, including water and forage.[4] Climate change and the widening variability of climatic conditions affect human-wildlife conflict through their effects on food and water supply for wildlife. In particular, recurrent droughts and rising temperatures cause frequent food and water shortages for herbivores, and contribute to greater wildlife mobilities and the likelihood of contacts between wildlife, people and livestock (Mukeka et al. 2019).

---

4   Editors' note: for more detail regarding the implications of unpredictably varying rainfall on the dryland ecologies and production systems of west Namibia, see Sullivan (1996) and Sullivan and Rohde (2002).

Livestock farming (pastoralism) is both culturally important and provides wealth for households. Although livestock have traditionally been predominantly owned by men, women in the two conservancies also own mainly small stock, primarily goats, and some also own donkeys used for transport. In our survey, most women indicated that farming helps to generate income for the school education of their children, and to provide for their needs and provisions through sales or bartering. Donkeys are especially important for transportation, assisting in fetching water, taking people to key services such as hospitals and pension grant pay-points, and enabling them to attend social gatherings such as community meetings, funerals and weddings. In one of the interviews, a respondent shared that

> an elderly woman lost her donkey to predators [...] and after that it was a challenge for her to travel to the centre for her pension (pers. comm. Otjimboyo 2019).

The frequency of droughts experienced in these two communities has led to poor pasture in these areas, meaning that livestock are forced to travel long distances from homesteads, exposing them as prey to predators. Donkeys in particular were subjected to high mortalities or becoming vulnerable to predators due to limited food sources. A large proportion of the impacts of HWC during 2018 and 2019, for example, was linked to the devastating country-wide drought, which left many poor households destitute, and especially those which were female-headed (MET/NACSO 2018; Shikangalah 2020). Human-wildlife conflict has been found to exacerbate female-headed households' vulnerabilities and to negatively affect household wealth, as livestock ownership in these households tends to be characterised by low numbers and the inability to afford herders (Kamupingene et al. 2016).

Kamupinge et al. (2016) additionally report that: the common type of HWC incidents experienced in these communities include livestock predation, damage to property and attacks on and loss of human lives. Severe drought increases HWC as herders experience difficulties with kraaling animals; and these impacts are set in a context wherein Erongo region is also reported to have inherent water problems which have been aggravated by poor rainfall resulting in a receding water table and drying up of water bodies. Respondents in our study narrated that during drought, cattle may travel long distances from villages (between 10 and 15 km) to where water can be found, and in most cases do not return home the same day. While men can go after the livestock or hire livestock herders, households headed by women are more affected by livestock losses as they are less likely to be able to leave their household to follow animals grazing large distances away (Sadie n.d).

Conversely, degradation of natural resources close to villages may mean that women and children must travel long distances and spend more time in the field in search of resources important to their livelihoods, such as firewood, mopane worms,[5] items for craft production products, and medicinal or edible plants. Drought enhances their fear of encountering dangerous animals such as elephants when travelling far from their homes.

These observations are set in a context wherein climate change is also believed to have disrupted wildlife movement patterns, resulting in different species frequently moving close to homesteads and their movement becoming unpredictable, either causing damage to property or livestock predation (MET/NACSO, 2018: 45). Farmers in Erongo Region are experiencing livestock predation, whilst those along the Ugab river area are more affected by elephant-related conflicts. Thus, "[p]eople in Erongo Region along the Ugab River landscape, had freedom of movement restricted to their homesteads, and they disappeared into their houses just at dusk for fear of their lives" (National Council of Namibia 2017).

Indeed, the presence of elephants is the HWC that communities fear the most. Although livestock predation is prevalent, predator conflict seemed to be more manageable. For example, not all women in our study expressed that predation is a problem, but most of them expressed that human-elephant conflict is a key concern, referring especially to the man killed by an elephant in Ohungu conservancy.

In addition to unpredictable wildlife movements, the presence of elephants makes women feel helpless and with little or no options, due to elephants moving up to homesteads in search of water and food. Human-elephant encounters may leave the poor more destitute as their property, including food storage facilities, gardens and infrastructure

---

5   Editors' note: caterpillars of the emperor moth *Gonimbrasia belina*.

such as fences and water points, is destroyed. Furthermore, women fear for their children and grandchildren when they go to school, look after livestock, or fetch water, in case they encounter wildlife.

Human-Wildlife Conflict Policy Compensation Scheme

The national HWC management policy compensation scheme is a control measure designed to mitigate against effects of human-wildlife conflict and is carried out by conservancies in collaboration with the MEFT and supporting NGOs. The scheme is perceived to be problematic, however, because in most cases compensation is delayed, does not provide market value of lost livestock, and does not reach all affected parties due to the nature of evidence required for compensation payments. To allow for compensation of lost livestock, the human-wildlife conflict policy regulations require evidence and reporting within twenty-four hours. For women in particular, to get this evidence within the timeframe is often impossible and a particular problem is that affected parties are often not able to provide evidence of the cause of death of their livestock, because incidents happen far from their home and at times evidence disappears before they discover the missing animal.

Further, whilst the HWC management policy makes provisions for cash compensation, this might not be the ideal form of compensation for affected parties who may prefer direct replacement of the lost animal(s). Services such as electricity and provisions such as household feeding schemes may instead go a long way to assist communities affected by climate-induced or -exacerbated HWC. Electricity helps to keep elephants away, leading people in the community to start using light-based deterrents to chase away elephants, especially at night (Shaffer et al. 2019). More protection and/or compensation in terms of repairs for properties and water infrastructure damaged by elephants are needed to prevent the reported situation that

> the little income we have left, is what we use to repair some damages caused by elephants to get water for households, livestock, and even to protect our people (female respondent, Ohungu Conservancy).

For women, fencing off villages or households to prevent elephants from invading villages and homesteads may be desired. In this context, women also remain disadvantaged in terms of accessing information to enable them to make informed decisions or benefit from opportunities. Men tend to be more mobile, networked, connected and are likely to access information quicker than women in these contexts. A situation of minimal information shared among women tends to create fear, affecting their day-to-day livelihood activities such as gathering firewood, fetching water, caring for and sending children to schools and/or harvesting foods and medicines.

Conclusion

In addition to changes in land use, climate change is among the key factors predicted to cause losses of natural resources during the coming decades in southern Africa (Biggs et al. 2008; Kupika et al. 2017; Khalife 2020). National and international efforts have provided crucial information for planning, formulating policies, and implementing programmes aiming to address climate change. For about twenty-five years, Parties to the United Nations Framework Convention on Climate Change have congregated to deliberate on the risks of climate change, but with limited substantial outcomes that make a difference on the ground for the most affected communities at risk in dryland countries such as Namibia. Communities at local levels experience climate change in different ways. For the Otjimboyo and Ohungu communal area conservancies of Namibia's Erongo Region, their tale is one of the intersecting effects of climate change, HWC, and now, COVID-19.

While the impacts of these phenomena are felt broadly, they are also gendered. Women in these communities are finding it difficult to adapt to the effects of climate change due to limited opportunities, combined with already existing inequalities and vulnerabilities. The combination of climate change and HWC presents complexity for women farmers by negatively impacting their farming and household economy. There is a need for regular support with gardening and income opportunities to supplement their vulnerable livestock farming. Regarding the fear of lack of awareness about the presence of problem animals and other relevant information, an early warning system using simple technology is needed. Such a system might use existing communication structures in the conservancies to bring forth information warning of the proximity of problem animals and HWC incidents. Diversified means of information sharing, such as using local leadership structures, social grouping, media (mainly the radio) and mobile phones, could assure that most people are reached, irrespective of gender, age, and other social categories.

As world leaders, scientists and civil societies gather at COP26 in the hope of finding the best solutions to climate change, let us remember the local communities in the remotest parts of our planet: communities such as those inhabiting the Otjimboyo and Ohungu conservancies in rural Namibia. Let us also remember that climate change impacts are differentiated and that the most vulnerable social groups—women, the poor, and others—tend not to be present at international negotiations such as the UNFCCC COPs to share their experiences of dealing and living with the impacts of climate change in their daily lives. This chapter is intended as a short communiqué to foreground the types of concerns women in rural dryland communities might wish to voice if they were able to be present at COP26.

References

Bandyopadhyay, Sushenjit, Michael N. Humavindu, Priya Shyamsundar, and Limin Wang, *Do Households Gain from Community-Based Natural Resource Management? An Evaluation of Community Conservancies in Namibia* (Washington, DC: The World Bank, 2004).

Bann, C., and S. C. Wood, 'Valuing Groundwater: A Practical Approach for Integrating Groundwater Economic Values into Decision Making—A Case Study in Namibia, Southern Africa', *Water SA*, 38(3) (2012), 461–66, https://doi.org/10.4314/wsa.v38i3.12.

Biggs, Reinette, Henk Simons, Michel Bakkenes, Robert J. Scholes, Bas Eickhout, Detlef van Vuuren, and others, 'Scenarios of Biodiversity Loss in Southern Africa in the 21st Century', *Global Environmental Change*, 18(2) (2008), 296–309, https://doi.org/10.1016/j.gloenvcha.2008.02.001.

Hartman, Adam, 'Elephant Kills Man in Erongo' (Namibian.com, 2017), https://www.namibian.com.na/161622/archive-read/Elephant-kills-man-in-Erongo.

Jacobsohn, Margaret, and Garth Owen-Smith, 'Integrating Conservation and Development: A Namibian Case-Study', *Nomadic Peoples*, 7(1) (2003), 92–109.

Kamupingene, Gift, Sikala Elma, Obong'o David, and Gabayi Princess, *Assessment of Impacts and Recovery Needs of Communities Affected by El Niño-Induced Drought in Kunene, Erongo and Omusati Regions of Namibia* (Rome: Food and Agriculture Organization of the United Nations, 2016), http://www.fao.org/3/a-i6604e.pdf.

Kapolo, I. N., 'Drought Conditions and Management Strategies in Namibia', *Windhoek: Namibia Meteorological Services* (2014), 1–9.

Khalife, Sawsan, 'Climate Change Impact on Natural Resources and Migration Across the Regions of Africa' (Thesecuritydistillery.org, 2020), https://thesecuritydistillery.org/all-articles/climate-change-impact-on-natural-resources-and-migration-across-the-regions-of-africa.

Kupika Olga L., Gandiwa Edson, Kativu Shakkie and Nhamo Godwell, 'Impacts of Climate Change and Climate Variability on Wildlife Resources in Southern Africa: Experience from Selected Protected Areas in Zimbabwe', *Intechopen* (2017), 1–23, https://doi.org/10.5772/intechopen.70470.

Lendelvo, Selma, Faith Munyebvu, and Helen Suich, 'Linking Women's Participation and Benefits within the Namibian Community Based Natural Resource Management Program', *Journal of Sustainable Development*, 5(12) (2012), 27–39, https://doi.org/10.5539/jsd.v5n12p27.

Lendelvo, Selma, Margaret N. Angula, Immaculate Mogotsi, and Karl Aribeb, 'Towards the Reduction of Vulnerabilities and Risks of Climate Change in the Community-Based Tourism, Namibia', in *Natural Hazards—Risk Assessment and Vulnerability Reduction*, ed. by José Simão Antunes do Carmo (IntechOpen, 2018), https://doi.org/10.5772/intechopen.79250.

Lendelvo, Selma, Mechtilde Pinto, and Sian Sullivan, 'A Perfect Storm? The Impact of COVID-19 on Community-Based Conservation in Namibia', *Namibia Journal of Environment*, 4 (2020), 1–15, http://www.nje.org.na/index.php/nje/article/view/volume4-lendelvo/43.

Leslie, Sam, Brooks Ashley, Jayasinghe Nilanga, and Hilderink Femke, 'Human Wildlife Conflict Mitigation: Lessons Learned from Global Compensation and Insurance Schemes', *HWC SAFE Series. WWF Tigers Alive* (2019), 1–50, https://wwfeu.awsassets.panda.org/downloads/wwf_human_wildlife_conflict_mitigation_annex.pdf.

MET/NACSO, *The State of Community Conservation in Namibia—A Review of Communal Conservancies, Community Forests and Other CBNRM Activities* (Annual Report 2017) (Windhoek: MET/NACSO, 2018), http://www.nacso.org.na/sites/default/files/State%20of%20Community%20Conservation%20book%20web_0.pdf.

Mukeka, Joseph M., Joseph O. Ogutu, Erustus Kanga, and Eivin Røskaft, 'Human-Wildlife Conflicts and Their Correlates in Narok County, Kenya', *Global Ecology and Conservation*, 18 (2019), e00620, https://doi:10.1016/j.gecco.2019.e00620.

NACSO (Namibian Association of CBNRM Support Organisations), *Namibia's Communal Conservancies: A Review of Progress and Challenges in 2003* (Windhoek: NACSO, 2004).

National Council of Namibia, *Report of the Standing Committee on Habitat on the Motion on Human-Wildlife Conflict to Zambezi, Oshikoto, Oshana, Ohangwena, Omusati, Kunene, Kavango East, Kavango West, Erongo Regions and Benchmark Study to Tanzania and Zimbabwe from 09 September—05 October 2017* (Windhoek: National Council of Namibia, 2017), pp. 1–38, http://parliament.na/index.php/archive/category/187-report-2017?download=8455.

Nhongo, Kaula 'Feature: Drought Exacerbates Human-Wildlife Conflict for Namibia Rural Communities' (Xinhuanet.com, 2019), http://www.xinhuanet.com/english/2019-03/05/c_137871079.htm.

Sadie, Yolanda, 'Human-Wildlife Conflict and Wildlife Conservation' (Accord.org.za, 2019), https://www.accord.org.za/conflict-trends/human-wildlife-conflict-and-wildlife-conservation/.

Schnegg, Michael, and Michael Bollig, 'Institutions Put to the Test: Community-Based Water Management in Namibia during a Drought', *Journal of Arid Environments*, 124 (2016), 62–71, https://doi.org/10.1016/j.jaridenv.2015.07.009.

Shaffer., L. Jen, Kapil K. Khadka, Jamon Van Den Hoek, and Kusum J. Naithani, 'Human-elephant Conflict: A Review of Current Management Strategies and Future Directions', *Frontiers in Ecology and Evolution*, 6 (2019), 1–12, https://doi.org/10.3389/fevo.2018.00235.

Shikangalah, Rosemary N., 'The 2019 Drought in Namibia: An Overview', *Journal of Namibian Studies: History Politics Culture*, 27 (2020), 37–58, https://namibian-studies.com/index.php/JNS/article/view/8635.

Silva, Julie, and Alfons Mosimane, '"How Could I Live Here and Not Be a Member?": Economic Versus Social Drivers of Participation in Namibian Conservation Programs', *Human Ecology*, 42 (2014), 183–97.

Sullivan, Sian, 'Towards a Non-equilibrium Ecology: Perspectives from an Arid Land', *Journal of Biogeography*, 23 (1996), 1–5.

Sullivan, Sian, '"How Can the Rain Fall in This Chaos?": Myth and Metaphor in Representations of the North-West Namibian Landscape', in *Challenges for Anthropology in the 'African Renaissance': A Southern African Contribution*, ed. by D. LeBeau and R. J. Gordon (Windhoek: University of Namibia Press, 2002), pp. 255–65.

Sullivan, Sian, 'Dissonant Sustainabilities? Politicising and Psychologising Antagonisms in the Conservation-development Nexus', *Future Pasts Working Paper Series*, 5 (2018), https://www.futurepasts.net/fpwp5-sullivan-2018.

Sullivan, Sian, and Rick Rohde, 'On Non-equilibrium in Arid and Semi-arid Grazing Systems', *Journal of Biogeography*, 29(12) (2002), 1595–618, https://doi.org/10.1046/j.1365-2699.2002.00799.x.

UNDP, *Integrated Approach to Proactive Management of Human-wildlife Conflict and Wildlife Crime in Hotspot Landscapes in Namibia* (Na.undp.org, 2020), https://www.na.undp.org/content/namibia/en/home/library/environmental-and-social-management-framework--esmf-.html.

Weaver, L. Chris, and Theunis Petersen, 'Namibia Communal Area Conservancies', in *Best Practices in Sustainable Hunting—A Guide to Best Practices from Around the World*, ed. by Rolf D. Baldus, Gerhard R. Damm, and Kai-Uwe Wollscheid (Budakeszi: International Council for Game and Wildlife Conservation, 2008), pp. 48–55.

Yavinsky, Rachel, 'Women More Vulnerable Than Men to Climate Change' (Prb.org, 2012), https://www.prb.org/women-vulnerable-climate-change/.

13. Environmental Change in Namibia: Land-Use Impacts and Climate Change as Revealed by Repeat Photography

Rick Rohde, M. Timm Hoffman and Sian Sullivan

This essay draws on repeat landscape photography to explore and juxtapose different cultural and scientific understandings of environmental change and sustainability in west Namibia. Change in the landscape ecology of western and central Namibia over the last 140 years has been investigated using archival landscape photographs located and re-photographed, or 'matched', with recent photographs. Each set of matched images for a site provides a powerful visual statement of change and/or stability that can assist with understanding present circumstances at specific places. The chapter shows in a practical way an innovative possibility for documenting and analysing environmental and social change, helping us to contextualise projected and predicted environmental futures, and sometimes offering complexity with regard to modelled climate change projections and scenarios.

https://doi.org/10.11647/OBP.0265.13

Historicising Environmental Change through Repeat Landscape Photography

Fig. 7. Repeat photos of Mirabib inselberg in the Namib Desert. Composite image © Rick Rohde, drawing with permission on images by John Jay, Frank Eckhardt, Rick Rohde and Timm Hoffman.

The above composite image illustrates a key method drawn on in exploring environmental change in west Namibia. The image depicts the Mirabib rocky outcrop in Namib-Naukluft National Park at three different moments in time. The first view (the small black and white photo) is a still from the film *2001: A Space Odyssey* which director Stanley Kubrick used for some of the opening scenes. The film's still photographer, John Jay, took this photo in 1965. This image was re-photographed in 1995 (the image on the clipboard) by University of Cape Town geomorphologist Frank Eckardt; followed by a third retake in 2015 by Rick Rohde and Timm Hoffman.

Repeat landscape photography can be used to explore and juxtapose different cultural and scientific understandings of environmental change and sustainability. Sometimes this method reveals ecological markers of historical events and climate change trends that contradict both popular and scientific assumptions. The material brought together through this method can tell stories of environmental change that are different to those assumed in popular imagination and scientific predictions alike.

In this essay we summarise some findings from repeat photography research for changes in the landscape ecology of western and central

Namibia for a longer time period than the effects of recent drought reported for the same area in Lendelvo et al.'s chapter, this volume. We have revisited and re-photographed sites that were originally photographed as long ago as 1876, analysing ecological changes these 'matched' images record. We select examples from a large dataset of repeat images brought together since the 1990s by Rick Rohde and Timm Hoffman from the University of Cape Town. Each set of matched images for a site provides a powerful visual statement of change and/ or stability that can assist with understanding present circumstances at specific places and across regions. They help us to contextualise projected and predicted environmental futures linked, for example, with understandings and assumptions about climate change.

Historical Ecology[1]

Given the dramatic events that have shaped the present socio-economic landscape of central and west Namibia—the establishment of colonial enterprise, a genocidal colonial war, seven decades of apartheid rule, and the ushering in of broadly neoliberal policies since independence in 1990—it is not surprising that traces of past impacts are inscribed on the landscape. These traces create layered landscape 'palimpsests'[2] in which past influences can be read and deciphered in the present.

In the archive image below, for example, we see the kraal of Maherero, the first Paramount Chief of the pastoralist ovaHerero, who was powerful in central Namibia prior to German colonisation in the 1880s. The photograph was made by the photographer who accompanied British colonial magistrate W. C. Palgrave as he travelled from Cape Town to central Namibia in 1876, seeking the possibility of establishing colonial 'Protection Treaties' with the diverse autochthonous peoples

---

1   For more detail on the research informing this first part of the chapter see Rohde and Hoffman 2012.

2   Human and cultural actions in relation to landscapes are sometimes framed metaphorically as akin to writing on a page, leaving behind signs that can be read by future inhabitants. The overlaying of multiple workings of the land thereby make a landscape something of a *palimpsest*—a text overlain by successive writings, the earliest writings never quite completely erased. In other words, landscapes are multi-layered and readings of them invite an unpicking of these layers, and an awareness of the influence and interplay of earlier inscriptions on those that follow (as explored more fully in, for example, Rohde 2010; Heatherington et al. 2019).

inhabiting the territory (Coates Palgrave 1877; Stals 1991). What was then a pastoral scene of grassland savanna with umbrella acacias (*Acacia tortilis*) providing some shade for dung-plastered dwellings and kraaled animals, is now a cluster of twenty-first-century settler houses built like Bavarian castles on high square stone or concrete plinths, surrounded by electric fences and barbed wire. The vegetation changes are astonishing. Where once there was a wide ephemeral river bordering the receding Namibian grasslands (centre left in the top image), there is now hardly an opening in the thornveld canopy. Some of the riverine Ana trees (*Fairdherbia albida*) in the bottom image are now gigantic—like old grandparents care-worn and haggard surrounded by a new generation of hungry dependents. The lush flowering grasses in the bottom image are breast-high in places.

Fig. 8. Maherero's kraal (1876) above, and present day Okahandja (2009) below. Sparse riverine Ana trees and thornveld savanna have been replaced by alien tree species such as eucalyptus (Australian) and prosopis (North American) that now obscure the view of the upper reaches of the Swakop River. These social and environmental changes are emblematic of the reshaping of central Namibian landscapes since colonial times. (Above) W. C. Palgrave Collection (National Library of South Africa), out of copyright, used with permission; (below) © R. F. Rohde and M. T. Hoffman.

The photographic evidence from many sites in central Namibia reveals a complex story of radical political, cultural and socio-economic change amounting to an ecological revolution associated with colonialism and an expanding global capitalist economy.

In 1876 the grazing lands of central Namibia were more grassy and less woody; rangelands were highly shaped by large pastoral herds, transhumance routes, temporary settlements and the use of fire to manage grasslands. During the 1890s, however, several cataclysmic events converged to bring about radical change. These included the rinderpest epizootic of especially 1897, smallpox, and a colonial war of the German state against autochthonous Namibians that intensified in 1904–1907, all of which effectively decimated indigenous peoples and their herds (Esterhuyse 1968; Olusoga and Erichsen 2010; Wallace 2011). As illustrated in Figure 9, the resulting hiatus in land use resulted in ecological changes that persist in the landscape today as a signature of events of a century ago.

Fig. 9. River Skaap, Hatsemas Central Highlands, central Namibia. Matched photographs spanning 1876 (left) to 2005 (right). Camelthorn individuals have colonised previously bare or sparsely vegetated river terraces, probably in a single recruitment event. The increase in small trees and shrubs on the rocky pediments and hill slopes is indicative of regional bush encroachment patterns. For scale, note the white tent and ox-wagon of the Palgrave expedition at the bottom left quadrant of the 1876 image. (L) W. C. Palgrave Collection (National Library of South Africa) out of copyright, used with permission; (R) © R. F. Rohde and M. T. Hoffman.

The change in riverine habitats lining the margins of an ephemeral river in the pastoral landscape shown in Figure 9 is immediately apparent when viewing these images alongside one another. Considering the large size of individual trees and the uniform population structure of these long-lived camelthorn trees (*Acacia erioloba*), it is likely that they originated as a seedling cohort sometime between 1876 and 1910. It is reasonable to assume that they recruited in response to the sudden release of pastoral grazing pressure when indigenous pastoral groups were effectively displaced by disease, war and subsequent German colonisation. This woodland thus endures as an ecological marker of historical events with political, social and environmental dimensions.

Other repeat image pairs from this collection show the development of towns and settlements, such as in the open uninhabited savanna of central Namibia, now dominated by the Presidential Palace and the extensive southern suburbs of the capital, Windhoek: as shown in Figure 10.

Fig. 10. Windhoek southern suburbs, looking south-west from Wassenberg. Matched photographs spanning 1919 (left) to 2014 (right). Bush thickening is evident throughout the landscape, not to mention the sprawl of urban residential housing and the new Presidential Palace built by a North Korean company, completed in 2008. (L) I. B. Pole-Evans (South African National Biodiversity Institute), out of copyright, used with permission; (R) © R. F. Rohde.

Bush thickening ('bush encroachment') has become an increasing problem in central Namibia, reducing the available grazing for cattle and in many instances closing the thornbush canopy to grazing altogether, as illustrated in the matched images in Figure 11. Palgrave's 1876 photo depicts a newly established mission station inhabited by several hundred indigenous pastoralists situated around a wetland in a tributary of the Swakop River. Intermittent conflict, cattle theft and violent confrontation took place between Nama and ovaHerero until the 1890s when a police station was established here by the German colonial administration. Otjisewa farm was bought from the local ovaHerero chief in the early 1900s and has remained a privately owned commercial cattle farm owned by a succession of German and Afrikaner families. Although a problem for cattle farmers, the thickening of thornbush savanna here forms a massive carbon sink that might be advantageous in a global context of increased anthropogenic greenhouse gases and atmospheric CO_2 fertilisation. At the same time, however, and illustrating the complex trade-offs involved between different production / protection choices, the management of bush encroachment often involves clearing woody plants for conversion to saleable charcoal (Dieckmann and Muduva 2010), or converting the "bothersome biomass" of bush encroachment into biomass fuel to power Biomass Industrial Parks (Heck 2021). Both strategies contribute significant emissions.

Fig. 11. Otjisewa, central Namibia. The change from the savanna landscape in 1876 (above) to the image of 2006 (below) illustrates how bush thickening typifies these more mesic (i.e. moist) highland areas of Namibia today. (Above) W. C. Palgrave Collection (National Library of South Africa), out of copyright, used with permission; (below) © R. F. Rohde and M. T. Hoffman.

In contrast, many of the more arid rangelands in the South of Namibia have remained stable and relatively unaltered over long periods of political, socio-economic and climatic change. The ephemeral Guireb River in the Karas Region of southern Namibia is a stunning example: see Figure 12. These more arid areas are less impacted by land use and have remained relatively stable over long periods of time. Patterns of little or no woody vegetation cover change are correlated with Mean Annual Precipitation (MAP) below a threshold of around 250 mm, regardless of land tenure system or land-use practice.

Fig. 12. Aub / Gurieb River, southern Namibia. In spite of the contrast of the dry, sparsely populated, pastoral landscape depicted in 1876 and the lush grass and flowing river during the exceptional rainy season of 2009, woody vegetation species and the extent of cover has hardly changed. (L) W.C. Palgrave Collection (National Library of South Africa), out of copyright, used with permission; (R) © R.F. Rohde and M.T. Hoffman.

Bush encroachment (as per Figures 10 and 11) is positively correlated with MAP above 250mm and associated with land-use practices such as commercial cattle ranching arising in the wake of colonialism, drought, and the epidemics and epizootics of the late-nineteenth and early-twentieth centuries. Legacies of demographic collapse, land-use change and landscape fragmentation are evident in these more mesic savannas. We see no evidence of recovery from bush-encroached to open grassland over the timescale of this study.

Repeat Photography and Assessing Climate Change

In order to assess the impacts of climate change in west Namibian landscapes we examined a dataset of one hundred repeat landscape images compiled over the past twenty-five years in the western desert landscapes of the Pro-Namib and Namib Desert. These arid and hyper-arid areas have been less impacted by human development and historical events than many of the more mesic parts of Namibia. A detailed vegetation survey was conducted and changes in dominant species cover was ascertained (Rohde et al. 2019).

One of the main obstacles to researching climate change in the Namib Desert is the paucity of historical climate data. Because woody vegetation in this relatively undisturbed environment is strongly influenced by climate, we regard long-term vegetation change as a proxy for changes in climate. These repeat photo sites cover an area of approximately 40,000 square kilometres and the average time between the date of the original image and the repeat photograph is seventy-eight years.

Illustrating the diversity of habitats characterising west Namibia, these sites were categorised into four broad groups: Large Ephemeral Rivers; Fogbelt, Grasslands & Shrublands, Savanna Transition—the latter three being largely determined by rain and fog. These groupings are shown in Figure 13, together with the locations of the repeat photographs used in our analysis.

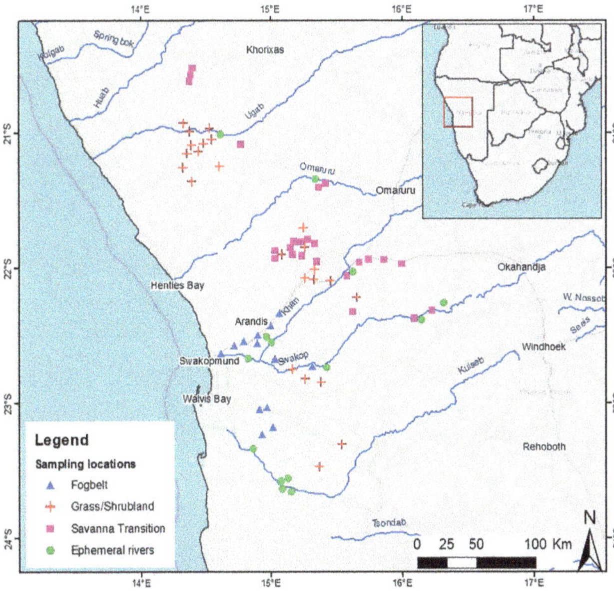

Fig. 13. Location of repeat photo sites in each of four vegetation types, west and central Namibia. Image by chapter lead author.

The region is characterised by a distinct environmental gradient from the hyper-arid coastal areas to the savannas of the plateau. Mean annual precipitation (fog and rain) is relatively low for sites in the coastal fogbelt and inland grass/shrubland sites, but increases in the savanna areas and sites within large ephemeral rivers farthest from the coast. Mean annual temperature is lowest in the first 60 km within the fogbelt zone and is higher further inland, although considerable variability

exists between sites. The number of fog days declines steeply away from
the coast and provides relatively little moisture input for sites further
than 100 km from the ocean.

Fogbelt

The sparse woody cover in the coastal western portion of the Fogbelt (9
km to 35 km from coast) increased by 0.36% per year[3] while sites between
40 km to 74 km from the coast showed very little change (+0.003% per
year) (Figure 14). Over the study period woody cover declined in only
two of the thirteen fogbelt sites, both of which were located towards the
eastern margin of this vegetation zone.

Fig. 14. Fogbelt site near Rössing uranium mine (33km from coast) showing
examples of the long-lived shrubs between 1919 (top) and 2016 (below) displaying
the same individuals in each image (white dots). The population of this species
has doubled during the last ninety-seven years (green dots) due to increased
fog. Mortality of individuals is shown as red dots. (Top) I. B. Pole-Evans (South
African National Biodiversity Institute), out of copyright, used with permission;
(below) © R. F. Rohde and M. T. Hoffman.

---

3   Woody cover change per year is calculated by dividing the total change over time by
     the total number of years between the original and repeat photo, recognising that
     vegetation growth increments are not necessarily smooth, i.e. they will not in fact be
     the same every year as they will be coupled with other dynamic factors.

Grasslands and Shrublands

The average percentage cover of woody vegetation in the grass/shrubland zone was more than twice the average value for the fogbelt zone. Unlike in the fogbelt zone however, woody cover in the majority of sites in the grass/shrublands did not increase over the sampling period (Figure 15). The average change in woody cover was only +0.007% per year. Variability in woody cover change was also relatively low in this vegetation zone. In seven of the eight sites located in the more eastern part of the grass/shrublands (i.e. 92 km to 125 km from the coast) woody plants, more typical of the savanna transition vegetation zone have increased in cover.

Fig. 15. Grass/shrublands (Aukas East). Matched photographs of *Euphorbia damarana, Calicorema capitata* shrubland (110km from coast) illustrating slight decrease in cover over ninety-seven years between 1919 (left) and 2016 (right). The replacement of *E. damarana* by savanna species (white dots) such as *Acacia reficiens, A. mellifera, Adenolobus pechuelii,* and *Maerua parvifolia* and *Commiphora spp.* is indicative of the westward expansion of savanna rainfall. (L) I. B. Pole-Evans (South African National Biodiversity Institute), out of copyright, used with permission; (R) © R. F. Rohde and M. T. Hoffman.

Savanna Transition

The savanna transition zone is comprised of a large number of woody species but is dominated by acacia and subtropical savanna tree and shrub species. In the repeat photographs, average values for woody plant cover were nearly twice those of the grass/shrubland zone (Figure 16). Woody plant cover declined in one site only and remained the same in four of the twenty-four sites in the savanna transition zone. The average rate of increase in woody plant cover was 0.07% per year.

Species composition in savanna transition sites varied in relation to latitudinal position where the hotter northern sites were dominated by *Colophospermum mopane* in contrast to southern sites where acacia species predominated.

Fig. 16. North Usakos Railway. Savanna transition showing significant increase in woody cover between 1919 (left) and 2014 (right), indicating the influence of a westward shift in summer rainfall. (L) I. B. Pole-Evans (South African National Biodiversity Institute); (R) © R. F. Rohde and M. T. Hoffman.

Large Ephemeral Rivers

The sites located within or adjacent to large ephemeral rivers are dispersed throughout the fogbelt, grass/shrubland and savanna zones. Woody plant cover was greatest in this vegetation zone with values as high as 75% woody cover recorded at one location. A wide range of woody plants dominated the ephemeral rivers and although changes in woody plant cover varied considerably between sites in this zone, it nearly doubled over the study period with average values of 0.26% increase per year. Only two sites exhibited a decline in woody cover with relatively modest reductions in cover. Each of the five ephemeral rivers in the study area present distinct traits in relation to vegetation change: e.g. the Khan River is the most stable in terms of species and percentage cover (although this may have been affected since by the opening of Husab Uranium Mine in the area), the Kuiseb appears to be the most dynamic (see Figure 17), and the Swakop the most impacted by anthropogenic disturbance and alien species.

Fig. 17. Large ephemeral river (Kuiseb). Matched photograph from Gobabeb (55km from coast) illustrating significant thickening and size increase of riverine woody cover over fifty years between 1965 (left) and 2015 (right). Dominant riverine species: *Faidherbia albida, Tamarix usneoides, Acacia erioloba, Salvadora persica.* (L) photographer unknown (Gobabeb Namib Research Institute), out of copyright, used with permission; (R) © R. F. Rohde.

Discussion

Past climate trends are not predictors of future environments but they do inform our understanding of the causes of present conditions. A tipping point might reverse a historical trend, but at present we see no evidence of such a shift from our analysis of woody vegetation change in the Namib and Pro Namib.

For example, we see no evidence for the expansion of desert and arid shrublands into higher rainfall savanna areas, nor do we find any evidence of a predicted decrease in groundwater or increased evaporation as a result of global warming. Rather than an expansion of more arid-adapted species into more mesic environments, our analysis documents the incursion of savanna species into more arid environments and an increase in woody plant cover in most localities.

Preliminary analysis of the changes in woody vegetation lead us to the following four possible explanations:

- That increased vegetation in the fogbelt is associated with a change to a colder, more intense Benguela Current upwelling in the Atlantic Ocean off the coast of Namibia. This cooling in turn generates more coastal fog, making more moisture available to plants in the coastal areas of west Namibia. This enhanced fog moisture has combined with recent increased occurrences of the Benguela Niño climate fluctuation (associated with desert rainfall events) that have supported population recruitment of coastal fog dependent woody vegetation species such as *Arthraerua leubnitziae* and *Zygophyllum stapffii*.

- Further inland, increased temperatures in the grasslands/shrublands areas have resulted in the reduced incidence of fog and the desiccation of woody vegetation apart from the ecotone bordering the savanna where increased summer rainfall has resulted in an expansion of more savanna species.

- The savanna transition areas show increases in vegetation consistent with increased rainfall and atmospheric CO_2 fertilisation in the more eastern landscapes.

- The large increases in woody vegetation along the azonal large ephemeral rivers may be due to upstream dams and fewer recent large flood events (which can uproot and wash trees and shrubs downstream).

These observations add to current debates regarding trends that are predicted or contradicted by various modelled future scenarios and that posit intensification (or weakening) of cold Benguela Current upwelling, leading to increased (or decreased) fog near the coast and desiccation further inland. Researchers have described hypothetical scenarios based on the modelling of the climatic gradient across central and western Namibia. None have done so, however, by showing the relationships between woody cover change, climate change and land use derived from historical sources in order to substantiate past and present trends across different kinds of vegetation communities.

The research presented here thus offers a summary of the empirical evidence for historical changes in Namibia's vegetation and climate. It has significant implications for understanding global phenomena such as the effects of historical events on contemporary landscapes and vegetation diversity, bush encroachment, rainfall and temperature trends, and hence climate change. It can also contribute to understandings of the effects of global warming on the intensity of the Benguela Eastern Boundary Upwelling System, and its associations with global synoptic moisture fluxes from the southeast Atlantic, southwest Indian Ocean, Sea Surface Temperature (SST) anomalies and the El Niño Southern Oscillation (ENSO).

References

Coates Palgrave, William, *Report of W. Coates Palgrave, Esq., Special Commissioner to the Tribes North of the Orange River, of his Mission to Damaraland and Great Namaqualand in 1876* (Cape Town: Saul Solomon and Co., 1877).

Dieckmann, Ute, and Theodor Muduva, *Namibia's Black Gold? Charcoal Production, Practices and Implications* (Windhoek: Legal Assistance Centre, 2010).

Esterhuyse, Jan Hendrik, *South West Africa 1800–1894: The Establishment of German Authority in South West Africa* (Struik: Cape Town, 1968).

Heatherington, Catherine, Anna Jorgensen, and Stephen Walker, 'Understanding Landscape Change in a Former Brownfield Site', *Landscape Research*, 44(1) (2019), 19–34.

Heck, Peter, *Road Map to a Biomass Industrial Park: Biomass Partnership with Namibia* (Dasnamibia.org, 2021), https://www.dasnamibia.org/?wpfb_dl=117.

Olusoga, David, and Casper Erichsen, *The Kaiser's Holocaust: Germany's Forgotten Genocide and the Colonial Roots of Nazism* (London: Faber and Faber, 2010).

Rohde, Rick F., 'Written on the Surface of the Soil: West Highlands Crofting Landscapes of Scotland During the Twentieth Century', in *Repeat Photography: Methods and Applications in the Natural Sciences*, ed. by R. H. Webb, D.E. Boyer, and R. M. Turner (London: Island Press, 2010), pp. 247–64.

Rohde, Rick, and M. Timm Hoffman, 'The Historical Ecology of Namibian Rangelands: Vegetation Change Since 1876 in Response to Local and Global Drivers', *Science of the Total Environment*, 416 (2012), 276–88.

Rohde, Rick F., M. Timm Hoffman, Ian Durbach, Zander Venter, and Sam Jack, 'Vegetation and Climate Change in the Pro-Namib and Namib Desert Based on Repeat Photography: Insights Into Climate Trends', *Journal of Arid Environments*, 165 (2019), 119–31.

Stals, Ernst L. P. (ed.), *The Commissions of WC Palgrave Special Emissary to South West Africa 1876–1885* (Cape Town: Van Riebeeck Society, 1991).

Wallace, Marion, *A History of Namibia: From the Beginning to 1990* (London: Hurst & Co, 2011).

14. On Climate and the Risk of Onto-Epistemological Chainsaw Massacres: A Study on Climate Change and Indigenous People in Namibia Revisited

Ute Dieckmann

On behalf of a Danish organisation (Charapa Consult), in 2012 the Legal Assistance Centre in Windhoek undertook a research study on climate change and indigenous people in Namibia. Charapa Consult had itself been commissioned by the World Bank Trust Fund for Environmentally and Socially Sustainable Development to undertake a regional research project in Africa, and parallel studies for Asia and Latin America had also been commissioned. As a researcher involved in the Namibia study, in this essay I critically assess its methodological challenges and dilemmas in relation to the global framework within which it was conducted. I place special emphasis on the predicament of *short-term* 'participatory' research with indigenous communities on climate change. I also outline the challenges arising from the necessity of squeezing indigenous environmental knowledge and experience into internationally acknowledged scientific frameworks, an approach which implies a subordination of indigenous peoples' ontologies to western ontologies. The compartmentalising necessitated by such a methodology risks

https://doi.org/10.11647/OBP.0265.14

the loss of the most important aspects of indigenous ecological knowledge related to climate change.

Introduction

Who better to lead during this time of dramatic climate change than peoples who know or can recollect in their indigenous traditions of TK [Traditional Knowledge] and/or TEK [Traditional Ecological Knowledge] practices of sustainability and indigenous ingenuity—Indigenuity? Can you imagine a world where nature is understood as full of relatives not resources, where inalienable rights are balanced with inalienable responsibilities and where wealth itself is measured not by resource ownership and control, but by the number of good relationships we maintain in the complex and diverse life-systems of this blue green planet? I can (Wildcat 2013: 515).

In this essay, I draw on a number of methodological challenges encountered during a study on climate change and indigenous people in Namibia as a starting point for a critique of climate change studies that attempt to integrate indigenous knowledge into dominant scientific frameworks. I was involved in the study as an anthropologist employed by the implementing organisation as part of a multi-disciplinary team. I illustrate what happens when we try to compartmentalise indigenous knowledge in order to fit it into our own conceptual frameworks.

Complementing Sullivan's Chapter 3 (this volume), I outline what we would gain from taking indigenous onto-epistemologies seriously, in the context of climate change and beyond. In short, I argue that avoidance of onto-epistemological chainsaw massacres, and the opening up of more possibilities for radical (re-)learning so as to avert ecological crisis, requires putting normalised 'western' frameworks aside in order to stop, listen and think carefully. I am drawing here on theoretical physicist Karen Barad's (2003) call for a revised "onto-epistem-ology", and borrow the term "chainsaw massacres" from Dianne Rocheleau's (2005: 339) analysis of the risks in cartography on fixing indigenous onto-epistomologies into the "iron grid of Descartes" (ibid: 328).

The Study

In 2012, the World Bank Trust Fund for Environmentally and Socially Sustainable Development (TFESSD) commissioned a Danish organisation (Charapa Consult) concerned with human rights and development[1] to undertake a regional research project on indigenous peoples and climate change in Africa, having commissioned similar regional studies for Asia and Latin America. The research in Africa, coordinated by Charapa with a number of implementing partners, looked at three ecological sub-regions of the African region: the tropical forest zone (Republic of Congo); arid/desert areas in southern Africa (Namibia); and lakes and wetlands (Kenya) (Charapa Consult 2012: 7). The overall research initiative had three main objectives: to analyse how indigenous peoples were affected by climate change; to identify indigenous peoples' local and traditional knowledge, practices and adaptation strategies; and to support strengthening of indigenous peoples' capacities for their engagement and direct participation in the formulation of public policies regarding climate change.

In Namibia, the Legal Assistance Centre (LAC)[2] in Windhoek, having an excellent record of research regarding marginalised/indigenous communities, was contracted to undertake the research for this project. Two indigenous Namibian communities were selected as case studies: the Topnaar (ǂAonin) and Haiǁom communities, both speaking Khoekhoegowab but living in different parts of the country (see Figure 18). These two communities were selected due to the difference of environmental circumstances in which they live, as well as the prior research experience of the lead author (for example, Dieckmann 2007, 2012). Both communities belong to the most marginalised people in Namibia (Odendaal and Werner 2020).

As requested by the organisations commissioning the study, the main components of the research were literature review, field research—including focus group discussions, household surveys, trend lines, ranking of livelihood strategies, mapping of well-being, knowledge

---

1   See http://www.charapa.dk/.
2   See http://www.lac.org.na/.

and political assets and semi-structured interviews—and data analysis (Charapa Consult 2012: 10).

Fig. 18. Locations of Topnaar and Hail̦lom research communities in Namibia. Dieckmann et al. (2013), http://www.lac.org.na/projects/lead/Pdf/climate_change.pdf, p. 35, CC BY 4.0.

On Ethics and Frameworks

Undertaking 'participatory' research regarding climate change in communities that are severely marginalised and struggling for daily survival felt extremely inadequate to me, especially given the limitations on participation caused by having to follow a pre-determined framework and methodology. The Topnaar and Hail̦lom communities today lack access to land and only have very limited access to natural resources. In post-colonial Namibia they experience high unemployment, low levels of education, very limited political representation and serious discrimination.

During the study people wanted to talk about their current situation and needs, rather than climate change and anticipated impacts. In addition, the limitations in access to land and natural resources meant that the direct impacts of climate change seemed to be minimal compared to other urgent threats to their livelihoods, at least in 2012.

These research observations and experiences led to fundamental questions being asked of the common conceptual framework being deployed so as to make the research consistent with the wider study: as shown schematically in Figure 19. This framework drew on the vulnerability concept developed by the Intergovernmental Panel on Climate Change (IPCC 2007), combined with the framework used for the World Bank study on Indigenous Peoples and Climate Change in Latin America and the Caribbean Region (Kronik and Verner 2010), itself adapted from the UK Department of International Development's (DFID) Sustainable Livelihood Framework as a tool to assess the vulnerability of different socio-economic groups and their adaptive capacity.

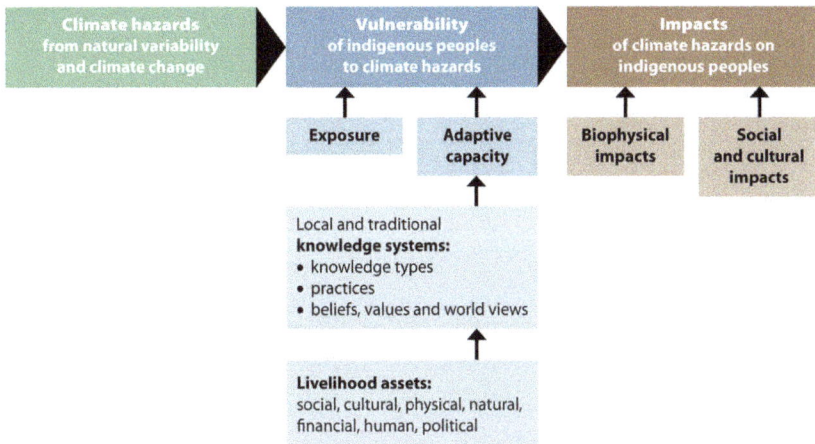

Fig. 19. Conceptual framework deployed in the World Bank Trust Fund for Environmentally and Socially Sustainable Development (TFESSD) study. Dieckmann et al. (2013), http://www.lac.org.na/projects/lead/Pdf/climate_change.pdf, p. 27, CC BY 4.0.

This conceptual framework analytically distinguishes between the impacts of climate change hazards and the conditions established by the contexts in which indigenous communities live. The framework is based on specific scientific assumptions operating within a specific scientific logic, although even in a western ontology, separating climate change-related impacts from other factors such as governance, access to land, and socio-economic status—which are interrelated and have a cumulative impact on indigenous peoples—seems highly problematic (also see Barnes et al. 2013: 543). This conceptual separation runs the risk of *de-politicising* and *re-naturalising* climate change in turning the attention away from the unevenly distributed anthropogenic/industrial causes of climate hazards.

Furthermore, merely providing a slot for "local and traditional knowledge systems" implies—as Mario Blaser (2009: 15) points out for the context of conservation—that "Indigenous environmental knowledges and practices" are "translated into discrete packages of knowledge that can be integrated into the toolkit of conservation practitioners, often as mere informational inputs". When applied in specific localities, the usefulness of the framework and its underlying assumption becomes highly questionable.

On Non-Existent and Non-Fitting Concepts

The issue of frameworks is closely connected to the question of concepts. Neither the Haillom nor the Topnaar had prior knowledge of the concept of climate change as such, an observation also reported for indigenous Baka and Babongo communities participating in the parallel Republic of Congo study (Charapa Consult 2012: 11). As Charapa (2012: 12, emphasis added) state in their final report: "[w]hen attempting to compare scientific and indigenous notions of climate change and related impact, *it becomes clear that these are not immediately comparable*".

It is not only that these notions may not be immediately comparable nor translatable, however, but that at times they may simply be incompatible (as also documented for Khoekhoegowab-speaking communities in north-west Namibia by Sullivan (2002), and more recently for Andean circumstances in Bolivia by Burman (2017)). To complexify matters further, this situation is not limited to the rather abstract concept of climate change—a concept whose definition even scientific experts disagree on—but relates also to additional associated terms such as drought, weather and environment.

During our fieldwork, Haiǁom participants came up with three terms for drought. Eventually, the community involved with the study agreed that |khurub should be used. This term also means hunger or no food. It is not only related to a lack of rain or dry environment but also includes impacts on the community. Complexifying matters, a frost that kills bushfood can also cause |khurub, meaning that the term and concept is not limited to low rainfall alone. The difficulty of comparing the concept of |khurub with 'drought' is thus evident: while 'drought' in science relates to a climate phenomenon, 'drought' for Haiǁom relates more specifically to associated broad spectrum impacts resulting in a loss of foods for humans that may have multiple climatic causes. Such complexities are also apparent elsewhere: Turkana and Maasai participants in the overall study had similar concepts combining drought and hunger (Charapa Consult 2012: 53; also see Goldman et al. 2016).

The concept of 'weather' is another telling example. Thomas Widlok (2017: 4) points out that the translation of the English term 'weather' in the Khoekhoegowab spoken by Haiǁom constitutes a compound of agentive forces: |nanutsiǁhaotsiǂôab literally translates as 'rains-and-clouds-and-wind', although this term is rarely used in everyday discourse. ǂNūkhoen (Damara), who like Haiǁom and Topnaar (ǂAonin) speak Khoekhoegowab, use the term ǂoab tsî |nanub (wind and rain) for weather (Schnegg 2019).

'Environment' is another instance where understandings do not fit. According to Widlok, translations such as ǂnamibeb and !ha!hais were originally coined by official language committees, but are also hardly used, and he suggests that for Haiǁom, 'environment' mainly refers to man-made environmental features (e.g. houses/huts or fire places) (Widlok 2017: 5). This understanding also points to other relevant concepts, especially the western dichotomy of natural and human/cultural, which do not exist in the same form in many indigenous understandings, Haiǁom (Widlok 2009, 2017) and other Khoekhoegowab-speaking peoples (Sullivan and Hannis 2016) included. The distinction between natural and supernatural agencies also seems to be non-existent or at least blurred in these contexts (Schmidt 2014; Sullivan and Low 2014; Widlok 2017: 5–6; Dieckmann 2021a).

In sum, central terms used for key concepts in climate change discourse either do not exist in, or do not seamlessly translate into, indigenous languages.

On Relationships and Agency

Arguably, then, events and developments that scientists place in the context of climate change and relate in certain ways (mostly causally) to each other, may be perceived differently by indigenous peoples through their distinct experiences of being-in-the-world and accompanying explanations of causality (also see Charapa Consult 2012: 66–67). During the studies in Namibia and the other African countries, it was thus a challenge "to directly relate and compare the perceptions and experiences of the indigenous communities participating in this study with, for example, the climate change phenomenon and first order impacts identified through the literature review" (Charapa Consult 2012: 52). Similarly, it was often impossible to provide sufficient room for the interpretations most meaningful to the participating communities.

Khoekhoegowab-speaking communities experience and establish relationships, including their drivers and effects, that may be different to scientific models. Some match with scientific explanations (cf. Sullivan 1999), while others do not. Some indicative fragments are provided below.

Haillom regard the pied crow (*!kha-nub*) as a protected bird, because according to Haillom tales, it brings back the rain after it is taken away from them by the animal "married to the rain", i.e. the elephant (Dieckmann 2012: 12–13), again indicating that there is no clear-cut distinction between the world of myth, legend, and the supernatural and the natural world.

Haillom, like other Khoe and San peoples, report the existence of 'water snakes' that protect waterholes, such that if the snake is killed or dies the water will dry up (Hoff 1997; Sullivan and Low 2014; Dieckmann 2021a).

‡Nūkhoen connect winds and rain, and moreover associate both with non-human agents, speaking of the power of winds, good and bad winds, and gendered winds (Low 2007; Schnegg 2019). The strongest

spirit-being (*ǁgamab*) for Haiǁom is the spirit of the rain (Dieckmann 2021a: 121).

Haiǁom look to the moon to see what will happen in the next season, as indicated schematically in Figure 20. When it is half-moon, there will be no rain in the season (1); when it is half-moon but one side is higher, the rain will start (2); and when the left side is even higher, it will be a sign of death and no rain (3).

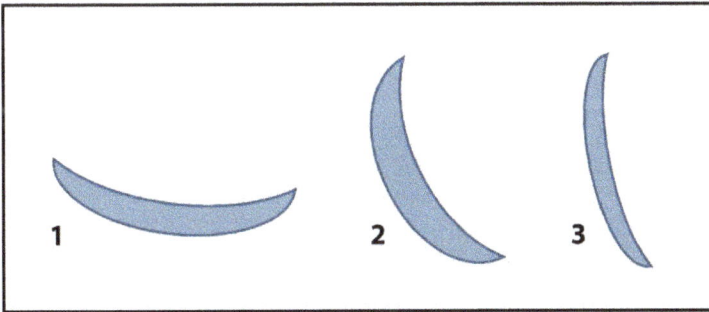

Fig. 20. Position of the moon in Haiǁom rain forecasts. Dieckmann et al. (2013), http://www.lac.org.na/projects/lead/Pdf/climate_change.pdf, p. 91, CC BY 4.0.

These are just a number of snippets connected to what scientists and most westerners would commonly call weather or climate, suggesting 'deviations' from scientific models based on western ontologies that assume dichotomies of nature vs. culture, human = animated vs. nature = unanimated and natural vs. supernatural. As brief and decontextualised examples, they nonetheless illustrate that rain and wind, celestial bodies, certain animals and more are regarded as agents. Khoekhoegowab-speaking communities, like other indigenous communities, seem to have an animistic understanding of the world (cf. Sullivan 2010; Low 2014) wherein the world is deemed "full of persons, only some of whom are human" (Harvey 2006: 11). As invoked above, the world is also inhabited by a variety of agential spirit-beings, connected—inter alia—to weather, animals and ancestors.

In these Namibian indigenous contexts, humans are an integral part of a wider ecology animated by other non-human agents, past and present. Relationships are thereby conceptualised in fundamentally different ways to the scientific framework of the Charapa study and most other scientific studies on climate change. This situation also appears true for other indigenous peoples worldwide (Yeh 2016; Goldman et al. 2018).

Why Does This Matter?

The above examples are provided as potentially puzzling onto-epistemological snippets to serve as illustrations of the epistemic violence—or chainsaw massacre—that can happen when knowledge is removed from its context. While the authors (both of the Namibian and the other African studies) tried to make space to mention at least some of these nuanced understandings and relationships, they remained odds and ends in their final reports. Because they were not overtly connected to scientific notions, they were also subsumed under headings such as 'beliefs' or 'culture' (see e.g. Charapa Consult 2012: 57), further undermining their relevance to the main business of understanding climate change. While certain aspects of indigenous 'ideas' may be called *knowledge*, namely those that can be made to correspond with scientific understandings, others are framed as *beliefs* unworthy as contributions to science.

A number of interrelated arguments—political, ethical, methodological, theoretical, philosophical—suggest that the above challenges should be taken seriously in the context of climate change.

Disempowering Indigenous People

The study discussed above is just one of many international studies which "attempt to combine 'expert assessment' with processes of 'stakeholder consultation'" (Scoones 2009: 548; also see Brosius 2006; contributions in Cameron, Leeuw and Desbiens 2014; Yeh 2016). Admittedly, the Charapa study (like many other studies) tried to do justice to indigenous peoples' needs and rights, and the partners explicitly agreed on general principles for the research to this end (see Charapa Consult 2012: 9): but did we meet these needs?

In retrospect, I would reply with a rhetorical question: whose ontology counts?

The study followed the familiar road of one ontology and the belief that this ontology can come to be known by different epistemologies.

Blaser calls this a "multiculturalist" perspective on indigenous knowledge, "according to which cultural differences are ultimately negotiable because they are mutually commensurable via what is common to all: a world or reality 'out there'" (Blaser 2009: 15; see also Goldman et al. 2016: 28).

This conceptualisation has been disputed in certain branches of academic thought (e.g. philosophy, post-colonial studies, feminism, science and technology studies, see e.g. Blaser 2013; Mol 2002). Relatedly, almost two decades ago, Karen Barad, a trained theoretical physicist, pointed to the Cartesian origin of the analytical separation of epistemology and ontology and stressed the analytical inseparability of the two:

> [t]he separation of epistemology from ontology is a reverberation of a metaphysics that assumes an inherent difference between human and nonhuman, subject and object, mind and body, matter and discourse. *Onto-epistem-ology*—the study of practices of knowing in being—is probably a better way to think about the kind of understandings that are needed to come to terms with how specific intra-actions matter (Barad 2003: 829, emphasis in original).

Whilst there are now attempts in academic argumentation to overcome these separations, it is important to acknowledge that many indigenous philosophies did not distinguish between them. Indeed, the inseparability of the two is an essential feature of so-called relational ontologies (Sidorkin 2002: 91), in which relationships constitute beings or persons (including non-human beings) rather than vice-versa. This perspective stands in stark contrast to the atomistic or substantivist ontology dominant in the western world.

I thus argue that by conceptually separating *how-we-know* from *what-we-know*, studies like the one described in this essay further disempower indigenous people by squeezing their knowledge into scientific conceptual straightjackets or subsuming it in a side note as 'beliefs' or 'culture', despite the stated intention to do otherwise (also see Muller et al. 2019: 402).

Preventing (Radical) Learning

As long as 'we' try to shoehorn indigenous knowledge into our onto-epistemological frameworks, we will never reach the roots of the problem. Climate change is the outcome of practices entangled with a specific western philosophical heritage, by the dominant 'western' onto-epistemology. With this acknowledgement, it would be wise to look *beyond* this tradition for possible paths forward as 'humanity' has much more to offer. Although many thinkers have already stressed this possibility (e.g. Rose 2005; Sullivan 2013, 2017; Umkeek 2014; Castree 2016), there is still a tendency in much climate change research to ignore these calls. To my point of view, indigenous ontologies offer a variety of interrelated aspects in response to the current climate and ecological crisis, a few of which are encouraged below.

Take relational onto-epistemologies seriously. As soon as we acknowledge the inseparability of *how*-we-know and *what*-we-know, we can stop the bizarre fighting about one nature/several cultures or one culture/several natures. We can also stop fighting about many other things, e.g. 'the truth', as 'truth' evolves in the field of relations between different beings. We have different, partly overlapping, onto-epistemologies which we need to consider holistically or, in Escobar's words, we have a "pluriverse" as "a world where many worlds fit" (Escobar 2011: 139).

'Dethrone' the human. Although many philosophers, posthumanist scholars and other academics have already pointed to the need to conceptually and practically re-integrate the human into ecology, many of them, based on Eurocentric scholarship, (still) tend to pay no or very little attention to indigenous knowledges, perspectives and ontologies (see also Bignall and Rigney 2019). Disclosing a western onto-epistemology as particular to a specific area and period and philosophising about new approaches to imagine the world are useful endeavours, but it might be less abstract and less theoretical to encourage more learning from concrete cases of existing or past alternatives of human-environment relationships as lived by particular groups of indigenous peoples.

Re-learn mutual respect and relatability by (re-)animating nature. Indigenous ways of being-in-the-world and onto-epistemologies epitomise what is needed in dealing with climate change (Wildcat 2013; Umeek 2014). The necessity to maintain ethical and mutual relationships to non-human others is a central part of their experience, an experience which appears lost in post-Enlightenment European thought. The objectification of nature is an important cause for the current ecological crisis and technology on its own will not bring salvation (see e.g. Umeek 2014: 7). What is needed is a 'relational turn' (e.g. Dépelteau 2015), not only in science but in the western approach to life.

Focus on local knowledge and acknowledge people's connection to/knowledge of the land. The points above refer to general principles connected to indigenous onto-epistemologies. The concept of onto-epistemology also stresses the importance of place, i.e. of locality with regard to knowledge evolution, and thus of the situatedness of knowledge. While the relational ontologies of indigenous peoples located in continents beyond Africa have been studied and compared extensively, case studies focusing on onto-epistemological issues of indigenous peoples in Africa have rarely been considered in comparative discussions within the field of 'new animism' (although see Sullivan 2010; Low 2014; Dieckmann 2021b: 25–26) or indeed linked to the ecological crisis (with few exceptions, e.g. Goldman et al. 2016; Sullivan 2017; Schnegg 2021). Khoekhoegowab-speaking communities in southern Africa, being severely affected by climate change, deserve special attention in this regard. These communities have lived for millennia with a harsh environment but due to their degree of marginalisation, their voices have hardly found their ways into official discourses. Muller et al. (2019: 405–07) provide a number of promising examples from other continents, where indigenous peoples' onto-epistemologies have been integrated into environmental management and legal provisions.

What if Topnaar or Haiǁom experiences of the world and their acknowledgement of the importance of mutual relationships between a variety of human and non-human actors (including winds, rain, animals and plants) could find their ways into the Namibian (and global) climate change discourse? What if these communities could be integrated into the management of the national parks established on parts of their ancestral lands? What if these national parks became legal persons? Would 'other' people around the world change their/our behaviour if they/we took these ways of engaging with their surroundings as our example? How might this unfold?

References

Barad, Karen, 'Posthumanist Performativity: Toward an Understanding of How Matter Comes to Matter', *Journal of Women in Culture and Society*, 28(3) (2003), 801–31.

Barnes, Jessica, M. Dove, M. Lahsen, A. Mathews, P. McElwee, R. McIntosh, et al., 'Contribution of anthropology to the study of climate change', *Nature Climate Change*, 3(6) (2013), 541–44, https://doi.org/10.1038/nclimate1775.

Bignall, Simone, and Daryle Rigney, 'Indigeneity, posthumanism and Nomad Thought Transforming Colonial Ecologies', in *Posthuman Ecologies— Complexity and Process After Deleuze*, ed. by Rosi Braidotti and Simone Bignall (London: Rowman & Littlefield International Ltd, 2019), pp. 159–81.

Blaser, Mario, 'The Threat of the Yrmo: The Political Ontology of a Sustainable Hunting Program', *American Anthropologist*, 111(1) (2009), 10–20.

Blaser, Mario, 'Ontological Conflicts and the Stories of Peoples in Spite of Europe: Toward a Conversation on Political Ontology', *Current Anthropology*, 54 (2013), 547–68.

Brosius, J. Peter, 'What Counts as Local Knowledge in Global Environmental Assessments and Conventions?', in *Bridging Scales and Knowledge Systems: Concepts and Applications in Ecosystem Assessment A Contribution to the Millennium Ecosystem Assessment*, ed. by Walter V. Reid, Fikret Berkes, and Thomas J. Wilbanks (Washington: Island Press, 2006), pp. 129–44.

Burman, Anders, 'The Political Ontology of Climate Change: Moral Meteorology, Climate Justice, and the Coloniality of Reality in the Bolivian Andes', *Journal of Political Ecology*, 14 (2017), 921–38, https://doi.org/10.2458/v24i1.20974.

Cameron, Emilie, Sarah de Leeuw, and Caroline Desbiens, 'Indigeneity and ontology', *cultural geographies*, 21(1) (2014), 19–26, https://doi.org/10.1177/1474474013500229.

Castree, Noel, 'Broaden research on the human dimensions of climate change', *Nature Climate Change*, 6(8) (2016), 731, https://doi.org/10.1038/nclimate3078.

Charapa Consult, '*Indigenous Peoples and Climate Change in Africa: Traditional Knowledge and Adaptation Strategies: Draft Report*' (Rainforest Foundation UK, Legal Assistance Centre, Namibia and Mainyoito Pastoralist Integrated Organization, 2012), http://www.charapa.dk/wp-content/uploads/Africa-IPs-CC-adaption.pdf.

Dépelteau, François, 'What Is the Direction of the "Relational Turn"?', in *Conceptualizing Relational Sociology: Ontological and Theoretical Issues*, ed. by Christopher Powell and François Dépelteau (Basingstoke: Palgrave Macmillan, 2015), pp. 163–85.

Dieckmann, Ute, *Hai||om in the Etosha Region: A History of Colonial Settlement, Ethnicity and Nature Conservation.* (Basel: Basler Afrika Bibliographien, 2007).

Dieckmann, Ute, *Born in Etosha: Living and Learning in the Wild* (Windhoek: Legal Assistance Centre, 2012).

Dieckmann, Ute, 'Hai||om in Etosha: Cultural maps and being-in-relations', in *Mapping the Unmappable? Cartographic Explorations with Indigenous Peoples in Africa*, ed. by Ute Dieckmann (Bielefeld: Transcript, 2021a), pp. 93–137, https://doi.org/10.14361/9783839452417.

Dieckmann, Ute, 'Introduction: Cartographic exploration with indigenous peoples in Africa', in *Mapping the Unmappable? Cartographic Explorations with Indigenous Peoples in Africa*, ed. by Ute Dieckmann (Bielefeld: Transcript, 2021b), pp. 9–46, https://doi.org/10.14361/9783839452417.

Dieckmann, Ute, Willem Odendaal, Jacquie Tarr, and Arja Schreij, *Indigenous Peoples and Climate Change in Africa: Report on Case Studies of Namibia's Topnaar and Hailom communities* (Windhoek: Legal Assistance Centre, 2013), http://www.lac.org.na/projects/lead/Pdf/climate_change.pdf.

Escobar, Arturo, 'Sustainability: Design for the Pluriverse', *Development*, 54(2) (2011), 137–40, https://doi.org/10.1057/dev.2011.28.

Goldman, Mara J., Meaghan Daly, and Eric J. Lovell, 'Exploring Multiple Ontologies of Drought in Agro-pastoral Regions of Northern Tanzania: A Topological Approach', *Area*, 48(1) (2016), 27–33, https://doi.org/10.1111/area.12212.

Goldman, Mara J., Matthew D. Turner, and Meaghan Daly, 'A Critical Political Ecology of Human Dimensions of Climate Change: Epistemology, Ontology, and Ethics', *WIREs Climate Change*, 9(4) (2018), e526, https://doi.org/10.1002/wcc.526.

Harvey, Graham, *Animism: Respecting the Living World* (New York: Columbia University Press, 2006), https://doi.org/10.1111/j.1748-0922.2008.00315_1.x.

Hoff, Ansie, 'The Water Snake of the Khoekhoen and |Xam', *South African Archaeological Bulletin*, 52(165) (1997), 21–37, https://doi.org/10.2307/3888973.

IPCC, *Climate Change 2007: Impacts, Adaptation and Vulnerability. Contribution of Working Group II to the Fourth Assessment Report of the Intergovernmental Panel on Climate Change*, ed. by M. L. Parry, O. F. Canziani, J. P. Palutikof, P. J. van der Linden, and C. E. Hanson (Cambridge: Cambridge University Press, 2007).

Kronik, Jakob, and Dorte Verner, *Indigenous Peoples and Climate Change in Latin America and the Caribbean* (Washington, DC: The World Bank, 2010), https://doi.org/10.1596/978-0-8213-8237-0.

Low, Chris, 'Khoisan Wind: Hunting and Healing', *Journal of the Royal Anthropological Institute*, 13 (2007), 71–90, https://doi.org/10.1111/j.1467-9655.2007.00402.x.

Low, Chris, 'Khoe-San Ethnography, "New Animism" and the Interpretation of Southern African Rock Art', *South African Archaeological Bulletin*, 69(200) (2014), 164–72, https://doi.org/10.1111/j.1467-9655.2007.00402.x.

Mol, Annemarie, *The Body Multiple: Ontology in Medical Practice* (Durham NC: Duke University Press, 2002).

Muller, Samantha, Steve Hemming, and Daryle Rigney, 'Indigenous Sovereignties: Relational Ontologies and Environmental Management', *Geographical Research*, 57(4) (2019), 399–410, https://doi.org/10.1111/1745-5871.12362.

Odendaal, Willem, and Wolfgang Werner (eds), *'Neither Here nor There': Indigeneity, Marginalisation and Land Rights in Post-independence Namibia* (Windhoek: Legal Assistance Centre, 2020).

Rocheleau, Dianne, 'Maps as power tools: locating communities in space or situating people and ecologies in place?', in *Communities and Conservation: Histories and Politics of Community-based Natural Resource Management*, ed. by J. Peter Brosius, Anna Tsing, and Charles Zerner (Walnut Creek, CA; Oxford: AltaMira Press, 2005), pp. 327–62.

Rose, Deborah Bird, 'An Indigenous Philosophical Ecology: Situating the Human', *The Australian Journal of Anthropology*, 16(3) (2005), 294–305, https://doi.org/10.1111/j.1835-9310.2005.tb00312.x.

Schmidt, Sigrid, 'Spirits: Some Thoughts on Ancient Damara Folk Belief', *Journal of the Namibian Scientific Society*, 62 (2014), 133–60.

Schnegg, Michael, 'The Life of Winds: Knowing the Namibian Weather from Someplace and from Noplace', *American Anthropologist*, 121(4) (2019), 830–44, https://doi.org/10.1111/aman.13274.

Schnegg, Michael, 'Ontologies of climate change', American ethnologist, 48(2) (2021), https://doi.org/10.1111/amet.13028.

Scoones, Ian, 'The politics of global assessments: the case of the International Assessment of Agricultural Knowledge, Science and Technology for Development (IAASTD)', *The Journal of Peasant Studies*, 36(3) (2009), 547–71, https://doi.org/10.1080/03066150903155008.

Sidorkin, Alexander M., 'Chapter 7. Ontology, Anthropology, and Epistemology of Relation', *Counterpoints*, 173 (2002), 91–102.

Sullivan, Sian, 'Folk and Formal, Local and National: Damara Cultural Knowledge and Community-based Conservation in Southern Kunene, Namibia', *Cimbebasia*, 15 (1999), 1–28.

Sullivan, Sian, '"How Can the Rain Fall in This Chaos?": Myth and Metaphor in Representations of the North-West Namibian Landscape', in *Challenges for Anthropology in the 'African Renaissance': A Southern African Contribution*, ed. by D. LeBeau and R. J. Gordon (Roma, Lesotho: Institute of Southern African Studies, National University of Lesotho, 2002), pp. 255–65.

Sullivan, Sian, 'Ecosystem service commodities'—a new imperial ecology? Implications for animist immanent ecologies, with Deleuze and Guattari', *New Formations: A Journal of Culture/Theory/Politics*, 69 (2010), 111–28.

Sullivan, Sian, 'Nature on the Move III: (Re)countenancing an Animate Nature', *New Proposals: Journal of Marxism and Interdisciplinary Enquiry*, 6(1–2) (2013), 50–71, https://ojs.library.ubc.ca/index.php/newproposals/article/view/183771.

Sullivan, Sian, 'What's ontology got to do with it? On nature and knowledge in a political ecology of the 'green economy'', *Journal of Political Ecology*, 24(1) (2017), 217–42, https://doi.org/10.2458/v24i1.20802.

Sullivan, Sian, and Mike Hannis, 'Relationality, Reciprocity and Flourishing in an African Landscape: Perspectives on Agency amongst ‖Khao-a Dama, !Narenin and ‖Ubun elders in west Namibia', *Future Pasts Working Papers*, 2 (2016), https://www.futurepasts.net/fpwp2-sullivan-hannis-2016.

Sullivan, Sian, and Chris Low, 'Shades of the Rainbow Serpent? A KhoeSan Animal between Myth and Landscape in Southern Africa—Ethnographic Contextualisations of Rock Art Representations', *Arts*, 3(2) (2014), 215–44, https://doi.org/10.3390/arts3020215.

Umeek, Richard A. E., *Principles of Tsawalk: An Indigenous Approach to Global Crisis* (Vancouver: UBC Press, 2014).

Widlok, Thomas, 'Where Settlements and the Landscape Merge', in *African Landscapes: Interdisciplinary Approaches*, ed. by Michael Bollig and Olaf Bubenzer (New York: Springer, 2009), pp. 407–27.

Widlok, Thomas, 'No Easy Talk about the Weather: Eliciting "Cultural Models of Nature" among Hai‖om', *World Cultures ejournal*, 22(2) (2017), https://escholarship.org/uc/item/69n0s73f.

Wildcat, Daniel R., 'Introduction: Climate Change and Indigenous Peoples of the USA', *Climatic Change*, 120(3) (2013), 509–15, https://doi.org/10.1007/s10584-013-0849-6.

Yeh, Emily T., 'How Can Experience of Local Residents be "Knowledge"? Challenges in Interdisciplinary Climate Change Research', *Area*, 48(1) (2016), 34–40, https://doi.org/10.1111/area.12189.

V

GOVERNANCE

15. Towards a Fossil Fuel Treaty

Peter Newell

We need a new approach to tackling climate change. We need to start using the 'f' word much more: fossil fuels. The Paris Agreement does not even mention fossil fuels. The deliberate neglect by the climate regime of the largest source of greenhouse emissions is as shocking as it is unsurprising in a world in which fossil fuel lobbies still wield such power and have delayed effective climate action for so long that climate chaos is now upon us. This chapter urges that it is time to rein in the power these actors have over our collective fate, through international agreements and law which effectively and fairly leave large swathes of remaining fossil fuels in the ground. A *Fossil Fuel Non-Proliferation Treaty* (FF-NPT) based, like the Nuclear Non-Proliferation Treaty, on the three pillars of non-proliferation, disarmament and peaceful use, could fulfil that purpose.

The 'F' Word

We need a new approach to tackling climate change. We need to start using the 'f' word much more: fossil fuels. The Paris Agreement does not even mention fossil fuels. The deliberate neglect by the climate regime of the largest source of greenhouse emissions is as shocking as it is unsurprising in a world in which fossil fuel lobbies still wield such power and have delayed effective climate action for so long that climate chaos is now upon us. These companies have long wielded such power (Newell and Paterson 1998; Kolk and Pinkse 2007)—as also documented by Wright and Nyberg, this volume. But if further evidence of their influence were needed, it is observable in the distribution of

https://doi.org/10.11647/OBP.0265.15

bailout funds in response to the COVID crisis where G20+ countries have pledged over $207 billion so far to fossil fuel projects, according to the Energy policy tracker.[1]

It is time to reign in the power these actors have over our collective fate. Just six of the largest listed oil and gas companies alone hold reserves that together would use up more than a quarter of the remaining 2°C budget (McKibben 2012). And, historically speaking, only ninety companies have caused two-thirds of anthropogenic global warming emissions, including companies such as Chevron, Exxon, Shell and BP, with half of the estimated emissions produced in the past twenty-five years when the scale of the climate threat was clear (Heede 2014). Governments are complicit in this situation by planning to produce about 50% more fossil fuels by 2030 than would be consistent with a 2°C pathway, and 120% more than would be consistent with a 1.5°C pathway (SEI et al. 2019).

The long-neglected supply-side needs to occupy a central place in collective efforts to address climate change (Erikson et al. 2018; Gaulin and Le Billon 2020), starting with the Glasgow COP. The IPCC Special Report on 1.5 degrees published in October 2018 makes clear that realising the ambition of the 2015 Paris Agreement to keep global warming below 1.5°C requires deep and rapid decarbonisation.

A crucial, yet neglected, aspect of this is the need for international agreements and laws which effectively and fairly leave large swathes of remaining fossil fuels in the ground. A *Fossil Fuel Non-Proliferation Treaty* (FF-NPT) could fulfil that purpose (Newell and Simms 2019).

Though there have been calls for a Coal Elimination Treaty (Burke and Fishel 2020), it is clear we now need a more general fossil fuel treaty since the majority of remaining oil and gas reserves also need to remain in the ground. Such a treaty could have three pillars, modelled on the Nuclear Non-Proliferation Treaty.

The first pillar is *non-proliferation*. This would imply a moratorium on further expansion in rich OECD+ countries, underpinned by a model-driven assessment of which reserves of fossil fuels are un-burnable carbon and need to stay in the ground to be Paris compliant. This would underpin negotiations about the sequencing of commitments regarding different fossil fuels and the point at which other groups of countries take on commitments.

---

1   https://www.energypolicytracker.org/.

The second pillar is *disarmament*, which here refers to the accelerated phaseout, and managed decline of, existing investments and infrastructures in fossil fuels. It would be underpinned by the principle of a just transition to address both historical responsibility and the capacity to diversify away from fossil fuels, providing support for countries to do so (Kartha et al. 2018; Le Billon and Kristoffersen 2019; Muttitt and Kartha 2020).

The third pillar is *peaceful use*. This pillar refers to the financial and technological support to developing countries that will be needed for the adoption of renewable energy pathways. This support could be achieved, in part, by redirecting finance from fossil fuels, both public and private, and including the US$10 million a minute the IMF calculates that the world spends on fossil fuel subsidies (Coady et al. 2015), into a global transition fund to finance technology, retraining and compensation (see the chapters by Bracking and by Kaplan and Levy, this volume, on the complexities of climate finance).

There is precedent for international treaties which ban or regulate particularly harmful substances—think of the WHO Framework Convention on Tobacco Control (WHO FCTC), the Ottawa Treaty to ban landmines and the Chemical Weapons Convention. Internationally, there are also precedents for bans on fossil fuels such as the moratorium in place for mining projects in Antarctica (Article 7 of the Environmental Protocol of the Antarctic Treaty). The International Council on Mining and Metals has committed its members (including the World Coal Association) to neither explore nor mine in World Heritage Sites and to "respect legally designated protected areas" (ICMM 2003). Likewise, there are calls for banning oil drilling in the Arctic Sea and to halt exploitation in protected areas and on indigenous lands. Meanwhile, the 2017 Lofoten Declaration, signed by over 500 organisations, highlights the need to put an end to fossil fuel development and manage the decline of existing production.

There is much to be worked out in terms of overarching principles, modalities and procedures to ensure a fair, workable and effective fossil fuel treaty. But criteria for allocating and sequencing responsibility might include that (i) the costs of action should be borne disproportionately by those who have the greatest ability to pay defined by per capita income levels and who are best placed to redirect finance, production and technology towards lower carbon alternatives; (ii) the greatest emitters of GHG emissions from the direct burning of their own fossil fuel reserves should act first; and (iii) cumulative emissions are assessed to take adequate account of historical responsibility and use of fossil fuels to date.

These three criteria would imply that OECD countries, plus the Russian Federation (OECD+), take the lead in the first instance with near-term targets and timetables for the phaseout of fossil fuels. Multilateral responses may be attractive to powerful countries wanting to ensure other states do not free-ride on commitments they are now making to leave fossil fuels in the ground. They would likely be supported in such an endeavour by the climate vulnerable groupings in the climate change negotiations such as the Least Developed Countries (LDCs) and the Small Island Developing States (SIDS) (Newell and Simms 2019). A universal treaty like the UNFCCC might not be required. Hence, even if major fossil fuel producers would not join a Fossil Fuel Non-Proliferation Treaty at first, there is still a strong rationale for initiating a treaty process led by a group of first movers who encourage others to join to avoid free-riding and problems of leakage. Supply-side policies adopted could also of course be included under countries' Nationally Determined Contributions under the Paris Agreement, providing a further incentive to participate in negotiations for a new treaty. Though negotiating the nuclear NPT took three years, this treaty would take longer and needs to be supplemented by other strategies aimed at keeping fossil fuels in the ground.

But there is momentum in this direction. Initial moves in this direction would include the formation of a first movers alliance, such as the Beyond Oil and Gas Alliance (BOGA), building on the precedent of the Powering Past Coal Alliance of countries. A number of countries in recent years have adopted bold supply-side policies in the form of moratoria, bans, production limits and so on, including most prominently Costa Rica, New Zealand, Denmark, Spain, France and Belize. France announced in December 2017 that it would phase out oil and gas exploration and production, a move then followed by Belize (which announced a moratorium on all offshore oil activity in late December 2017), Denmark (which implemented a ban on onshore oil and gas exploration in February 2018), New Zealand (which banned new offshore oil exploration licences in April 2018), and Ireland (which enacted a ban on future oil exploration licences in September 2019) (Carter and MacKenzie 2020). Gaulin and Le Billon (2020), drawing on a fossil fuel cuts database, found that 1302 initiatives were implemented between 1988 and 2017 in 106 countries across seven major types of supply-side approaches. This demonstrates both a rapid growth in the number of supply-side initiatives taken during the past decade, but also their highly uneven adoption across the world, underscoring the need for a multilateral approach.

There is no underestimating the scale of the challenge of deliberately and legally calling time on the fossil fuel era that has provided such riches for some of the world's most powerful actors. Although it can appear daunting, it is worth recalling that many of the world's largest and most powerful private fossil fuel companies have their home base in OECD+ countries. This is key to avoid problems of carbon leakage and to improve compliance. An important move in this direction, and around which there is already some support, would be a public transparent registry of existing and planned sites of fossil fuel extraction that would form the basis of negotiations about *which* and *whose* reserves would be put beyond limits for reasons of avoiding further climate chaos.

An FF NPT is clearly not the only way forward. Any multilateral agreement to restrict the supply and production of fossil fuels will take many years to be negotiated. The urgency of the climate crisis and the need to improve the speed and depth of action in the way called for in the IPCC SR15[2] means that other routes to action must also be pursued in the meantime or alongside a multilateral endeavour. If an international agreement is to be achieved, it will likely only come about due to a confluence of political and economic factors favouring more ambitious action and a new approach to the issue. With regard to supply-side policies, this might include changes in the price and availability of alternatives to fossil fuels, particularly renewable energies such as wind and solar whose prices have fallen dramatically in recent years (notwithstanding the problems associated with industrial renewable energy production identified by Dunlap, this volume), and improvements in battery storage capacity. For many countries, further

---

2   i.e. the UN Intergovernmental Panel on Climate Change Special Report 15, see https://www.ipcc.ch/sr15/.

investments in a fossil-based infrastructure could lock in a higher cost fossil energy path and lead to stranded fossil fuel infrastructure assets and decreased competitiveness in a global energy market moving in the opposite direction and where 'peak demand' is also a growing consideration (Van de Graaf 2018).

Momentum is also likely to come from social movements and pressure groups both in terms of resistance to new sites of exploration at fossil fuel frontiers involving environmental defenders and other groups and advocacy around specific proposals for new mines and airport expansions, for example. Temper et al.'s (2020) analysis finds, for example, that over a quarter of fossil fuel projects encountering social resistance have been cancelled, suspended or delayed. Another source of pressure comes from the recent waves of litigation targeted at fossil fuel producers in recent years. The Urgenda case in the Netherlands stands out as the first case that successfully enforced the implementation of stricter national emission targets, followed up by the ruling in May this year in the Netherlands against Shell demanding that the oil company reduce its emissions within a more ambitious timeframe.

Proposing a new fossil fuel treaty is a bold thing to do. Let us not be naïve about the prospects that any such treaty will emerge in the very near future. Opposition will be immense. But really, if not this, then what? It is clear the vast majority of fossil fuels need to remain in the ground. Activism and resistance aimed at cutting off finance and resisting new infrastructures on the ground is vital. But we also need a multilateral approach to fairly agree who leaves which resources in the ground and helps poorer countries meet their energy needs in a lower-carbon way. This would complement, not replace, the Paris Agreement, but has the advantage of getting to the root of the problem. As cities, NGOs, citizens and even some businesses, as well as leading figures, such as former Irish President Mary Robinson, lend their support to this proposal,[3] it may be an idea whose time has come.

3　See https://www.fossilfueltreaty.org/.

References

Burke, Anthony, and Stefanie Fishel, 'A Coal Elimination Treaty 2030: Fast Tracking Climate Change Mitigation, Global Health and Security', *Earth System Governance* (3) (2020), 1000462, https://doi.org/10.1016/j. esg.2020.100046Get.

Carter, Angela V., and Jannette McKenzie, 'Amplifying "Keep It in the Ground" First-Movers: Toward a Comparative Framework', *Society & Natural Resources*, 33(11) (2020), 1339–58, https://doi.org/10.1080/08941920.2020.1772924.

Coady, David, Ian Parry, Louis Sears, and Baoping Shang, 'How Large Are Global Energy Subsidies?', *IMF Working Paper*, WP/15/105 (Washington: IMF, 2015), https://www.imf.org/external/pubs/ft/wp/2015/wp15105.pdf.

Erickson, Peter, Michael Lazarus, and Georgia Piggot, 'Limiting Fossil Fuel Production as the Next Big Step in Climate Policy', *Nature Climate Change*, 8 (2018), 1037–43, https://doi.org/10.1038/s41558-018-0337-0.

Gaulin, Nicolas, and Philippe Le Billon, 'Climate Change and Fossil Fuel Production Cuts: Assessing Global Supply-side Constraints and Policy Implications', *Climate Policy*, 20(8) (2020), 888–901, https://doi.org/10.1080 /14693062.2020.1725409.

Heede, Richard, 'Tracing Anthropogenic Carbon Dioxide and Methane Emissions to Fossil Fuel and Cement Producers, 1854–2010', *Climatic Change*, 122 (2014), 229–41, https://doi.org/10.1007/s10584-013-0986-y.

ICMM, 'Mining and protected areas position statement. ICMM position statement', https://www.icmm.com/engb/members/member-commitments/ position-statements/mining-and-protected-areas-position-statement.

Kartha, Sivan, Simon Caney, Navroz Dubash, and Greg Muttit, 'Whose Carbon is Burnable? Equity Considerations in the Allocation of a "Right to Extract"', *Climatic Change*, 150 (2018), 117–29, https://doi.org/10.1007/ s10584-018-2209-z.

Kolk, Ans, and Jonathan Pinkse, 'Multinationals' Political Activities on Climate Change', *Business & Society*, 46 (2007), 201–28, https://doi. org/10.1177/000765030730138.

Le Billon, Phillippe, and Berit Kristoffersen, 'Just Cuts for Fossil Fuels? Supply-side Carbon Constraints and Energy Transition', *Environment and Planning A: Economy and Space*, 52(6) (2020), 1072–92, https://doi. org/10.1177/0308518X18816702.

McKibben, Bill, 'Global Warming's Terrifying New Math' (Rollingstone. com, 2012), http://www.rollingstone.com/politics/news/ global-warmings-terrifying-new-math-20120719.

Muttitt, Greg, and Sivan Kartha, 'Equity, Climate Justice and Fossil Fuel Extraction: Principles for a Managed Phase Out', *Climate Policy*, 20(8) (2020), 1024–42, https://doi.org/10.1080/14693062.2020.1763900.

Newell, Peter, and Matthew Paterson, 'A Climate for Business: Global Warming, the State and Capital', *Review of International Political Economy*, 5 (1998), 679–704, https://doi.org/10.1080/096922998347426.

Newell, Peter, and Andrew Simms, 'Towards a Fossil Fuel Non-proliferation Treaty', *Climate Policy*, 20(8) (2020), 1043–54, https://doi.org/10.1080/1469 3062.2019.1636759.

SEI, IISD, ODI, Climate Analytics, CICERO, and UNEP, 'The Production Gap: The discrepancy between countries' planned fossil fuel production and global production levels consistent with limiting warming to 1.5C or 2C' (Productiongap.org, 2019), http://productiongap.org/.

Temper, Leah, Sofia Avila, Daniela Del Bene, Jennifer Gobby, Nicolas Kosoy, Philippe Le Billon, et al., 'Movements Shaping Climate Futures: A Systematic Mapping of Protests Against Fossil Fuel and Low-carbon Energy Projects', *Environmental Research Letters*, 15 (2020), 123004, https://iopscience.iop.org/article/10.1088/1748-9326/abc197/meta.

Van de Graaf, Thijs, 'Battling for a Shrinking Market: Oil Producers, the Renewables Revolution, and the Risk of Stranded Assets', in *The Geopolitics of Renewables*, ed. by D. Scholten (Cham: Springer, 2018), pp. 97–121, http://hdl.handle.net/1854/LU-8544872.

16. How Governments React to Climate Change: An Interview with the Political Theorists Joel Wainwright and Geoff Mann

Joel Wainwright and Geoff Mann (Interviewed by Isaac Chotiner)

In *Climate Leviathan: A Political Theory of Our Planetary Future* (2018), Joel Wainwright, Professor of Geography at Ohio State University, and Geoff Mann, Professor of Geography and Director of the Center for Global Political Economy at Simon Fraser University, consider how to approach the global politics of climate change. They look at several different potential futures for our warming planet, and argue that a more forceful international order, or "Climate Leviathan," is emerging, but unlikely to mitigate catastrophic warming. An edited and condensed version of our conversation about the book follows.

Chotiner: Does global warming fundamentally change how you evaluate international politics and sovereignty and the idea of the nation-state, or is it more evidence of a crisis that already existed?

Wainwright: One of the arguments in our book is that, under pressure from the looming challenges of climate change, we can expect changes in the organisation of political sovereignty. It's going to be the first major change that humans have lived through in a while, since the emergence of what we sometimes think of as the modern period of sovereignty, as theorised by Thomas Hobbes, among others. We should expect that after, more than likely, a period of extended conflict and real problems

    https://doi.org/10.11647/OBP.0265.16

for the existing global order, we'll see the emergence of something that we describe as planetary sovereignty.

So, in that scenario, we could look at the current period with the crisis of liberal democracies all around the planet and the emergence of figures like Bolsonaro [President of Brazil] and Trump [former US President] and Modi [Indian Prime Minister] as symptoms of a more general crisis, which is simultaneously ecological, political, and economic. Maybe this is quibbling with your question, of trying to disaggregate the causal variable. Which comes first—is it the ecological or the political and economic?—is a little bit difficult because it's all entangled.

Mann: I think we're going to witness and are already witnessing, in its emergent form, lots of changes to what we think of as the sovereign nation-state. Some of that change right now is super-reactionary—some groups are trying to make it stronger and more impervious than it's been in a long time. Then, other kinds of forces are driving it to disintegrate, both in ways we might think of as pretty negative, like some of the things that are happening in the E.U., but also in other ways that we might think of as positive, in the sense of international cooperation (also see Newell, this volume). There's some discussion about what to do about climate migration, at least.

I think one of the interesting things that's happening right now is that we have so few political, institutional tools, and, I would say, conceptual tools to handle the kinds of changes that are required. Everyone knows climate change is happening and it's getting worse and worse, and everyone's trying to fight off the worst parts of it, but we're not really getting together as everyone thinks that we need to.

I think that the nation-state is one of the few tools that people feel like they have and so they're wielding it in crazy ways. Some people are trying to build walls. Other people are trying to use their powers to convince others to go along with their plans. I think we have so few tools to deal with this problem that the nation-state is kind of being swung around like a dead cat, with the hope that it'll hit something and help.

One of the most depressing and scary parts of this is that global warming is exacerbating economic problems, and migration- and refugee-related problems, that are actually making the political dynamics within these countries worse and opening up a window for people like Trump.

Wainwright: I think your hypothesis, of a cyclical undermining of the global liberal order, is potentially valid. In fairness, it's not exactly what Geoff and I are saying in the book. You may be right and you may be wrong. If you wanted to strengthen that hypothesis, you'd have to clarify in exactly what way the authoritarian, neoliberal, climate-denialist position that we see represented by those diverse figures—again Modi, Bolsonaro, Trump, et cetera—represents the opposite of something else.

Part of the reason we wrote the book is because—I think Geoff and I would both say—there's a lot of talk right now in places like Canada and the United States about what we have and what we need, that when it comes to climate change is pretty vague on the political, philosophical fundamentals. What exactly do Trump and Modi represent? Where does it come from, and why is it so clearly connected to climate denialism, and in what way is that crazy ensemble—or what appears to us as crazy and new—connected to the liberal dream of a rational response to climate change that's organised on a planetary basis?

Chotiner: This gets to some of the scenarios you lay out in the book, and why you are so pessimistic about the current order. What are those scenarios?

Mann: In the book, we lay out what we think of as possible futures. They're really, really broad, and there's lots of room for maneuvers in them and they could blur a bit.

One of them, which we think is quite likely, is what we call 'Climate Leviathan'. Another one is 'Climate Mao'—that would be a sovereign, but it would operate more on the principles of what we might think of as a Maoist tradition, a quasi-authoritarian attempt to fix climate change by getting everyone in line. Then there's the 'Behemoth' [their term for a reactionary order]. We, at the time we started to work on the book, had in our heads the caricature of Sarah Palin, because that was the moment of "Drill, baby, drill."

The last thing we call 'Climate X', and that's the hopeful scenario. That is the sense we both have that the way to address climate change is definitely not international meetings that achieve nothing over and over again, in big cities all over the world. The attempts by liberal capitalist states like Canada or the US to regulate tiny bits here and there, implement tiny little carbon taxes, to try to get people to buy solar panels. This is not anywhere near enough, nor coordinated in any meaningful way to actually get us out of this problem.

I think Joel and I really feel strongly that Climate X describes a whole array of stuff that isn't attached to this completely failing set of institutions. So, with Climate X, we're going to see activity happening at local levels, bridges across boundaries that you don't think about now, institutions refuting the state entirely, like so many indigenous people from Canada going ahead and doing things on their own, building new alliances, discovering ways of managing the collapsing ecosystems and political institutions in creative ways. We don't see a map to this and the attempts to map it thus far have been a total and complete failure. Our hope is that we reinforce what is already happening in so many communities.

Chotiner: Climate change has caused me to think not just about what kinds of action are needed but also about whether our whole moral framework should change. I don't want people in Bangladesh to start blowing up Chinese coal plants, but I also wonder whether we need to start thinking about what is and is not O.K. differently because this is so dire.

Wainwright: We agree with you completely. What's notable is the disjuncture between what any clear-eyed observer will see really needs to happen fast and the depth of the seeming incapacity in the world's political and economic arrangements to move beyond even the first basic steps. So, the masses as well as many élites are realigning in all these strange combinations and producing figures like Trump and Bolsonaro.

As far as refugees go, the world has a large number of people who are sometimes called climate refugees today. There is still no international definition of a climate refugee that is generally accepted. If we take a reasonably capacious definition of a climate refugee, it's someone who has been displaced, at least in part, because of climate change. There are probably already tens of millions of climate refugees in the world today, including a pretty significant number of people from places like Honduras and Guatemala and Mexico, who have come to the United States, although we don't tend to talk about them that way.

Some estimates are as high as two hundred million climate refugees by 2050 or so, although that's really speculation because no one really knows (on the complexities of these numbers, see Durand-Delacre et al., this volume). It could easily creep into [several] hundreds of millions if the expectations of flooding in places like Bangladesh and the Caribbean and Indonesia come to pass.

In the face of all that, the present liberal-capitalist international order has utterly failed, as we've all said, and we can't expect people to just do nothing. They're going to look elsewhere for answers to their problems. To make a huge generalisation, they're not turning toward the mainstream ideological resources of liberal modernity. They're turning to variations on religious metaphysics and often, unfortunately, forms of ethnic and religious exclusion. Hence the desperate need for us to develop a new political theory of this moment and new utopian ideas.

I don't think that's entirely wrong, but, at least in the United States, people say they don't believe in climate change because there's been a systematic campaign to lie to them. Exxon documents are coming out in lawsuits all the time. It is one thing to say, "Well, this is a failing of the liberal order," and people looking for alternatives, which I think is true, but it's also true that people are being taken advantage of and lied to, and maybe the critique of capitalism is that it allows people like Rupert Murdoch to shape the perceptions of large chunks of the country.

Mann: You're right, there's tons of media flying around, there's all sorts of efforts to hide the truth, to hide the science, to twist things to get people to naïvely take up positions that are not only against everyone's interests but against their own as well, and in the interests of the most powerful.

It's also the case that these are generally characterised, and accurately so, as class issues. One aspect of the critique of capitalism that you mentioned is the way in which capitalism produces and reinforces class divides that lead to a situation in which, to some extent, we're seeing different factions of the élite struggle over the support of the masses. So, in many ways, the problem can be attributed to the fact that so many voters don't believe in climate change, but in actual fact, I would say that the problem really is a failure of the liberal order that can produce a situation in which, for one thing, that can occur, but secondly, in which the élites who control the state water down all its attempts to confront climate change.

Even here in Canada, where of course the problems are bad, but not as bad as they are in the US, we have a state that says it's fully committed to addressing climate change, but it actually is doing no more than Trump. So we're in a situation where it's hard to believe that it's only conspiracy theory that has prevented us achieving anything. I really do think it's much more systematic than that.

Chotiner: How do you want people to think and respond to something like what Bolsonaro is proposing with the rain forest?

Mann: I think both Joel and I would say that the most effective mechanisms are supporting those in Brazil who oppose Bolsonaro, and there are millions and millions. We sometimes forget that a lot of leaders are in power with the support of far less than half their population, just because of the way that the elections work. So it's not like there's not an enormous part of Brazil that is terrified of Bolsonaro and doing everything they can to stop him. I think that our reaction from far away, of course, should take into account the fact that we can't restart imperialism in the interest of climate change, but we can figure out ways to support those who are doing their best to stop this from happening.

Some of that, of course, could be something as simple as a consumer boycott, but I think that, fundamentally, it's going to require alliances and support that reach much further down in the political, economic strata of Brazil. Figuring out how to get in there and help those people, that's a challenge in and of itself.

Chotiner: We've heard a lot about how Western countries industrialised at a time when we didn't really know climate change was happening, and we here in the West got really rich. Now countries in the rest of the world want to go through the same process to raise the standard of living for their people, but at the same time we know that climate change is happening. I'm curious how you, as leftists, think about a situation where rich countries start telling poor ones what they can and can't do and enforcing that in some way, even if it's in the service of an end that we all think is beneficial to the planet.

Mann: That scenario you just described is a pretty big part of what Joel and I call Climate Leviathan. That's not what we're hoping for, but we think it's very likely.

Wainwright: I would say that, right now, the core powerful capitalist societies are in fact telling developing and poor countries what to do about all kinds of things. But their general encouragement—whether it's through financial policy or trade policy or military bases or what have you—tends to be in the direction of locking in fossil-fuel extraction and consumption. There is no way around the fact that the US government has played a major role in building, reinforcing, and protecting the global oil industry—Saudi Arabia is just the best-known illustration. What

Geoff and I would point to instead, as an alternative to imperialism, is a lot more old-fashioned transnational solidarity on behalf of ordinary people all over the world, in the name of climate justice (on which, see Harris, this volume). That's what we desperately need.

On this point about transnational, trans-class solidarity and climate justice, it might be worth taking a look at Pope Francis's encyclical *Laudato Si* (2015), which has probably been, to my mind, the most important book on these questions in my lifetime. In a series of statements that Pope Francis makes in that text, he reconfigures Catholic theology as a process of forging a planetary solidarity for humanity, in a world still to come. O.K., we're not Catholics. Geoff and I aren't directly quoting Francis and saying, "You see, the Pope has it all figured out," but we're basically stretching and pointing in the same direction.

17. Inside Out COPs: Turning Climate Negotiations Upside Down

Shahrin Mannan, Saleemul Huq and

Mizan R. Khan

By now it is known that COP25, the latest UNFCCC conference of the parties (COP) and the longest in history, could not achieve its intended outcomes, as negotiators failed to agree on the core issues, thus pushing further away the implementation of the Paris Agreement. COPs that overrun, since it is now a standard practice to drag negotiations into overtime, appear an extremely inefficient process, which is not helped by the arcane language of the adopted texts. We argue that it is time to rethink the entire process and propose the concept of 'inside out COPs'. This proposal affirms that actions on the ground to implement the Agreement should be given greater prominence than political negotiations agreeing to a patchwork of compromises over its rulebook for implementation. The many actors, including civil society, private companies, cities, universities, indigenous communities, youth and others pressing for action, should be put at centre-stage, which will allow for space to deliver results on the ground, as opposed to fetishising the skilful weaving of texts run through with constructed ambiguities.

https://doi.org/10.11647/OBP.0265.17

COP Negotiations: A History of Inequitable and Inefficient Process

The climate change negotiations during the UNFCCC Conference of Parties (COPs) portray the underlying inequalities between industrialised and developing nations (Rennkamp 2020). One aspect of such inequality is the overrun of COPs beyond the schedule. COPs that overrun, since it is now a standard practice to drag negotiations into overtime, appear inefficient and unfair to the delegates of vulnerable countries such as Bangladesh, who have to return home after the official, prearranged timescale is over (Huq 2019). Invariably, the decisions made in the extra time are not in favour of vulnerable countries.

Figure 21 depicts the top 6 COP negotiations which were stretched beyond the schedule. COP25, the latest conference of the parties and the longest in history, could not achieve the intended outcomes with the negotiators failing to agree on the core issues, thus pushing further away the implementation of the Paris Agreement (Huq 2020).

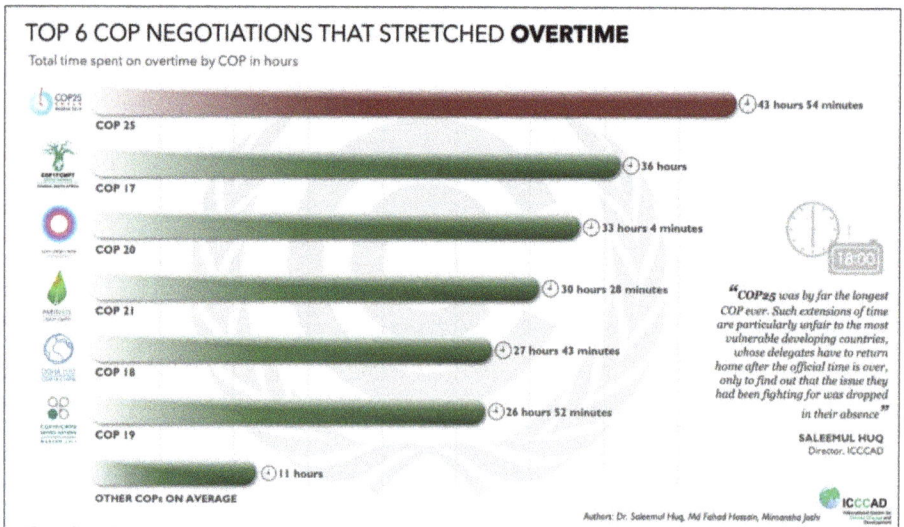

TOP 6 COP NEGOTIATIONS THAT STRETCHED OVERTIME
Total time spent on overtime by COP in hours

COP 25 — 43 hours 54 minutes
COP 17 — 36 hours
COP 20 — 33 hours 4 minutes
COP 21 — 30 hours 28 minutes
COP 18 — 27 hours 43 minutes
COP 19 — 26 hours 52 minutes
OTHER COPs ON AVERAGE — 11 hours

"COP25 was by far the longest COP ever. Such extensions of time are particularly unfair to the most vulnerable developing countries, whose delegates have to return home after the official time is over, only to find out that the issue they had been fighting for was dropped in their absence"

SALEEMUL HUQ
Director, ICCCAD

Authors: Dr. Saleemul Huq, Md Fahad Hossain, Mimansha Joshi

Fig. 21. Top 6 Negotiations That Stretched Overtime. Image by Saleemul Huq, Fahad Hossain and Mimansha Joshi (2019), Dhaka, Bangladesh. ©ICCCAD, CCBY 4.0.

Despite having plenty of time to reach an agreement over the course of two weeks, the COP often runs into overtime in order to reach a result, depriving the vulnerable developing countries from participating (Huq

2020). While negotiators continue mulling over texts and punctuation and arguing over agenda disputes (as detailed for COP21 by Sullivan Chapter 3, this volume), local communities are left to suffer. There is evidence that this is a deliberate tactic of the developed countries to water down agreements by getting rid of vulnerable country delegates from the negotiating process (Vihma 2014). The rhetorical facade of hope from the rich and powerful emitting nations to achieve the targets of the COP21 Paris Agreement has been replaced by this blatant game played on behalf of fossil fuel companies (Huq 2019).

Global Climate Agenda is Transforming and UNFCCC Negotiations Should Follow

The lack of progress in negotiations in recent years, and the failure to obtain a universal agreement on emission targets, have made the UNFCCC negotiation process questionable (Mfitumukiza et al. 2020). The recent failure of COP25 to allow the Warsaw International Mechanism on loss and damage to have an implementation and financing arm has made the vulnerable countries lose all hope (Mfitumukiza et al. 2020).

The current top-down approach of the UNFCCC negotiation process only deals with state actors, prioritising national demands over local needs questioning the efficacy of the process. But the potential of non-state actors (civil society organisations, communities, cities, businesses) to take on-the-ground climate actions is really high (Biasillo 2019).

Non-State Actors Transforming the Climate Agenda

While the negotiators continue to argue over agenda disputes, global greenhouse gas emissions have increased rapidly, making local communities more vulnerable (Huq 2020). But despite institutional evasion and government inaction, local climate actions are taking place around the world (Thew et al. 2020). Breaking from the popular notion of them as vulnerable victims to climate change-induced disasters, grassroot communities, together with local government and civil society organizations, are demonstrating locally-led adaptation measures (also see Lendelvo et al. and Sandover, this volume).

Researchers, academics and non-government organisations are also backing such initiatives by co-creating ground-level data and publishing the results.

While the global negotiations continue arguing over connotation and punctuation, local communities are addressing climate risks through collective actions (Mfitumukiza et al. 2020). Given the insufficient action by major emitting countries to combat climate change, youth movements have also intensified over the past year (Moosmann et al. 2020). Inspired by Greta Thunberg, young people around the world are engaging in school strikes drawing attention to the climate emergency and demanding justice and rapid political action. The 'Fridays for Future' movement, initiated by Greta Thunberg in 2018, brought thousands of students to the streets every Friday to demand intergenerational climate change justice. 'Extinction Rebellion' has also made a mark as an attempt to halt mass extinction and minimise the risk of social collapse (Moosmann et al. 2020)—as also documented in the chapters by North, Paterson and Gardham, this volume.

This community of practice, including grassroot organisations, youth networks, indigenous groups, researchers, civil society and NGOs—sharing a common concern, and the passion to work on the same set of problems and enhance their knowledge and expertise by interacting with each other—is promising in terms of a "whole of society approach" towards solving the climate crisis (Iyalomhe et al. 2013).

Given the increasing evidence of climate-related stresses, the current negotiation process must be rectified. In the face of successful implementation of community-led climate actions and movements carried out by different groups, there is a need for a paradigm shift in the UNFCCC negotiation process from top-down and state-centric to a bottom-up, people-centric approach.

Reimagining COPs: From a Conference of Parties towards a Community of Practice

To increase the efficiency of the UN climate negotiations, it is time to reimagine the entire process and come up with alternative ways such big events are run. One way to do so is to introduce the concept of 'inside out COPs', which calls for giving greater prominence to the climate actions taken on the ground than via the political negotiations. In these 'Action COPs' as opposed to the 'Negotiators' COPs', actors such as civil society, indigenous communities, private companies, cities and universities who are reducing greenhouse gas emissions as well as building climate resilience will take centre-stage, leaving the government negotiators on the periphery (Mfitumukiza et al. 2020). Instead of skilfully weaving compromises, the action COPs will demonstrate results on the ground.

Over the course of two weeks of a COP, only around 5,000 technical government officials conduct the negotiations, later joined by the ministers. At the same time, thousands of people from different backgrounds attend different adjacent events taking place both inside and outside the COP venue (Wei 2017). From COP23 the world has started witnessing how the green zone, which features on-the-ground action, brings more energy than the closed-door blue zone where government negotiations take place (Huq 2020). These events provide opportunities to both practitioners sharing experiential knowledge, and researchers to tailor research according to local needs (Corcoran 2020).

COP26: A Pilot Action COP?

While the global climate change discourse cannot bypass the government negotiation process, it is also important to start piloting the potential of alternatives. The COP26 to be held in Glasgow in November 2021 provides the perfect opportunity to host a parallel 'Action COP' (COVID-19 constraints notwithstanding). The 'Negotiators' COP' can be hosted by the British government while the 'Action COP' can be held elsewhere in Glasgow by the Scottish government. The aim of the action COP will be to bring together the communities of practice and facilitate peer learning.

In order to avoid repeating the mistakes of COP25, the organisers of the UK COP26 should confirm ahead of the event that the negotiations will finish on time, and that the remaining agenda will be dealt with at COP27. If overnight negotiations are required, these should be conducted in the middle of the week, rather than at (or beyond) the end of the official schedule.

References

Biasillo, Roberta, 'Grassroots Initiatives in Climate Change-Adaptation for Justice and Sustainability' (Undisciplinedenvironments.org, 2019), https://undisciplinedenvironments.org/2019/02/28/grassroots-initiatives-in-climate-change-adaptation-for-justice-and-sustainability.

Corcoran, Teresa, 'Local Voice Must Take Pride Of Place: D&C Calls For "Inside-Out COP"' (Iied.org, 2016), https://www.iied.org/local-voice-must-take-pride-place-dc-calls-for-inside-out-cop.

Huq, Saleemul, 'For The Vulnerable, UN Climate Talks Are No Longer Fit For Purpose' (Climatechangenews.com, 2019), https://www.climatechangenews.com/2019/12/17/vulnerable-un-climate-talks-no-longer-fit-purpose/.

Huq, Saleemul, 'COP25 Mistakes Must Not Be Repeated In Glasgow' (Thedailystar.net, 2020), https://www.thedailystar.net/opinion/politics-climate-change/news/cop25-mistakes-must-not-be-repeated-glasgow-1866664.

Iyalomhe, Felix, Anne Jensen, Andrea Critto, and Antonio Marcomini, 'The Science–Policy Interface for Climate Change Adaptation: The Contribution of Communities of Practice Theory', *Environmental Policy and Governance*, 23(6), 368–80, https://doi.org/10.1002/eet.1619.

Moosmann, Lorenz, Cristina Urrutia, Anne Siemons, Martin Cames, and Lambert Schneider, 'International Climate Negotiations—Issues At Stake In View Of The COP25 UN Climate Change Conference In Madrid' (Europarl.europa.eu, 2019), https://www.europarl.europa.eu/thinktank/en/document.html?reference=IPOL_STU(2019)642344.

Mfitumukiza, David, Arghya Roy, Belay Simane, Anne Hammill, Mohammad Rahman, and Saleemul Huq, *Scaling Local and Community-based Adaptation* (Gca.org, 2020), http://www.gca.org/global-commission-on-adaptation/report/papers.

Rennkamp, Britta, 'Negotiating Climate Change between Unequal Parties: COP25 From an African Perspective' (Africaportal.org, 2020), https://www.africaportal.org/features/negotiating-climate-change-between-unequal-parties-cop25-african-perspective/.

Thew, Harriet, Lucie Middlemiss, and Jouni Paavola, '"Youth Is Not a Political Position": Exploring Justice Claims-Making In The UN Climate Change Negotiations', *Global Environmental Change*, 61 (2020), e102036, https://doi.org/doi:10.1016/j.gloenvcha.2020.102036.

Vihma, Antto, 'How to Reform the UN Climate Negotiations? Perspectives From the Past, Present and Neighbour Negotiations' (Finish Institute of International Affairs, 2014), https://www.files.ethz.ch/isn/184844/wp82.pdf.

Wei, D., 'COP23: The 'inside-out' approach to delivering climate action' (Bsr.org, 2017), https://www.bsr.org/en/our-insights/blog-view/cop23-the-inside-out-approach-to-delivering-climate-action.

18. Local Net Zero Emissions Plans: How Can National Governments Help?

Ian Bailey·

Since 2016, nearly 2,000 local government authorities in thirty-four countries have declared climate emergencies and begun initiating plans to reduce emissions in their areas. Local governments have the potential to make a major contribution to achieving global climate mitigation goals but they need greater support from their national governments. Assistance is particularly needed through the provision of supportive national climate policy environments, greater empowerment of local governments, and enhanced finance for local net zero transitions.

Introduction

In December 2016, Darebin Council in North Melbourne became the world's first local government body to declare a climate emergency. Since then, over 1,948 local authorities across thirty-four countries have made similar declarations and initiated plans to reduce or achieve net zero emissions by or before 2050, and the number is still rising (Climate Emergency Declaration 2021). Local action on climate change is far from a new phenomenon. Initiatives like the *Cities for Climate Protection* campaign and *ICLEI—Local Governments for Sustainability* network have already demonstrated the potential for urban and regional initiatives to mobilise actors, catalyse capacity building and knowledge exchange, and promote policies and practical actions to address climate change

https://doi.org/10.11647/OBP.0265.18

and other sustainability issues (Bulkeley 2012). However, the current wave of local emergency declarations offers a once-in-a-generation opportunity to unleash the capabilities of local actors to promote flexible, place-based, and democratically-informed approaches to achieving net zero or negative emissions (Davidson et al. 2020). In South Korea, for example, 226 local governments issued a joint climate emergency declaration at its National Assembly in June 2020, one of the largest climate declarations by any country in the world. Even in countries like the United States and Australia, whose federal governments have yet to declare climate emergencies at the national level, 12% and 36% of their respective populations now live in local government areas that have declared climate emergencies.

Local government powers in areas like land-use planning and infrastructure development provide vital tools for the development of practical strategies to reduce emissions. Local climate emergencies nevertheless remain a predominantly Global Northern phenomenon with just a handful of sub-national authorities from five countries in the Global South having declared climate emergencies. Equally, local governments' capacity to act is heavily influenced by factors over which they often have limited control. Of particular importance is the support—or lack thereof—they receive from national governments. As such, the approach of national governments is likely to be critical in determining whether local climate emergencies become a driving force for reducing emissions or another false dawn for effective climate mitigation (Ghag 2019).

This chapter examines three areas where support from central governments is essential for the future of local net zero plans: the creation of supportive national policy environments; ensuring local governments are granted and can exercise delegated powers to influence emissions; and the provision of finance to support emissions-reduction activities.

Supportive National Policy Environments

A growing number of countries have adopted framework climate legislation establishing emissions targets, carbon budgets, and other institutional arrangements aimed at giving clarity on the long-term direction of climate policy. Many governments have also introduced

'Green Deal' policies to stimulate green industrialisation, investment and public procurement, particularly in Asia, North America and Europe, and many have introduced national carbon taxes and trading schemes to target emissions more directly (Eskander and Fankhauser 2020). However, although countries like the UK and Germany have made encouraging progress towards their decarbonisation goals, more concerted action is needed before the majority of countries can claim to have coherent and durable policy landscapes to support net zero transitions at either the national or sub-national levels.

National climate policies are also likely to be significantly influenced in the short-to-medium term by economic and fiscal responses to the COVID-19 pandemic. 'Build back better' has become a clarion call for ambitions to embed low-carbon investment and nature protection into COVID recovery programmes and was further underlined by the launch of the Build Back Better World (B3W) Partnership at the 2021 G7 meeting in June 2021 (The White House 2021). Yet governments also face strong pressures to regrow their economies as rapidly as possible, by both low- and high-carbon means (Hepburn et al. 2020). It is imperative that recovery programmes do not short-circuit climate mitigation and that investment and policy continue to flow towards climate goals in addition to addressing short-term economic and social concerns. The United Nations Summit on Biodiversity in September 2020 offered further cause for optimism about governments' commitments to integrate environmental protection into COVID-19 responses (General Assembly of the United Nations 2020).

However, such pledges will need to be supported by convincing and durable policies. One of the most important things national governments can do is to declare country-wide climate emergencies and develop action plans to achieve net zero, as New Zealand did recently with the publication of its Climate Change Commission's first advice on carbon budgets and emissions reduction plan (Climate Change Commission 2021). Other national policies to meet climate-change and green-deal goals need to extend beyond national concerns to provide active support for local and regional net zero initiatives (also see Whitmarsh's chapter, this volume). The widespread application of full-cost pricing to emissions generating activities will be crucial to ending subsidies for climate liabilities in areas like waste management and transport.

Financial support mechanisms for renewable and other low-carbon energy sources equally need to focus consistently on providing levels of reward and policy stability that will attract investment in infrastructure with long-term payback periods (Liu et al. 2019). This is especially the case for high potential but less commercially advanced technologies, such as tidal, wave and deep geothermal power, which will continue to require support for some years to realise their contribution to decarbonising national energy systems. Regulatory reforms are also needed to enable local authorities to capitalise on interest in local generation tariffs, peer-to-peer options, and other local energy supply models as part of efforts to create a supportive policy environment for local climate initiatives (Regen/Scottish and Southern Electricity 2020). Whatever approaches are adopted, central government policies need to provide clear and stable signals of their commitment to net zero transitions and avoid undermining the direction or stability of local government net zero plans.

Empowering Local Governments

Each country apportions governing responsibilities between central, regional and local government in different ways. However, areas like planning, transport, housing, and land use typically fall within the remit of sub-national governments and provide important levers for influencing emissions. Transport, housing and planning feature prominently in many local net zero carbon strategies (Davidson et al. 2020) but, even here, local governments rarely operate independently. Local planning decisions in the United Kingdom, for example, need to demonstrate compliance with goals of the National Planning Policy Framework and sector-specific planning guidance issued by the central government, where local planning authorities are obliged to follow nationally determined interpretations of climate action and sustainable development. Local plans in the UK similarly require approval from the Ministry of Housing, Communities and Local Government—which also controls approvals for national infrastructure projects—while rejected planning applications may be appealed to the Planning Inspectorate, another central government executive agency (Berisha et al. 2021).

Such checks and balances are commonplace and necessary for coordinating central and local government policy but reinforce the need for central governments to provide local governments with clear policy direction and powers to pursue net zero emissions. One example of ways to confer these powers involves ensuring planning policy allows local governments to demand that all new developments achieve zero or negative emissions criteria and do not compromise other critical objectives, such as nature protection and combating poverty (Friends of the Earth 2020). Another is to ensure local transport authorities have sufficient powers and coordinate with service providers, users and neighbouring areas to develop integrated public transport networks operating single ticketing systems that promote user-friendly ways of linking with non-motorised transport networks (Buehler et al. 2019).

Vertical coordination in the planning system is also vital to safeguarding against successful appeals—in other than exceptional circumstances—against development proposals local planning authorities reject as incompatible with net zero emissions. Removing ambiguity over when emissions-intensive developments, such as fossil-fuel extraction and airport expansions, should be rejected is especially important to achieving climate-coherent planning systems. More generally, kneejerk planning-policy responses from central governments to the COVID crisis have the potential to lock in carbon-intensive infrastructure for decades, and may determine whether or not carbon neutrality by 2050 remains possible. Put simply, national land-use policies and regulations in areas like building construction and transport need to have zero emissions at their core and to be applied with conviction and consistency over the next thirty years if local net zero ambitions are to survive and thrive (also see Dyke et al. and Lankford, this volume, on these net zero ambitions and complexities).

Empowering local governments in the ways described above will require a combination of approaches and cannot happen just by central governments deciding which powers to grant and in what measure. More dynamic and creative relationships are only likely to emerge through open dialogue with local government representative bodies and other concerned groups, such as Canada's *Climate Emergency Unit*, Australia's *Council Action in the Climate Emergency* and the US's *The Climate*

Mobilization, to share ideas on how to develop local climate emergency plans and capacities (Council Action in the Climate Emergency 2021).

Financing Local Net Zero Transitions

Studies of local climate initiatives frequently stress their role as crucibles for experimentation in technological, governance and social innovations, and their potential to provide 'learning-by-doing' in emissions reduction (Bulkeley and Castán Broto 2013). However, a large proportion of innovative approaches falter in their early stages and scaling-up successful experiments remains a major challenge. Local governments in many countries have additionally suffered sizable cuts in funding as a result of responses to the 2008–2009 global financial crisis, while their resources have been stretched further by the need to provide healthcare and other services during the COVID pandemic (Anand 2020). Additional and secure funding will be critical to whether local governments can support the low-carbon innovation and infrastructure development needed to stimulate a genuinely green recovery (on climate finance, see the chapters by Bracking, and Kaplan and Levy, this volume).

Friends of the Earth (2020) estimates that £7 billion-£10 billion is needed per year in the UK alone to fund urban public transport and cycling and argues that substantial further investment is needed in low-carbon skills development in housing retrofits and heat-pump installation. Retrofitting programmes for existing buildings require significant funding commitments, though they are more cost-effective and quicker to achieve than many state-of-the-art building programmes (Zuo and Zhao 2014). Analysis of low-carbon transportation policy scenarios in California, for example, indicates initial additional investment of around $4 billion but potential long-term savings in the region of $23 billion by 2045, in addition to $28 billion in health benefits (Brown et al. 2021).

The ability of councils to stimulate low-carbon employment will be severely restricted without greater direct funding from central governments, measures to encourage private low-carbon finance, and enhanced powers for local governments to raise more funds within their areas, for example, through road-user charging and local levies on waste, single-use plastics and other emissions creating activities in their areas. In many countries, the lion's share of environmental tax revenues flow to the central government and either contribute to general finances or are hypothecated for environmental or social investments (Cadoret et al. 2020). Although the latter (if targeted suitably) can increase overall expenditure on climate mitigation, these funds could gain greater and more nimble leverage by directing a greater proportion of funding towards local government to empower climate emergency response plans. Even where funds remain controlled by central government, higher direct investment in emissions reduction and green infrastructure (e.g. sustainable travel, renewable energy, and ecological restoration) could provide assistance to local authority net zero plans, cost-effective job creation and the generation of co-benefits, such as reducing fuel poverty, improving health, and flood prevention.

Conclusions

The recent surge in climate emergency declarations offers one of the clearest indications to date of the strength of grassroots concern about climate change. The simple—and uncontroversial—plea in this chapter is for national governments to be more energetic in supporting local and regional net zero initiatives. Three main priorities have been identified: a clearer and more reliable orientation of national policies towards net zero emissions; empowerment of local governments in planning, transport and other areas of delegated responsibility; and enhanced financial and practical assistance for local net zero initiatives. If governments fail to support the current enthusiasm for local net zero strategies, there is no guarantee this momentum will be regained in time for local governments to make a meaningful contribution to climate mitigation efforts. National governments need the support of local administrations as much as local governments need support from their national governments for net zero to become a reality.

Alongside practical considerations, how national governments approach climate politics in the future is also likely to have a direct bearing on the fortunes of local net zero initiatives. A number of recent studies have explored how experiences from the COVID crisis might inform responses to climate change (Howarth et al. 2020; Manzanedo and Manning 2020). A common conclusion is that the pandemic has redrawn the boundaries of acceptable central government interventions and limits on personal freedoms to protect health and employment. However necessary these actions have been, an ethos of democratic deficit—where national governments bypass or overrule local governments—must not become pervasive. Responses to the climate crisis will need to be sustained over a protracted period and require long-term social mandates that are only likely to be achieved through dialogue rather than prescription. The need for zero-carbon strategies to reflect the emissions profiles and needs of individual regions adds further weight to arguments for dialogue within regions and between central and local actors, rather than overreliance on top-down approaches.

A further challenge is to broaden the geographical range of local net zero initiatives beyond their present concentration in a small number of affluent countries. At the time of writing, local governments from just three Latin American and three Asian countries (excluding Japan and South Korea) had declared climate emergencies according to the *Climate Emergency Declaration* database, although Bangladesh and Maldives had adopted national declarations. Relatively few climate emergency declarations had been made by local governments in Central and Eastern Europe, in contrast with 104 in Germany, 113 in Italy, and 510 in the UK. Soberingly given their exposure to climate risks, no African national or local governments had declared climate emergencies (Climate Emergency Declaration 2021). Local net zero initiatives in Asian, African and Latin American countries are likely to have different goals, action programmes, and working relationships with their national governments from those in the Global North, but this in no way diminishes their importance or the urgency of encouraging communities across the world to be actively involved in debating and shaping their climate futures.

Of equal importance is the need for all tiers of governance to avoid the partisan and even tribal climate politics that often has dominated discussions on climate change. One common feature of countries that have adopted national climate change acts is cross-party support for ambitious long-term action, even where disagreements persist on targets and implementation (Nash and Steurer 2020). A return to—or the failure to break free from—ideological partisanship and short-term politicking on climate change risks undermining not only national climate policy but also the consensuses that enabled local politicians to agree on climate emergency plans in the first place. One of the greatest contributions national governments can make to local emergency response initiatives is to defend the idea that climate change is too important for party politics and demands new and more cooperative forms of political leadership.

References

Anand, Paul, 'Economic Policies for COVID-19', *IZA Institute for Labour Economics Policy Paper*, 156 (2020), http://ftp.iza.org/pp156.pdf.

Berisha, Erblin, Giancarlo Cotella, Umberto Janin Rivolin, and Alys Solly, 'Spatial Governance and Planning Systems and the Public Control of Spatial Development: A European Typology', *European Planning Studies*, 29(1) (2021), 181–200, https://doi.org/10.1080/09654313.2020.1726295.

Brown, Austin, Daniel Sperling, Bernadette Austin, J. R. DeShazo, Fulton Lew, Timothy Lipman, et al., *Driving California's Transportation Emissions to Zero* (Institute of Transportation Studies, University of California, 2021), https://escholarship.org/uc/item/3np3p2t0.

Buehler, Ralph, John Pucher, and Oliver Dümmler, 'Verkehrsverbund: The Evolution and Spread of Fully Integrated Regional Public Transport in Germany, Austria, and Switzerland', *International Journal of Sustainable Transportation*, 13 (2019), 36–50, https://doi.org/10.1080/15568318.2018.1431821.

Bulkeley, Harriet, *Cities and Climate Change* (London: Routledge, 2013).

Bulkeley, Harriet, and Vanesa Castán Broto, 'Government by Experiment? Global Cities and the Governing of Climate Change', *Transactions of the Institute of British Geographers*, 38 (2013), 361–75, https://doi.org/10.1111/j.1475-5661.2012.00535.x.

Cadoret, Isabelle, Emma Galli, and Fabio Padovano, 'How Do Governments Actually Use Environmental Taxes?', *Applied Economics*, 52 (2020), 5263–81, https://doi.org/10.1080/00036846.2020.1761536.

Climate Change Commission, *Ināia Tonu Nei: A Low Emissions Future for Aotearoa* (Climate Change Commission, 2021), https://ccc-production-media.s3.ap-southeast-2.amazonaws.com/public/Inaia-tonu-nei-a-low-emissions-future-for-Aotearoa/Inaia-tonu-nei-a-low-emissions-future-for-Aotearoa.pdf.

Climate Emergency Declaration, *Climate Emergency Declarations in 1,948 Jurisdictions and Local Governments Cover 844 million Citizens* (Climateemergencydeclaration. org, 2021), https://climateemergencydeclaration.org/ climate-emergency-declarations-cover-15-million-citizens/.

Council Action in the Climate Emergency, *Council and Community Action in the Climate Emergency* (Caceonline.org, 2021), https://www.caceonline.org/ about.html.

Davidson, Kathryn, Jessie Briggs, Elanna Nolan, Judy Bush, Irene Håkansson, and Suse Moloney, 'The Making of a Climate Emergency Response: Examining the Attributes of Climate Emergency Plans', *Urban Climate*, 33 (2020), 100666, https://doi.org/10.1016/j.uclim.2020.100666.

Eskander, Shaikh, and Sam Fankhauser, 'Reduction in Greenhouse Gas Emissions from National Climate Legislation', *Nature Climate Change*, 10 (2020), 750–56, https://doi.org/10.1038/s41558-020-0831-z.

Friends of the Earth, *How Can Government Help English Councils Act on Climate Breakdown?* (Friendsoftheearth. uk, 2020), https://policy.friendsoftheearth.uk/insight/ how-can-government-help-english-councils-act-climate-breakdown.

General Assembly of the United Nations, *United Nations Summit on Biodiversity* (Un.org, 2020), https://www.un.org/pga/75/ united-nations-summit-on-biodiversity/.

Ghag, Jasbinder, 'A ticking time bomb? Liverpool declares a climate emergency: what next?' *Environmental Law Review*, 21(3) (2019), 169–72, https://doi. org/10.1177/1461452919872012.

Hepburn, Cameron, Brian O'Callaghan, Nicholas Stern, Joseph Stiglitz, and Dimitri Zenghelis, *Will COVID–19 Fiscal Recovery Packages Accelerate or Retard Progress on Climate Change? Smith School Working Paper 20–02* (Lagone. it, 2020), https://www.lagone.it/wp-content/uploads/2020/05/STUDIO-STIGLITZ-ART4.pdf.

Howarth, Candice, Peter Bryant, Adam Corner, Sam Fankhauser, Andrew Gouldson, Lorraine Whitmarsh, and Rebecca Willis, 'Building a Social Mandate for Climate Action: Lessons from COVID-19', *Environmental and Resource Economics*, 76 (2020), 1107–15, https://doi.org/10.1007/ s10640-020-00446-9.

Kern, Kristine, 'Cities as Leaders in EU Multilevel Climate Governance: Embedded Upscaling of Local Experiments in Europe', *Environmental Politics*, 28 (2019), 125–45, https://doi.org10.1080/09644016.2019.1521979.

Liu, Wenfeng, Xingping Zhang, and Sida Feng, 'Does Renewable Energy Policy Work? Evidence from a Panel Data Analysis', *Renewable Energy*, 135 (2019), 635–42, https://doi.org/10.1016/j.renene.2018.12.037.

Manzanedo, Rubén, and Peter Manning, 'COVID-19: Lessons for the Climate Change Emergency', *Science of The Total Environment*, 742 (2020), 140563, https://doi.org/10.1016/j.scitotenv.2020.140563.

Nash, Sarah Louise, and Reinhard Steurer, 'Taking Stock of Climate Change Acts in Europe: Living Policy Processes or Symbolic Gestures?', *Climate Policy*, 19 (2020), 1052–65, https://doi.org/10.1080/14693062.2019.1623164.

Regen/Scottish and Southern Electricity, *Local Leadership to Transform our Energy System* (Regen.co.uk, 2020), https://www.regen.co.uk/publications/local-energy-leadership-to-transform-our-energy-system/.

The White House, *President Biden and G7 Leaders Launch Build Back Better World (B3W) Partnership* (Whitehouse.gov, 2021) https://www.whitehouse.gov/briefing-room/statements-releases/2021/06/12/fact-sheet-president-biden-and-g7-leaders-launch-build-back-better-world-b3w-partnership.

Zuo, Jian, and Zhao, Zhen-Yu, 'Green Building Research–Current Status and Future Agenda: A Review', *Renewable and Sustainable Energy Reviews*, 30 (2014), 271–81, https://doi.org/10.1016/j.rser.2013.10.021.

19. Reversing the Failures of Climate Governance: Radical Action for Climate Justice

Paul G. Harris

Addressing climate change effectively will require focused attention on the most vital sources of failure in climate governance. Much, if not most, of that failure can be attributed to a lack of climate justice—a lack of ecological and environmental justice, a lack of social and distributive justice, and a lack of international and global justice. Demands for justice began decades ago. Had they been listened to and acted upon then, radical action would not be required now. To avert climate catastrophe, climate governance must embrace and implement all forms of climate justice.

Introduction

The governance of climate change has failed (Harris 2021). Apart from a temporary decline due to the worldwide economic fallout of the COVID-19 pandemic, greenhouse gas emissions are still *increasing* globally, with concentrations of carbon dioxide in the atmosphere reaching 420 parts per million (ppm) in 2020—compared to the pre-industrial level of 280 ppm—the highest since measurements began (Monroe 2021). Global warming is continuing apace, already reaching 1.1°C above pre-industrial norms (World Meteorological Association 2020: 6) and likely to exceed 3°C, even if all of the promises arising from past climate negotiations, specifically pledges (i.e., Intended Nationally Determined

https://doi.org/10.11647/OBP.0265.19

Contributions) toward the Paris Agreement, are implemented (United Nations Environment Programme 2019: 8). The impacts of climate change—wildfires, storms, droughts, pestilence and much more—are being felt with greater intensity, with the only prospect being that things will grow much worse in the years ahead (Intergovernmental Panel on Climate Change 2015). What is more, very little has been done to help the most vulnerable nation states, communities and individuals adapt to the inevitable, potentially existential, impacts of climate change. Recent events, notably widespread bushfires in Australia and forest fires in the United States, demonstrate that even the world's affluent societies and individuals will have trouble avoiding the impacts of climate change. Many millions, perhaps even billions, of people in poor societies have little to no hope of doing so.

Failed Climate Governance

After several decades of international negotiations, nation states still cannot agree to take concrete actions that will reverse climate change. Indeed, that lack of agreement helps explain the existence of the climate crisis, what many are now calling—justifiably—a climate emergency. The best that negotiations among governments have achieved—the 2015 Paris Agreement on Climate Change—is a step in the right direction (on the Paris Agreement, also see Hannis, this volume). But it has followed many other steps resulting from international negotiations that have neither stemmed global greenhouse gas pollution nor mitigated climate change. Indeed, pledges by states to implement the Paris Agreement are a recipe for continued global warming. Certainly, international negotiations have failed to achieve the objectives of the 1992 UN Framework Convention on Climate Change (UNFCCC). That is, one cannot seriously argue that the negotiations have achieved anything akin to the "stabilization of greenhouse gas concentrations in the atmosphere at a level that would prevent dangerous anthropogenic interference with the climate system" (United Nations 1992: Article 2). All of the science and on-the-ground reporting about the impacts of climate change prove that 'dangerous interference' is manifestly upon us already.

Many of those involved in international climate negotiations will acknowledge their failures up to now (see Dyke et al., this volume). This underlying realisation is probably one factor motivating those involved in upcoming negotiations to strengthen national pledges toward the Paris Agreement and to agree on the means by which those pledges will be fulfilled and verified. Indeed, many hold out hope for the twenty-sixth Conference of the Parties (COP26), scheduled to convene in Glasgow in November 2021, to produce much more robust agreement among the world's governments to finally get to grips with the causes and consequences of the climate crisis—much as they held out hope for twenty-five previous conferences. It is likely that some progress will be made, but it is more likely that, as in all previous conferences, progress will still not be enough to prevent yet more dangerous interference with Earth's climate system.

Nationally, more governments are pledging to reduce greenhouse gas emissions, or at least limit increases in them, and a growing number have promised to achieve net carbon neutrality by mid-century, although few are on target to do so, and it is anyone's guess whether governments that succeed them will implement those promises. Most significantly, the greenhouse gas emissions of all of the countries that have made pledges of carbon neutrality add up to a minority of total global emissions. What is more, the materially consumptive and energy-hungry lifestyles and ways of doing business that have caused climate change remain largely unchanged; indeed, despite calls for 'green' growth, those lifestyles are being *advocated* by governments and businesses as a means by which to grow the world economy out of the COVID-19 slowdown that took hold in 2020. Even as most of the world was wallowing in a COVID-19-induced recession, China's millions of upper-class and hundreds of millions of middle-class consumers returned to shops in force, snapping up new cars and luxury products. At the time of writing in mid-2021, those Chinese are already being joined by consumers in other countries as their national economies start to turn back toward growth.

In short, the pollution causing the climate crisis, the behaviours causing that pollution and the impacts of climate change arising from it are all going in the wrong direction. All of these trends need to be acknowledged if there is to be much hope for climate negotiations to make substantial progress.

Holistic Climate Justice

What is to be done? More of the same just will not be enough. The momentum of climate change and the pollution that causes it mean that efforts to address the climate crisis must be stepped up by orders of magnitude. Radical action is needed to avert and cope with the most dangerous consequences of climate change. Doing that will require focused attention on identifying the most vital sources of failure in climate governance and overcoming them. Much of the failure of climate governance can be attributed to a lack of climate justice (a.k.a. climate equity)—a lack of ecological and environmental justice, a lack of social and distributive justice, and a lack of international and global justice.

Demands for climate justice began several decades ago and have been reaffirmed in climate negotiations ever since. That bears repeating: demands for climate justice have been ongoing for *several decades*. Poor states have repeatedly called for wealthy ones to reduce their greenhouse gas emissions and to provide robust aid to assist the poor states to adapt. Poor individuals and their advocates have for decades called for action to prevent climate change and to help the poor avoid suffering that was expected to arise from greenhouse gas pollution. Advocates for nature have called over and over again for action to protect ecosystems, biospheres, landscapes, seas, species and animals that are already suffering from climate change. Had these calls for different aspects of climate justice been heeded and acted upon to any substantial extent, radical action would probably not be required now. However, apart from mostly lip service—such as invocation of the principle of 'common but differentiated responsibility' in international climate change agreements—the calls were, in effect, mostly ignored. This pattern will have to change dramatically (also see Bond, this volume, for some suggestions in this regard).

To avert climate catastrophe, climate governance must wholeheartedly embrace *and robustly implement* climate justice. Viewed holistically, climate justice would, by definition, entail all actors around the world doing what they can to prevent the climate crisis from becoming a climate catastrophe. This would include decarbonising the global economy very rapidly, thereby limiting global warming as much as possible, and helping everyone to adapt to unavoidable climate change. It would mean implementing environmental, ecological, social, distributive, international and global justice, as outlined below.

Ecological Justice

Addressing the climate crisis effectively will require coming to terms with the inextricable connection between healthy societies and a healthy environment. As long as the non-human world is treated as merely a source of resources and a depository for society's pollution, there is little hope of mitigating the worst effects of climate change. From this perspective, realising climate justice would entail putting the needs of the nonhuman world alongside—not beneath—the needs of humanity (Schlosberg 2019; Wienhues 2020). Such a view has gained traction among philosophers, and it has been advocated by Green parties in several countries (as well as being integral to indigenous cosmologies, as gestured towards in the chapters by Dieckmann and Sullivan this volume). Now and in the future, it needs to become a top priority of all actors—political, economic and social—and more explicitly a part of climate negotiations. Because the growth model of capitalism as practiced up to now is premised on perpetual extraction of resources from nature, doing this will inevitably require an alternative global economic paradigm. Climate negotiations intended to address the climate crisis effectively cannot avoid this challenging reality.

Environmental Justice

Addressing the climate crisis effectively must also involve concerted and successful efforts to eliminate environmental injustices experienced by people within states. This means that climate negotiations will have to work out agreement on how to facilitate this for real. Climate justice from this perspective would, at least, aim to prevent and alleviate the impacts of climate-changing pollution on disadvantaged communities (Bullard 2000; Nesmith et al. 2021). Much as poor and minority communities have always been the dumping grounds for society's waste and the places where the most-polluting industries have been located, climate change and the pollution that causes it are already harming poor communities, whether they be favelas that are washed away by landslides and storm surges, or rural regions and coastal villages that must endure the impacts of fossil fuel extraction.

Social and Distributive Justice

Addressing the climate crisis effectively will require that other injustices in societies be alleviated (Preston et al. 2014; Chancel 2020). Social injustice and economic inequity, with poor, minority and underrepresented members of society suffering the consequences when power and resources are concentrated in the hands of the wealthy and the well-connected, create the conditions for perpetuating climate injustice and inequity. Thus, climate negotiations must do more not just to pay lip service to, for example, the plight of the world's poor, racial and ethnic minorities, and indigenous peoples; the negotiations must also reach agreement to genuinely end their plight and to more fairly share the world's resources (including financial ones) so that they can have an active role in addressing the climate crisis. This sort of thing has been advocated by socialists and others for a very long time, but of course it has not been implemented widely because it challenges accepted notions of power and supposedly free-market economics. Achieving social and distributive justice, as part of realising holistic climate justice, would also require the world's affluent individuals to temper their passions for consuming large quantities of stuff they do not need. None of this will be easy, not least because the climate negotiations themselves lack the official remit to deal with many of these issues. Nevertheless, negotiators ought not to shy away from them if they are to make real progress.

International Justice

Addressing the climate crisis effectively will require a fair and equitable international distribution of the burdens and benefits associated with climate change (Shue 2014; Okereke 2018). For many states, climate change is as much a matter of international (in)justice as it is one of environmental change. From this perspective, affluent states need to act aggressively to implement the principle of common but differentiated responsibility that was codified in the seminal climate convention three decades ago and which has been invoked at each conference of the parties in word—and largely ignored in practice—ever since. International climate justice requires a fairer distribution of power and resources internationally so as to give the least well-off states the ability to negotiate on a relatively equal footing, and it requires that the injustices associated with climate change suffered by states and their citizens, most obviously the costs and suffering that accompany the impacts of climate change, be ended (or, more realistically, greatly mitigated) and adequately compensated. This conception of climate justice is largely an extension of ideas about the need for international justice among states that have been argued by leaders, diplomats and scholars for most of the post-war period. International injustices that have existed for a century and more are multiplied by climate change.

Global Justice

There is another form of climate justice that is less commonly discussed in climate negotiations. It is a type that might be labelled as cosmopolitan in the sense that it is about the needs, rights, responsibilities and obligations of *all actors around the world*, encompassing state actors (as with the forms of climate justice within and among states mentioned above) but also involving non-state actors, including individuals. After all, the proximate cause of the climate crisis was, is and will be the behaviours of individuals, whether those behaviours result in greenhouse gas pollution directly, such as when driving a car, or indirectly, such as when consuming things whose production, transport and/or disposal result in such pollution. The notion here is that *all capable actors, regardless of where they are* (whether in rich countries or poor), ought to be acting to address climate change effectively. This means, for example, that people who consume more things than they need ought to refrain from doing so, and those who are capable of aiding others affected by climate change ought to do so. Put another way, in addition to requiring action on the common but differentiated responsibilities (bearing in mind respective capabilities) of states, holistic climate justice also requires action on the common but differentiated responsibilities (bearing in mind respective capabilities) of individuals (among other actors), regardless of where they live, too. Thus, while affluent people in the Global North have obligations to act on climate change, so do affluent people living in the Global South. For climate justice to be truly holistic and to aid in effectively addressing the climate crisis, it needs to include this *global* form of justice (Harris 2016; Dietzel 2019; Moss and Umbers 2020).

Prospects for Radical Action

None of the forms of climate justice outlined above, except, possibly, the last one, are unusual in any way. They have either been discussed since the earliest climate negotiations or at least promoted by activists and described by scholars for just as long. But all of these forms of climate justice have been practiced in the breach. Consequently, while propounding them is not at all revolutionary, to *implement* them, and perhaps even to approach negotiations in such a way as to assume that they should and must be implemented on the ground as soon as possible, would be. Just such a revolution is essential if coming climate negotiations are to achieve what must be, given the scale of the climate crisis, their supposed objectives: at minimum, to quickly reverse trends in greenhouse gas emissions, rapidly decarbonise the global economy, and, finally, take robust action without further delay to enable adaptation everywhere to the impacts of climate change that cannot be avoided. Negotiating holistic climate justice would help to create the conditions for making all of that possible.

Holistic climate justice amounts to making the world fairer and more equitable. That might seem to be idealistic, even fanciful, and certainly historical precedent would support such a sceptical view. But historically the world has not faced a threat as grave as climate change. The time has finally come when the fates of societies will depend on whether there are serious attempts to implement varieties of justice that remove incentives to live unsustainably and enhance the conditions that allow people easily to live in ways that mitigate climate change and its impacts. Action to achieve these forms of climate justice, and thereby to mitigate the climate crisis, must be, by definition, radical—that is, thorough and far-reaching. The need for radical action derives, very simply, from the failure to take modest action over the past several decades as governments and other actors have responded to climate change at a glacial pace while global warming and climate change, not to mention the pollution causing them, have sprinted toward catastrophe.

While rapid movement toward realising all forms of climate justice is essential (the fact that they can never be fully realised is no excuse for not doing all that is possible to make them so), a good place to focus is on ecological justice. That is because everything arguably arises from it.

If we treat the environment, including non-humans, justly, we thereby protect the environment upon which humans depend. But ecological justice cannot be achieved without treating humans justly, too. People trying to survive, and even people who just want their fair share of the economic pie, are not about to prioritise environmental stewardship if they do not perceive a very clear and present stake in doing so. Likewise, nation states are not about to protect the environment if they feel the same way vis-à-vis other states, so we need international justice, too. We need global justice as well because it captures the realities of climate change: ultimately, people cause it through their behaviours (to wit: if all humans dropped dead this week from a pandemic, new anthropogenic greenhouse gas pollution would immediately cease).

Conclusion

Climate change is an environmental problem, obviously. It is a political problem, locally, nationally and internationally. It is a social problem and, at its foundation, a human problem. But perhaps most of all it is a problem of justice, of treating others—other states, other communities, other races and genders, other individuals and other creatures (and their ecosystems) more fairly and equitably. The climate crisis is a crisis of injustice. Solving the climate crisis requires implementing climate justice (see Harris 2019). Whether climate negotiations can make that happen is an open question. Whether they try much harder to do so cannot be.

References

Bullard, Robert D., *Dumping in Dixie: Race, Class and Environmental Quality* (New York: Routledge, 2000).

Chancel, Lucas, *Unsustainable Inequalities: Social Justice and the Environment* (Cambridge, MA: Belknap/Harvard University Press, 2020).

Dietzel, Alix, *Global Justice and Climate Change: Bridging Theory and Practice* (Edinburgh: Edinburgh University Press, 2019).

Harris, Paul G., *Global Ethics and Climate Change* (Edinburgh: Edinburgh University Press, 2016).

Harris, Paul G. (ed.), *A Research Agenda for Climate Justice* (Cheltenham: Edward Elgar, 2019).

Harris, Paul G., *Pathologies of Climate Governance: International Relations, National Politics and Human Nature* (Cambridge: Cambridge University Press, 2021).

Intergovernmental Panel on Climate Change, *Climate Change 2014: Synthesis Report* (Geneva: Intergovernmental Panel on Climate Change, 2015), https://www.ipcc.ch/report/ar5/syr/.

Monroe, Robert, 'Coronavirus Response Barely Slows Rising Carbon Dioxide', Scripps Institution of Oceanography (Scripps.ucsd.edu, 2021), https://scripps.ucsd.edu/news/coronavirus-response-barely-slows-rising-carbon-dioxide.

Moss, Jeremy, and Laclan Umbers (eds), *Climate Justice and Non-State Actors: Corporations, Regions, Cities and Individuals* (London: Routledge, 2020).

Nesmith, Ande A., Cathryn L. Schmitz, Yolanda Machado-Escudero, Shanondora Billiott, Rackel E. Forbes, M. C. F. Powers, et al., *The Intersection of Environmental Justice, Climate Change, Community and the Ecology of Life* (Cham, Switzerland: Springer, 2021).

Okereke, Chukwumerije, 'Equity and Justice in Polycentric Climate Governance', in *Governing Climate Change: Polycentricity in Action*, ed. by Andrew Jordan, Dave Huitema, Harro van Asselt, and Johanna Forster (Cambridge: Cambridge University Press, 2018), pp. 320–37.

Preston, Ian, Nick Banks, Katy Hargreaves, Aleksandra Kazmierczak, Karen Lucas, Ruth Mayne, et al., *Climate Change and Social Justice: An Evidence Review* (York: Joseph Rowntree Foundation, 2014), https://www.jrf.org.uk/report/climate-change-and-social-justice-evidence-review.

Schlosberg, David, 'An Ethic of Ecological Justice for the Anthropocene', *ABC Religion & Ethics* (Abc.net.au, 2019), https://www.abc.net.au/religion/an-ethic-of-ecological-justice-for-the-anthropocene/11246010.

Shue, Henry, *Climate Justice: Vulnerability and Protection* (Oxford: Oxford University Press, 2014).

United Nations, *United Nations Framework Convention on Climate Change* (Bonn: United Nations Framework Convention on Climate Change Secretariat, 1992), https://unfccc.int/process-and-meetings/the-convention/what-is-the-united-nations-framework-convention-on-climate-change.

United Nations Environment Programme, *Emissions Gap Report 2019* (Nairobi: United Nations Environment Programme, 2019).

Wienhues, Anna, *Ecological Justice and the Extinction Crisis: Giving Living Beings Their Due* (Bristol: Bristol University Press, 2020).

World Meteorological Association 'WMO statement on the state of the global climate in 2019', WMO-No. 1248 (2020), https://library.wmo.int/doc_num.php?explnum_id=10211.

VI

FINANCE

20. Climate Finance and the Promise of Fake Solutions to Climate Change

Sarah Bracking

This essay explores how promises of money from global institutions and governments have financialised people's hopes and expectations of government action to adapt to climate change and slow the emission of greenhouse gases. Because of the cultural power of money in our understanding of the world, climate finance has had the particular job of signifying action while delivering very little. In order to move forward with the actual material changes to energy, infrastructure, production and income distribution that lie at the heart of an effective response to climate change, we need to accept that largely fictional promises of money that 'can change things' are a phantasmagorical expression of meaning—a firewall that prevents real change. In making this point, the essay traces the small disbursement figures for the main pots of climate finance and in doing so offers a stringent critique of the obfuscating power of the language of finance.

https://doi.org/10.11647/OBP.0265.20

Introduction

Finance is a key contemporary mediator of the relationship between humans, more-than-human natures and Nature.[1] This chapter explores how promises of money from global institutions and government have financialised people's hope and expectations of government action to adapt to climate change and slow the emission of greenhouse gases. Because of the cultural power of money in our understanding of the world, climate finance signifies extensive action. In practice, however, it is small and delivers even less (as also articulated by Kaplan and Levy, this volume).

Material and foundational changes to energy regimes, infrastructure, production and income distribution lie at the heart of an effective response to climate change. In order to progress with these changes, we need to discard the largely fictional promises of money that 'can change things' which act as a phantasmagorical expression of meaning: becoming a 'firewall' or barrier that prevents real change. We are being offered a financialised spectacle of climate change action which obscures both the empirical reality of ecosystem and biodiversity loss, and the uncomfortable imperative of how our ways of living need to change (as also foregrounded by Halme et al. and Harris, this volume). This essay is intended as a plea to give up on the idea of money as our conduit for action in favour of real shifts in production and in human and more-than-human relations.[2]

I proceed by exploring the definition, amounts and governance of climate finance that we currently have through a set of eleven propositions and their evidenced negation.

---

1   'More-than human' refers to the subset of the whole of nature that is not human—all other animals, trees, plants and so forth. For definitions of this term, and other related terms such as 'beyond-human', 'other-than-human' nature(s) or 'nature-beyond-the-human' I draw on Sullivan (2015: 3). For an extended ontological discussion see also Sullivan (2017).

2   This chapter updates an earlier version published in 2011 as the Green Climate Fund was being brought into existence in Durban, South Africa. See 'Climate Change: Beware, large-sounding-sum-of-money approaching!', https://www.theafricareport.com/7959/climate-change-beware-large-sounding-sum-of-money-approaching/.

Proposition 1: Climate Finance is Big and Expanding

The Paris outcome (COP21 2015) urged developed nations to mobilise US$100 billion per year by 2020 for climate action in developing nations. Partly as a consequence, many commentators believe the volume of public and private finance addressing climate change is slowly rising in aggregate toward this number—particularly at the sub-national level and by non-state actors—but that there remains a significant and large investment gap (UNCTAD 2020). In this world view, the Organisation for Economic Co-operation and Development (OECD) (2017) has estimated an 'infrastructure gap' of US$95 trillion globally in the investment required for energy, transportation, water and telecommunications decarbonisation transitions by 2030 to address climate change, of which 60–70% will be needed in developing countries.

These and similarly large-sounding numbers have inspired a wide body of work discussing the merits of blended finance and climate congruent activities of non-state and sub-state actors, such as corporations and cities, in order to meet the financing challenge in a climate crisis that is multi-scalar. Many academics and the public have also been mesmerised by this idea that we are discussing large numbers—but we are not. Current climate finance for adaptation that is unique, additional, and concessionary is approximately, on a generous interpretation, US$29 billion per year *globally* (Buchner et al. 2019). But even this figure is inflated. The NGO Care International recently analysed the details of reporting and wrote that official figures were hugely exaggerated, arriving at $9.7 billion globally as a corrected figure for 2018 (Care International, 2021). Paltry at $0.0097 trillion.

Whether estimated in billions or trillions, however, money matters in context, and in relation to how you count. For example, whilst the 'infrastructure gap' of US$95 trillion mentioned above evokes an emergency, it is in fact similar to 'normal' levels of investment that would be made anyway in a global economy of a ballpark $170 trillion. At best, these figures remind us of the real need to switch investments *in type and purpose* to decarbonisation pathways. Unfortunately, this switch is slow, and so far has been market-led as the price of energy generated from renewable technologies falls below the cost of energy generated by burning fossil fuels. The role of regulation and government action has contributed very little to the speed of this shift (as also noted by Wright and Nyberg, and Newell, this volume). Few governments have forcibly closed coal mines or oil fields.

Meanwhile, although the 'billions' figure for 'climate finance' from Paris sounds big, US$100 billion equates to only $0.1 trillion, and has not been implemented in practice. Indeed, the main purpose of the 'huge gap + large-sounding commitment' rhetoric appears to be to legitimise the next fashionable tinkering and boutique products of the climate finance market, and to privilege the private sector as a trustworthy handmaiden of change.

Proposition 2: Climate Finance is Innovative and Bespoke

This fore-grounding of the private sector in climate finance fits a wider pattern as capitalist development faces a legitimacy crisis, which has in turn generated a green-washing or 'green halo' effect (Sörqvist et al. 2015), involving constant rebranding of 'brown investments' and the lauding of finance, technology and innovation as components of a growing green economy (Bracking 2012, 2019; Sullivan 2012, 2018a). The depiction of 'greenness', complexity and novelty within environmental finance products appears to hold its academic and wider audiences in awe. This is despite the continuation in practice of both the environmental injustices born of centuries' old private property relations (see Lave 2018; Bigger and Millington 2020), and the salience of traditional metrics for calculating financial return, such as the discounted cash flow model, where 'green' is still a poor add-on.

Alongside grants, debt-based instruments have grown in type and apparent dedication, such as municipal bonds, habitat bonds, conservation bonds, species bonds, climate bonds, green bonds and more latterly transition bonds and sustainability bonds (Sullivan 2013, 2018b; Bracking 2019). These last two are the latest products, saluted and enthroned as comprising a spectacularly growing asset class in the UNCTAD 2020 *World Investment Report*. Private sector involvement is also growing in insurance-based instruments: climate risk insurance and securitisation (Taylor 2020), catastrophe bonds, hazard and disaster risk insurance (Surminski and Architesh 2020), and even humanitarian and pandemic health insurance (World Bank 2017; Erikson and Johnson 2020), although many products are faltering without public sector involvement to subsidise the cost of risk and artificially create 'demand' from a body who can afford to pay (see InsuResilience 2020). Many of these instruments promise the incorporation of modern innovations in algorithmic and artificial intelligence, weather and risk modelling, earth observation and even blockchain and cryptocurrency technologies as providing efficiency gains in what is basically debt finance.

These convivial sounding bonds and insurance products, however, also act both as a firewall and fetish to protect against encroaching reality, and provide a new means of providing debt-based finance to entities often already in ecological and financial deficit (see Jones et al. 2020). They largely fund incremental shifts in industrial emphasis, rather than the seismic shifts needed for meaningful infrastructure decarbonisation.

Proposition 3: Climate Finance Is a Distinct and Additional Source of Finance

Although an internationally-agreed definition of 'climate finance' has been elusive, the United Nations Framework Convention on Climate Change (UNFCCC) now refers to it expansively as "local, national or transnational financing—drawn from public, private and alternative sources of financing—that seeks to support mitigation and adaptation actions that will address climate change" (UNFCCC 2019: online). This definition signals the move in conception away from more traditional ideas of climate finance as principally flows of public development aid, concessional loans and grants, to a polycentric mix of public and private capital leveraged using financial technologies and institutions, governed by a range of actors in various combinations (Pattberg and Widerberg 2015: 685). Put more critically, what is envisioned is a New Washington Consensus[3] which subsidises investors in order to leverage and reward private capital (Mitchell and Sparke 2016).

---

3   The 'Washington Consensus' refers to the agreed set of conditionalities structuring lending to states by International Financial Institutions, post-1989, which thus shape flows of finance directed towards reform and structural adjustment.

In terms of the private green bond market where money to alleviate the effects of climate change (adaptation finance) or slow down and reduce the things that cause it (mitigation finance) is raised as 'climate bonds', 'green bonds' or 'transition bonds' (to help dirty or 'brown' industries change to be cleaner and more 'green'), the classification of what is 'green' is decided by the issuer in a 'self-labelled' action. Or it is classified according to what the money will be spent on—'use of proceeds'—with some reference to either the issuers' narrative or to a common 'standard' such as the Climate Bond Principles. This is kept deliberately vague, apparently so that market entrants are not deterred by too much regulation.

In terms of the public sector, climate finance is different, or additional, to market-based loans only because of the provenance of the issuer and the context of the lending. Climate finance is a part of a bigger pool of money generically called concessionary finance from governments, which includes grants, loans, and more recently 'blended finance'—a mixture of public and private money. Some call all of these categories 'aid'. The sums quoted are directly related to how it is counted and categorised, rather than to any actually growing amount of money or, technically, 'liquidity'. When public money is joined with private money as 'blended finance', the claim to be green or developmental, or both, is decided by the issuer and the regulator of official development assistance (ODA, or 'aid'), the OECD. In relatively new statistical rules implemented by the OECD *Development Assistance Committee* (OECD 2020a, 2020b), classification criteria were made more expansive, and reclassification of commercial flows as concessionary spiked, while actual grants have shrunk from most major countries. Now, anything looking vaguely developmental or climate-related can be added into the data as 'blended finance', even if it transfers from seller to buyer (or donor to beneficiary) at market rates and above. In other words, blended finance can be more expensive than private finance, but can be seen as 'green' or 'developmental' just because of who is issuing it and the authority of their claim to be 'green', within the technical rules of classification for overseas development assistance. Actual global climate finance in the form of grants were a measly $27 billion per year for 2017/18 (Climate Policy Initiative 2019: 12). The OECD estimates climate finance grants from the 'developed' to 'developing' countries at only $12 billion per annum from 2016–2018 (OECD 2020c: 9). Even here, we are including in the totals the salaries and overheads of the organisations delivering the money. On the ground, climate finance adaptation resources for the most vulnerable are as rare as an endangered species.

Proposition 4: Climate Finance Can Be Better as Blended Finance

The OECD *Development Assistance Committee* (OECD 2020b) argues that blended finance is the answer to drops in bilateral and multilateral public finance and offers synergies for increased efficiency, augmentation and the alignment of public and private ambition. Blended finance refers to public funds pooled with private funds, largely under private fund management. It forms the centrepiece of the 'billions to trillions' narrative (World Bank 2015; UNCTAD 2019) of mobilising private finance to meet the Sustainable Development Goals (SDGs) 'financing gap' of USD2.5 trillion per annum in developing countries (UNCTAD 2014). Often the public money is used to 'de-risk' the investment, which means that if it fails, the public sector takes the loss, and if the investment succeeds, the public sector has the last and worst dividend. It is a bonanza for private investors who enjoy highly competitive market rates on their 'tranches'. Within the blended finance realm, development and climate change management have morphed and merged into new categories depicting synergies and mutual co-benefits, often hiding contradictions in practice inherent to decarbonisation pathways.

In the context of climate finance, the official and hegemonic position dates from the Kyoto Protocol and sees an unproblematic synergy between market logic and public sector policy (Andrade and de Oliveira 2015). Current international climate governance thus emphasises partnerships, synergy with private actors, blended finance and leverage of private funds, alongside consensus-oriented governance driven by "[m]arket-oriented rationales" (Kuyper et al. 2018: 9). The *Climate Policy Initiative* compiles data on climate finance for their Global Landscape of Climate Finance report (Buchner et al. 2019). Their data for 2017–2018 show, for example, that finance for mitigation far outweighs adaptation, with the latter constituting only 5% of total flows. The former is paid to companies to clean up industrial processes to emit less carbon, often quite incrementally, such as by putting in sulphur capture chimneys at coal-fired power stations. The 5% for adaptation is intended to help people 'adapt' and become 'resilient' to climate change as it undermines their livelihoods and ecosystems. For example, it might be a grant for drought-resistant seeds. Of the US$30 billion of climate change adaptation finance, grants from governments totalled US$29 billion, reflecting that there is little money from the private sector to fund adaptation—there is no profit in it. By comparison, private sector actors contributed loans (debt) at market rates to mitigation projects worth US$223 billion; equity investment to projects worth US$44 billion; and balance sheet financing (debt and equity) worth US$219 billion, with these latter categories largely contributing to the US$537 billion for mitigation overall. All of this private sector climate finance used to be called (normal) debt and equity investment, made up of finance expecting a (normal) market rate of return. Counting this finance as 'climate finance' involves the self-labelling of climate-related 'improvements', which in practice can be just about anything. At best it is funding alternatives to fossil fuel energy generation (with due regard to surrounding people, animals and ecosystems). At worst, it is financing such oxymorons as 'clean coal'.

Proposition 5: Climate Finance Can Be Found in Private Debt Products

The illusion of money solving a problem is also maintained by the private markets in climate finance's sibling products—the green bonds, transition bonds and sustainability bonds—all of which are apparently enjoying a boom (Sullivan 2018b; Bracking 2019). According to the UNCTAD *World Investment Report* (2020: v) "investment in the SDGs show that sustainability themed funds in global capital markets are growing rapidly. [... But] they show these finances are not yet finding their way to investments on the ground in developing countries". The boom in green finance can be attributed to both classification issues and to trends in the immanent market. In terms of classification, a number of features wildly inflate the sense of 'greenness', including: that any investment can be 'self-labelled' green by its issuers; generally only just more than 50% of the principal needs to be 'green' for the whole bond to be classified that way; and because 'green' can be applied on the basis of a 'use of proceeds' narrative which may inflate the climate change mitigation/adaptation potential of the investments. In terms of the market, there has been a shift in the underlying cost of energy generated from renewable sources versus energy generated from fossil fuels, and many climate bonds and green bonds are simply following this market shift and investing in renewables because of better returns. This is a good thing, but giving these debt instruments the 'climate bond' or 'green bond' name makes it seem that investors are doing more than that; that they are in some way giving up a profit margin for the greater good. This is generally not the case.

In short, the private sector has been successful in continuing investments in existentially dangerous production practices, while simultaneously green-washing and reclassifying investments as green when the underlying asset and context has largely stayed the same. Meanwhile, all bond finance is still debt, and bonds issued in the Global South, particularly by municipal or sovereign authorities, ultimately extract from those least able to pay, and least likely to have historically caused planetary warming.

Proposition 6: Climate Finance Can be Found in Insurance

Climate finance also includes climate insurance, which is depicted as having several 'benefits' over other approaches to managing climate change. A loss and damage approach accepts that some people need compensation for losses that others have caused. Similarly, ecological debt and climate justice approaches endorse a variant of the 'polluter pays' principle where the victims of climate change are owed redress from the historical polluters (nations or companies). But climate insurance does not rest on these philosophical foundations, and for some this is seen as a benefit. For example, Horton (2018: 285) summarises in the Harvard-based *Carbon and Climate Law Review*, that climate insurance: "does not require that causation be demonstrated [...] is oriented toward the future rather than the past, [... and is] contractual, rather than adversarial". These three aspects make it look fair, based on the freedom of exchange that people widely associate with market-based solutions—while obscuring all the responsibility and culpability that could otherwise lie with the historic polluters.

But the limits to climate insurance are that if you are rich, and making profits fast and first, you can ignore the need for it by shifting costs to others, normally by effectively moving them into a time in the future. In Florida or Miami, for example, real estate investors build new towers by the waterside and then sell their stakes within a few years with no continued flood liability (Taylor 2020). Conversely, if you are poor, and in the absence of any other investments in basic goods or welfare, insurance is often not available, and weather and disaster prediction technologies are of limited use. For example, in some parts of Mozambique, Zimbabwe and Malawi advanced warning of the tropical cyclone 'Idai' in 2019 was ineffective as persons had few options to mitigate the outcome of the disaster. Another scenario where insurance is not an option is when the risk is certain and not probabilistic. For example, where inundation by the sea is already happening in small islands, and where it is seen as a certain outcome, insurance is not available to protect these first victims of climate catastrophe.

Theoretically, risk insurance algorithms and complex hazard and weather modelling, appropriately commoned, could assist the poor and vulnerable, if structured through a huge democratic risk-management and governance panopticon. This would only arise if action follows knowledge, in this case advanced modelling software of the likely weather. But under capitalism, action follows money, and it is more likely that these technologies will remain market edge and proprietary, allowing the owners of new complex predictive knowledge about the weather a financial advantage in futures trading. On the other hand, and metaphorically if not literally a world away, the poor and vulnerable may not get access to news about a pending hazard, or the resources to mitigate their risk.

This inequality reproduces itself when risk is used to manage resources. As we saw above, one benefit of climate insurance for the privileged has already been collected: using insurance as a way of managing a changing climate applies a future-looking resetting of the clock on who will pay, while discarding calculations of ecological debt. In this re-setting, risk pools for climate insurance bake in intersectionalities and hierarchies of economic inequality, postcoloniality, race and gender.

Proposition 7: Climate Finance Is Managed by People with Expertise Using Modern Technologies

Within financial products, particularly in climate insurance and disaster risk insurance, the insertion of calculative devices is common, to affect probabilistic calculation, but also to perform worth and expertise, helping to legitimise the central role of financial managers in our everyday lives (cf. Munden Project 2011). As Larry Lohmann (2020) suggests, however, the effort to use automation and technology to entrain humans and other species in actual processes of accumulation is constantly fraught with confrontation, a push and pull between capitalist asset making and peoples' resistance and acts of commoning. Some technologies end up working for capital, while others prove dysfunctional, and this depends largely on the class and power relations within the marketisation process. In particular, if a conservation, development or climate change project 'on the ground' seeks finance from a climate finance institution, its workers or 'beneficiaries' are then caught up in arrangements which make them subject to calculative technologies deciding risk and price. The product could involve earth observation and weather modelling, for example, with both or either of these becoming locked into parametric triggers for insurance pay-outs. Once climate change insurance becomes securitised and sold on as climate change catastrophe bonds, their risk will also be traded in markets using algorithmic 'sniffers' to check on the trading prices of the bonds, in the face of changing environmental conditions.

These exotic tools of investment management are not the norm, however. Old technologies remain the most common. For example, while the Green Climate Fund is home to the 'paradigm shift' to 'transformatory change' involving the co-production of climate change, environment, conservation and development co-benefits, it is also home to very orthodox calculative technologies, and extremely well-paid fund and project managers (Bracking 2015).

Consider, for example, a very recent Green Climate Fund project worth over $1 billion, about one tenth of all its committed funds:

The High Impact Programme for the Corporate Sector (GCF 2020). This 'High Impact Programme' is managed by the European Bank for Reconstruction and Development (EBRD) and addresses what it sees as deficits in corporate capacity in respect of climate change planning. It supports the "integration of risk analysis and gender-responsive climate change consideration into strategic decision making, target setting and investment planning", and aims to improve corporate climate governance and management by using apparently innovative 'High Impact Loans', which incorporate flexible interest rates and link these to "financial performance, the innovation being the link to climate and corporate governance performance" (GCF 2020: 3). Governance performance here is evaluated by the EBRD itself, using its own matrix and governance scorecard, a climate change governance (CCG) assessment tool that performs a gap analysis, finds entry points for low carbon strategy and then builds "low-carbon roadmaps" (GCF 2020: 5–13). The project will additionally promote "private-public sector dialogue [...] sector-level decarbonisation roadmaps... [and] collaborative knowledge exchange" in a two-step approach: "shift 1—uptake of high climate impact technologies; and shift 2—behavioural change at corporate governance and management levels" (GCF 2020: 5). In other words, the fund uses orthodox 1990s performance management of roadmaps and impact assessment. It re-packages these slightly for the 2020s by using more recent signifiers for "impact investment" (cf. Chiapello and Godefroy 2017; Sullivan 2018b), like "[p]aradigm shift potential: [where] The concessional loan has the potential to trigger behavioural change at corporate sector management level to incorporate climate change targets and corporate climate governance principles into strategic decision making" (GCF 2020: 13).

But behind this signalling of modernity and radicalism—the paradigm shift—is a stalwart mediocrity: the EBRD is spending $1 billion to ask managers to consider climate change. This is hardly novel, but it is insulting that the grant component, small as it is, appears to fund the technical assistance that the EBRD provides for its own loan, i.e. its own management costs (GCF 2020: 15).

Proposition 8: Climate Finance Is Spent with Due Accountability

When climate finance is being dispersed largely through the private sector as blended finance, aspects of its accountability, authority and legitimacy are handed to financiers to determine, framed using privatised metrics and calculations. In this form of governance, the public and private sectors join in what Asiyambi (2018: 533–36) so cogently analyses, for the green economy more broadly, as *spaces of mutuality,* where durable processes *of becoming* generate new green assets.

Asiyambi (2018) uses Foucault's idea of organising actions in his account of REDD+ (Reducing Emissions from Deforestation and Forest Degradation in Developing Countries) to explore how environmental financialisation is constructed. For climate finance, public finance is authorised globally by multilateral development banks (MDBs) and bilateral aid and development finance institutions (DFIs), and is then combined with private equity or used to leverage debt with diverse non-state actors. The mutuality is then a co-dependence. Public finance is critical to non-state actors in their contribution to aspects of climate finance governance: to the underwriting of risk and debt (reducing costs for private actors), the legitimising of the mode of implementation and the authority ascribed to the venture. In turn, private financiers contribute to climate finance governance, in that products are increasingly operated, implemented and governed by them, using market-based logics and profits-based rates of return.

The accountability of blended finance *ex ante* relates to contracts signed between the investors and the fund managers which are largely private as they contain 'commercially sensitive' data. There is also a process-based accountability found in corporate social responsibility monitoring and economic, social and governance scoring. Since fund managers themselves largely do this paperwork for their own investors, however, it is not a convincing exercise. It is self-reporting, as outlined above. Accountability *ex post* is largely financial and is indicated by the outcomes of the investment against the contract commitments on closure and any ESG and CSR scores attached. This again is largely private. In effect, given the opacity of all the metrics, peoples' trust in climate finance is largely a spectacle based on their trust in bankers and investors and the moral universe that they present. A spectacle of money.

Authority is inscribed by the status and reputation of the fund managers and banks involved. The legitimacy of the fund is built by the 'narrative authority' it produces (Leins, 2020), an account of itself which includes voluntary standards, disclosure, rankings, and ultimately financial performance. Thus the weakest area of research on climate finance governance is what happens once finance is co-invested and blended within the private financial sector, in this space of apparent mutuality. It is weak because researchers are rarely granted access to analyse these private transactions. This matters because scientific knowledge and climate justice concerns, and the civil society, government and academic actors who voice them (who are not mutually exclusive groups), are consigned to a weak power to comment on and influence how climate finance is spent. Without transparency there can be little accountability.

In climate finance provided through risk-based insurance, the opacity is a combination of conventional secrecy excused by 'commercial confidentiality' combined with the opacity of the automated machine of parametric insurance triggers, which few persons can see or understand. It is hard for the buyers of a product to work out how or why it may or may not pay out. Despite this uncertainty, insurance products use risk to socialise costs and privatise profits. In an interesting shifting calculation, risk shifts costs to sovereign states who pay premiums to access the insurance on behalf of their citizens. As the case of the Malawi drought in 2016 demonstrated, even in a famine a glitch in the model (in this case it being programmed on the basis of the wrong type of maize) might stop the insurance paying out (ActionAid 2017: 9–10, citing research from Lilongwe University of Agriculture and Natural Resources). But the insurance must be paid for, and its cost is a sovereign liability which means it is passed on to citizens through the tax relationship. Often, the poor pick up this bill, particularly so in regressive tax systems where the burden of tax falls disproportionately on them, despite their being least culpable for climate change. In many countries this is not an accountable relationship as increasing sovereign liabilities is effectively a privilege of the political class (see Pogge 2007), rather than subject to democratic process.

Proposition 9: Climate Finance Is a Public Good

The mainstream position on climate finance delivery revolves around the efficiency of the private sector within a business model and its contribution to climate change governance (Figueres et al. 2017). Correspondingly, the dominant model for providing climate finance is lending through equity funds, which are often domiciled in secrecy jurisdictions, which is sometimes called the indirect or 'fund-of-funds' model (Bracking et al. 2010). This has several consequences for efficacy and morality at the supranational level.

The first is that a significant amount of climate finance is used to pay for the management and service costs of the accrediting and implementing entities (DFIs, MDBs and so forth), and then again for the remuneration of fund managers if these are in the private sector (few are kept 'in house'). The supply chain of climate finance is skewed in favour of the suppliers who claim most of the value, which represents an unacceptable loss to the finance available for work with climate-affected persons (Bracking et al. 2010, 2015). This problem is compounded by the opacity of the indirect investment and lending model itself. Specifically, the secrecy jurisdiction domiciles of much public development finance compromises transparent reporting and makes evaluation of value-for-money challenging (NOU 2009; Bracking et al. 2010). In short, being a fund manager of climate bonds, even when issued nominally by a public institution, can be extremely lucrative. By comparison, many workers and 'project participants' at the site of the investment are very poorly remunerated and adversely incorporated, while their sovereign state may additionally become responsible for paying the loan back if the 'business model' for extracting an income stream from the project itself fails.

Proposition 10: Climate Finance Is Global and Inclusive

Citizens also become entrained in the representational language of climate finance, as 'beneficiaries' who are counted in order to express a figure for the worth and benevolence of the 'donor' financier. These narratives of climate finance and climate products are a ghostly reinvention of development power, where climate finance has inherited, largely intact, the intersectional, race, gender and postcolonial signifiers from within the international development discourse. The Global South is represented as 'lacking' and 'failing' on a number of counts, including in expertise, resources and in the generation of 'bankable projects' and 'governance standards'. By contrast, the MDBs, DFIs and their private partners can 'de-risk', make 'bankable', and insist on 'qualifying governance standards' from their self-assigned positions of expertise. When this binary world connects in an issuance of climate finance, whether it be equity, bond or insurance, the economic outcome is also similar to that generated in the political economy of development: it can be five times as expensive as a commercial loan (Africa Climate Resilience Investment Summit 2021). Of course, access to even these loans is not given to the riskier, often poorer, nations without the handmaiden 'leadership' and imposed governance of the MDGs and DFIs.

Climate finance projects and 'interventions' have thus inherited the same institutions and sometimes people who were the 'experts' in the age of development. This is because the structures of global power and political economy through which climate finance now travels, are inherited from a past that was justified and legitimised through ideas and practices of development expertise, knowledge and power, despite the amazing post-development (Rahnema and Bawtree 1997; Crush 1997; Escobar 1995; Ashish et al. 2019) and postcolonial critiques (Spivak 1988) that punctured development discourse from the late 1980s.

In sum, climate finance is managed within power structures which conditioned, and continue to do so, the political economy of development, through the institutional reproduction of economic inequality and vectors of race, coloniality and patriarchy (Bracking 2009). It might be tempting to see the climate crisis as a wider or bigger 'crisis' than the development crisis, which has arguably become normalised in the eyes of the privileged as an 'everyday' structural violence of poverty and premature death. After all, the climate crisis is an existential planetary crisis of the whole more-than human biosphere. But this might not be helpful as humanity is now facing both—and they are closely connected. Perhaps if the development challenge had been equitably addressed—by changing the foundational structures of power and political economy globally—the newer climate crisis might have been of a different disposition. The relationship between the two crises is complex, but the contributing underlying political economy of capitalism is the same. Also similar, is that the institutional arrangements currently directed toward the climate crisis are those that have already failed us in the development domain, and we can extrapolate that they will do the same again. In short, whatever its effect on climate change, the current arrangements for delivering climate finance mean a forecast of continued inequality, oppression and exclusion.

Climate financiers have replicated and extended the very old game of the development industry, where development, conservation and now climate change are marketised to suit the interests of northern financial institutions. This old game relies on projects with full operating costs recovery where a large proportion of funds are spent on consultancy, planning and management using northern-based firms or DFIs. Overpaid consultants make excessive claims for their own knowledge products while ignoring domestic capacities. Employment is generated in Europe, and the contribution of research money spent in Europe is double counted as Overseas Development Aid—but there is still no relief for the climate-stressed. The financiers make logframes and 'roll out road maps' that reproduce historical inequalities, while simultaneously retreating from the possibilities that new technology could be owned in common and democratised to produce outcomes in favour of the vulnerable. Instead, the application of risk calculation, folded into apparently 'radical' new concepts of 'resilience', 'adaptation', and 'just transitions', financialises nature at an abstract scale in order to provide dividends to people who own money and lend it out.

These concepts are synergistic in style and design to a superstructure of eco-cybernetics, eco-modernism and biopolitics (see Braun 2014). In Europe we hear of sustainability-linked loans (SSL), or performance-based financing (PBF), or the Task Force on Climate-Related Financial Disclosures (TCFD), the Carbon Disclosure Project (CDP) and the Principles for Responsible Investing as if they were revolutionising the future. Changing the behaviour of directors through High Impact Loans with flexible interest rates is still a 'paradigm shift'! The problem is that these initiatives, promoted as the most 'advanced international standards' are not working in Europe, and shouldn't be 'rolled-out at scale'.

These acronyms and other 'inventions', such as blended finance, transition bonds and sustainability bonds, will make up the (non) signifiers, firewalls and black boxes in discussion at the upcoming COP26. But they have very little substance, and certainly no high science. Rolling out metaphorical roads and road maps hides inaction, and even the continued financing of actual roads and fossil-fuel infrastructure.

Proposition 11: Climate Finance Works!

Unlikely.

References

Actionaid, *The Wrong Model for Resilience: How G7-backed Drought Insurance Failed Malawi, and What We Must Learn from It* (Actionaid.org, 2017), https://actionaid.org/sites/default/files/the_wrong_model_for_resilience_final_230517.pdf.

Africa Climate Resilience Investment Summit, Remote Conference, Panel 4, Innovative financing for resilience in the era of COVID-19 and beyond in Africa, 17 June 2021. Quoted by Senior UNECA figue, time tag 15.23 BST.

Andrade, Jose C. S., and José A. Puppim de Oliveira, 'The Role of the Private Sector in Global Climate and Energy Governance', *Journal of Business Ethics*, 130(2) (2015), 375–87, https://doi.org/10.1007/s10551-014-2235-3.

Ashish, K., Salleh, A., Escobar, A., Demaria, F., and Acosta, A., *Pluriverse: A Post-Development Dictionary* (New Delhi: AuthorsUpFront, Tulika Books 2019).

Asiyanbi, Adeniyi P., 'Financialisation in the Green Economy: Material Connections, Markets-in-the-making and Foucauldian Organising Actions', *Environment and Planning A: Economy and Space*, 50(3) (2018), 531–48, https://doi.org/10.1177/0308518X17708787.

Bigger, Patrick, and Nate Millington, 'Getting Soaked? Climate Crisis, Adaptation Finance, and Racialized Austerity', *Environment and Planning E: Nature and Space*, 3(3) (2019), 601–23, https://doi.org/10.1177/2514848619876539.

Bracking, Sarah, *Money and Power: Great Predators in the Political Economy of Development* (London: Pluto, 2009).

Bracking, Sarah, 'How do Investors Value Environmental Harm/care? Private Equity Funds, Development Finance Institutions and the Partial

Financialization of Nature-based Industries', *Development and Change*, 43(1) (2012), 271–93.

Bracking, Sarah, 'The Anti-politics of Climate Finance: The Creation and Performativity of the Green Climate Fund', *Antipode*, 47(2) (2015), 281–302.

Bracking, Sarah, 'Performativity in the Green Economy: how far does climate finance create a fictive economy?', *Third World Quarterly*, 36(12) (2015), 2337–57, https://doi.org/10.1080/01436597.2015.1086263.

Bracking, Sarah, 'Financialisation, Climate Finance, and the Calculative Challenges of Managing Environmental Change', *Antipode*, 51(3) (2019), 709–29, https://doi.org/10.1111/anti.12510.

Bracking, Sarah, David Lawson, Kunal Sen, and Danture Wickramasinghe, *Future Directions for Norwegian Development Finance* (Oslo: Norwegian Agency for Development Cooperation. Official document no: 0902364–55, 2010).

Braun, Bruce P., 'A New Urban Dispositif? Governing Life in an Age of Climate Change', *Environment and Planning D: Society and Space*, 32(1) (2014), 49–64, https://doi.org/10.1068/d4313.

Buchner, Barbara, Alex Clark, Angela Falconer, Rob Macquarie, Chavi Meattle, Rowena Tolentino, and Cooper Wetherbee, *Global Landscape of Climate Finance 2019* (Climatepolicyinitiative.org, 2019), https://www.climatepolicyinitiative.org/publication/global-landscape-of-climate-finance-2019/.

Care International, 'Developed Nations Hugely Exaggerate Climate Adaptation Finance for Global South' (Care-international.org, 2021), https://www.care-international.org/news/press-releases/developed-nations-hugely-exaggerate-climate-adaptation-finance-for-global-south.

Chiapello, Eve, and G. Godefroy, 'The Dual Function of Judgment Devices: Why Does the Plurality of Market Classifications Matter?', *Historical Social Research*, 42(1) (2017), 152–88.

Climate Policy Initiative, *The Global Landscape of Climate Finance: An Update* (2019), https://www.climatepolicyinitiative.org/wp-content/uploads/2020/12/Updated-View-on-the-2019-Global-Landscape-of-Climate-Finance.pdf.

Crush, Jonathan, *The Power of Development* (London: Routledge, 1997).

Erikson, Susan L., and Leigh Johnson, 'Will financial innovation transform pandemic response?', *The Lancet*, 20(May) (2020), 529–30, https://doi.org/10.1016/ S1473–3099(20)30150-X.

Escobar, Arturo, *Encountering Development: the Making and Unmaking of the Third World* (Princeton: Princeton University Press, 1995).

Figueres, Christiana, Schellnhuber, H. J., Whiteman, G., Rockström, J., Hobley, A., and Rahmstorf, S., 'Three Years to Safeguard Our Climate', *Nature*, 546(7660) (2017), 593–95, https://doi.org/10.1038/546593a.

Green Climate Fund Funding Proposal, *FP140: High Impact Programme for the Corporate Sector* (Greenclimate.fund, 2020), https://www.greenclimate. fund/sites/default/files/document/funding-proposal-fp140.pdf.

Horton, Joshua. B., 'Parametric Insurance as an Alternative to Liability for Compensating Climate Harms', *Carbon and Climate Law Review*, 12(4) (2018), 285–96, https://doi.org/10.21552/cclr/2018/4/4.

InsuResilience, *InsuResilience: Solutions Fund Annual Report 2020* (Isf-annual-report_final-web, 2020), isf-annual-report_final-web.pdf.

Jones, Ryan, Baker, T., Huet, K., Murphy, L., and Lewis, N., 'Treating Ecological Deficit with Debt: The Practical and Political Concerns with Green Bonds', *Geoforum*, 114 (2020), 49–58, https://doi.org/10.1016/j.geoforum.2020.05.014.

Kuyper, Jonathan W., Linnér, Björn-Ola, and Schroeder, Heike, 'Non-state Actors in Hybrid Global Climate Governance: Justice, Legitimacy, and Effectiveness in a Post-Paris Era', *Wiley Interdisciplinary Reviews: Climate Change*, 9(1) (2018), e497, https://doi.org/10.1002/wcc.497.

Lave, Rebecca, 'Not so Neo. Reflecting on Neoliberal Natures: An Exchange', *Environment and Planning E: Nature and Space*, 1(1–2) (2018), 25–75, https://doi.org/10.1177/2514848618776864.

Leins, Stefan, 'Narrative Authority: Rethinking Speculation and the Construction of Economic Expertise' *Ethnos*, 20 May 2020, https://doi.org/10.1080/00141 844.2020.1765832.

Lohmann, Larry, 'Interpretation Machines: Contradictions of 'Artificial Intelligence' in 21st-Century Capitalism', *Socialist Register*, 57(2020), https://socialistregister.com/index.php/srv/article/view/34947.

Mitchell, Katharyne, and Matthew Sparke, 'The New Washington Consensus: Millennial Philanthropy and the Making of Global Market Subjects', *Antipode*, 48(3) (2016), 724–49, https://doi.org/10.1111/anti.12203.

Munden Project, *REDD and Forest Carbon: Market-Based Critique and Recommendations* (The Munden Project, 2011), http://www.redd-monitor. org/wp-content/uploads/2011/03/Munden-Project-2011-REDD-AND-FOREST- CARBON-A-Critique-by-the-Market.pdf.

Norwegian Official Report (NOU), Government Commission, Norway, *Tax Havens and Development: Status, Analyses and Measures* (Oslo: Official Norwegian Reports, no 19, 2009).

OECD, *Investing in Climate, Investing in Growth* (OECD, 2017), https://www.oecd.org/environment/investing-in-climate-investing-in-growth-9789264273528-en.htm.

OECD, *DAC methodologies for measuring the amounts mobilised from the private sector by official development finance interventions* (OECD, 2020a), https://www.oecd.org/dac/financing-sustainable-development/development-finance-standards/DAC-Methodologies-on-Mobilisation.pdf.

OECD, *OECD DAC Blended Finance Principle 2: Guidance* (OECD, 2020b), http://www.oecd.org/dac/financing-sustainable-development/blended-finance-principles/principle-2/Principle_2_Guidance_Note_and_Background.pdf.

OECD, *Climate Finance Provided and Mobilised by Developed Countries in 2013-18: Climate Finance and the USD 100 Billion Goal* (Paris: OECD Publishing, 2020c), https://www.oecd.org/environment/climate-finance-provided-and-mobilised-by-developed-countries-in-2013-18-f0773d55-en.htm.

Pattberg, Philipp, and Oscar Widerberg, 'Theorising Global Environmental Governance: Key Findings and Future Questions', *Millennium*, 43(2) (2015), 684–705, https://doi.org/10.1177/0305829814561773.

Pogge, Thomas, *World Poverty and Human Rights: Cosmopolitan Responsibilities and Reforms* 2nd edn. (Cambridge: Polity, 2007).

Rahnema, Majid, and Rebecca Bawtree, *The Post-Development Reader* (London: Zedbooks, 1997).

Sörqvist, Patrik, Haga, Andreas, Langeborg, Linda, et al., 'The Green Halo: Mechanisms and Limits of the Eco-label Effect', *Food Quality and Preference*, 43 (2015), 1–9, https://doi.org/10.1016/j.foodqual.2015.02.001.

Spivak, Gayatri Chakravorty, 'Can the Sabaltern Speak?' *Die Philosophin*, 14 (27) (1988), 42–58.

Sullivan, Sian, *Financialisation, Biodiversity Conservation and Equity: Some Currents and Concerns* (Penang Malaysia: Third World Network Environment and Development Series 16, 2012), http://twn.my/title/end/pdf/end16.pdf.

Sullivan, Sian, 'Banking Nature? The Spectacular Financialisation of Environmental Conservation', *Antipode*, 45(1) (2013), 198–217.

Sullivan, Sian, 'Wild game or soul mates? On humanist naturalism and animist socialism in composing socionatural abundance', Paper for the conference Landscape, Wilderness and the Wild, Newcastle University, 26–29 March 2015, https://siansullivan.files.wordpress.com/2011/09/wild-game-or-soul-mates-sullivan-310315.pdf.

Sullivan, Sian, 'What's Ontology Got to do With It? On Nature and Knowledge in a Political Ecology of the "green economy"', *Journal of Political Ecology*, 24(1) (2017), 217–42, https://doi.org/10.2458/v24i1.20802.

Sullivan, Sian, 'Making Nature Investable: From Legibility to Leverageability in Fabricating "Nature" as "Natural Capital"', *Science and Technology Studies*, 31(3) (2018a), 47–76, https://doi.org/10.23987/sts.58040.

Sullivan, Sian, 'Bonding nature(s)? Funds, Financiers and Values at the Impact Investing Edge in Environmental Conservation, in *Valuing Development, Environment and Conservation: Creating Values that Matter*, ed. by Sarah Bracking, Aurora Fredriksen, Sian Sullivan, and Philip Woodhouse (London: Routledge, 2018b), pp. 101–21.

Surminski, Swenja, and Panda, Architesh, 'Disaster Insurance in Developing Asia: An Analysis of Market-Based Schemes', *ADB Economics Working Paper Series* (2020), http://dx.doi.org/10.2139/ssrn.3644910.

Taylor, Zac J., 'The Real Estate Risk Fix: Residential Insurance-linked Securitization in the Florida Metropolis', *Environment and Planning A: Economy and Space*, 52(6) (2020), 1131–49, https://doi.org/10.1177/0308518X19896579.

UNCTAD—United Nations Conference on Trade and Development, *World Investment Report 2014: Investing in the SDGs* (New York and Geneva: United Nations Conference on Trade and Development, 2014), https://unctad.org/system/files/official-document/wir2014_en.pdf.

UNCTAD, *Trade and Development Report: Financing a Green New Deal* (Geneva: UNCTAD, 2019), https://unctad.org/webflyer/trade-and-development-report-2019.

UNCTAD, *World Investment Report 2020: International Production Beyond the Pandemic.* (Geneva: UNCTAD, 2020), https://unctad.org/system/files/official-document/wir2020_en.pdf.

UNFCCC, *What Is Climate Finance?* (Unfccc.int, 2019), https://unfccc.int/topics/climate-finance/the-big-picture/introduction-to-climate-finance.

World Bank, *From Billions to Trillions: Transforming Development Finance. Document Prepared Jointly by the AfDB, ADB, EBRD, EIB, IADB, IMF and World Bank Group for the 18 April Development Committee Meeting* (Washington, DC: World Bank, 2015), https://thedocs.worldbank.org/en/doc/622841485963735448-0270022017/original/DC20150002EFinancingforDevelopment.pdf.

World Bank, *World Bank Launches First-Ever Pandemic Bonds to Support $500 Million Pandemic Emergency Financing Facility* (Worldbank.org, 2017), https://www.worldbank.org/en/news/press-release/2017/06/28/world-bank-launches-first-ever-pandemic-bonds-to-support-500-million-pandemic-emergency-financing-facility.

21. The Promise and Peril of Financialised Climate Governance

Rami Kaplan and David Levy

A recent development in climate governance has been the rise of investor-driven, or 'financialised governance' of corporate practices in relation to the natural environment. Investors and investment managers are demonstrating greater concern that the value of assets, from stock markets to real estate, are increasingly subject to climate risks. Financialised climate governance (FCG) puts investors and fund managers at the centre of efforts to limit greenhouse gas emissions, which suggests both the promise and peril of this advanced form of 'climate capitalism'. We describe these developments and point towards the peril that relying on investors and business self-interest is unlikely to result in the rapid structural shifts needed for full decarbonisation.

The Rise of Financialised Climate Governance

A notable recent development in climate governance has been the rise of investor-driven, or 'financialised governance' of corporate practices in relation to the natural environment (as also invoked by Bracking, this volume). Investors and investment managers are demonstrating greater concern that the value of assets, from stock markets to real estate, are increasingly subject to climate risks. These include physical risks from rising sea levels, storms, wildfires, and disease, together with financial risks, such as the loss of 'stranded assets' and product obsolescence, due to technological and regulatory changes, which are inducing a rapid shift toward renewable energy and other low-carbon products and processes.

https://doi.org/10.11647/OBP.0265.21

In January 2020, Larry Fink, CEO of BlackRock, the largest private investment company in the world with more than $8 trillion in assets under management, warned that "Climate change is different. Even if only a fraction of the projected impacts is realized, this is a [...] structural, long-term crisis. Companies, investors, and governments must prepare for a significant reallocation of capital."[1] In an even sharper letter in early 2021, Fink urged CEOs to take the COVID-19 pandemic as "a stark reminder of our fragility" and warned that companies that fail to quickly prepare for the net zero transition "will see their businesses and valuations suffer"[2] (also see Böhm and Sullivan, this volume).

Alongside this rhetoric, BlackRock joined Climate Action 100+, a rapidly growing consortium of more than 500 asset owners and managers with over $50 trillion under management. The initiative's strategy is to promote the greenhouse gases (GHG) reduction goals of COP21's Paris Agreement by leveraging the financial power of signatory investors into reforming the practices of 160 corporate "systemically important emitters" that account for two-thirds of global industrial emissions.[3] A hub of investor activism, Climate Action 100+ employs tactics ranging from formal appeals to boards, to filing shareholder resolutions, and action to remove uncooperative directors. The initiative claims to have already triggered a wave of commitments to adopt advanced disclosure standards and carbon reduction targets (Herd and Hillis 2019; Mooney 2020). For example, British Petroleum has committed to cut its fossil fuel production by 40% by 2030 and substantially increase its investment in renewable energy and electric transportation (British Petroleum 2020). Shell has declared its "ambition" to halve its carbon footprint by 2050 and stated that it will soon link executives' pay to short-term carbon goals. Many other major companies have committed to achieve net zero emissions by 2050 (the complexities of which are traced by Dyke et al. and Bailey, this volume), and to move to 100% renewable energy (see also Wright and Nyberg, this volume).

Financialised climate governance (FCG) puts investors and fund managers at the centre of efforts to limit GHG emissions, which suggests both the promise and peril of this advanced form of "climate capitalism"

---

1   https://www.ft.com/content/57db9dc2-3690-11ea-a6d3-9a26f8c3cba4.
2   https://www.blackrock.com/us/individual/2021-larry-fink-ceo-letter.
3   https://www.climateaction100.org.

(Newell and Paterson 2010). The promise lies in the centrality of financial mechanisms within capitalism; if climate indeed enters calculations of risks, returns, and asset pricing (Sullivan 2018), then FCG could have considerable leverage over corporate practices and strategies. Investors would be a major force in the low-carbon transition; operating with existing mechanisms and ideologies of corporate governance and shareholder value, FCG could be more effective than pressure from stakeholders or governmental and multilateral action. The peril is that relying on investors and business self-interest is unlikely to result in the rapid structural shifts needed for full decarbonisation, which will not always be profitable for individual companies and will require regulation to shape markets and large-scale government funding for new infrastructure. Moreover, relying on FCG shifts the balance of power in climate governance away from environmental activists and governmental agencies, with potentially dire long-term consequences.

The nexus between the financial world and climate change is not new. Funds specialising in 'socially responsible investment' (SRI) have proliferated since the 1990s, in parallel to the emergence of disclosure-based governance frameworks, such as certification schemes and sustainability disclosure initiatives (Bartley 2007, Levy et al. 2010; Depoers et al. 2016). From the 2000s, SRI and disclosure governance intersected around the emergence of 'environmental, social, and governance' (ESG) indices designed to inform investment decisions. According to several estimates, global assets under management integrating ESG considerations multiplied from roughly $10 trillion in 2010 to $40 trillion in 2020, which is close to half of the world's total assets under management (Social Investment Forum Foundation 2010; Basar 2020). The increasing concentration of the asset management industry—the top ten asset managers hold 34% of externally managed assets (Eccles and Klimenko 2019)—implies substantial pressure on corporate emitters. This concentration increases the leverage of activist consortia such as Climate Action.

Initiatives such as the Climate Disclosure Project and Ceres' Investor Network on Climate Risk explicitly sought to leverage investor pressure to change corporate practices (Knox-Hayes and Levy 2011). However, these were widely perceived as activist rather than investor-led projects and hence had little impact on capital flows or corporate emissions. The phenomenon of FCG is fundamentally different in that it represents a growing recognition of climate risks by investors and the mobilisation of the capitalist class more broadly, rather than just in response to external pressure. The original 'values-based SRI' has been displaced by 'profit-seeking SRI,' which asserts that ESG investment is more profitable. ESG-specialised investment management firms, indices, and professional associations have proliferated worldwide, and ESG strategies have diffused rapidly among general-purpose investment funds (Waddock 2008; Meyer et al. 2015; Yan et al. 2019).

The mobilisation by elite organisational investors has been global in scope and coordinated with governmental and multilateral organisations. The Asset Management Working Group, representing a dozen major investors organised by the UN Environment Programme Finance Initiative, pioneered the development and diffusion of ESG standards worldwide (UNEP-FI 2004, Asset Management Working Group 2009). Another key vector has been Bloomberg's Task Force on Climate-Related Financial Disclosures (TCFD), which was launched in 2016 by the Financial Stability Board, a coordinating body of national financial bodies and international standards organisations. The TCFD has legitimised and disseminated standardised climate risk management and disclosure internationally. Recently, the Big Four global accounting firms unveiled a unified reporting framework for ESG.

Investors are increasingly engaging in shareholder activism to pressure companies over climate change. For example, a coalition of seven Climate Action 100+ members narrowly passed a shareholder resolution in 2019 at Chevron, against management's opposition, to require the company to report on its climate lobbying expenditures and their alignment with Paris goals. A similar resolution was passed in May 2021 at the annual shareholder meeting of Phillips66, while a resolution passed the same month at the ConocoPhillips' shareholder meeting called for the company to set Scope 3 emission reduction targets, in other words, to take responsibility for the consumption of oil downstream.[4] The most surprising upset of 2021 was the successful effort by a relatively small activist hedge fund, Engine No. 1, to nominate and elect three new directors on to Exxon's twelve-person board. The

---

4   https://www.ceres.org/news-center/press-releases/
historic-votes-shareholders-demand-strong-climate-action-us-oil-and-gas.

hedge fund only held a 0.02% stake in Exxon but succeeded in winning the support of large state pension funds.

While some of these shareholder resolutions are non-binding, such open conflict between capitalist investors and fossil fuel companies is unprecedented and constitutes a marked shift from the prior use of shareholder activism by labour or church groups. The investor activists have claimed that corporate lobbying threatens governments' commitment to the Paris goals, which in turn threatens economic stability (BNP Paribas et al. 2019). This approach breaks strikingly from the traditional corporate preference for voluntarism (cf. Kaplan and Kinderman 2019, 2020; Kaplan and Lohmeyer 2020) and acknowledges a governmental role in addressing systemic financial and economic risks of climate disruption.[5] The activists also argue that the target companies need more visionary leadership to develop the comprehensive and far-reaching strategies required to survive and prosper in the low-carbon future.

Another remarkable development is the contestation around the status of ESG as a legitimate risk management criterion. In its final year, the Trump administration moved to restrict the use of ESG criteria in pension plans by requiring proof that ESG enhances profitability, and the investor community mobilised against this (Umpierrez 2020; Quinson 2020). The administration's action apparently responded to pressure from the fossil fuels sector, which was concerned about carbon divestment campaigns amongst activists and organisational investors (Quinson 2020). The contestation between the Trump administration and the asset management industry was remarkable because it centred on questions of shareholder value and risk calculation rather than the environmental and social impact of corporations. The Trump administration argued that ESG-informed investment reflected non-financial objectives and thus violated the fiduciary obligation of money managers; investment managers countered that ESG risk was fundamental to evaluating the long-term performance of investments. The Biden administration has since announced that it will not enforce the Trump rules restricting retirement investments and will revisit the issue. These developments express how the struggles over climate change are reframed and

---

5   https://www.ft.com/content/e6ad62f2-a9f3-4aec-b359-b662a07f5d01.

translated into the financial terrain, and the growing commitment of investment managers to ESG-informed financial strategies.

Critics of FCG will be quick to point to the historical failures of corporate self-regulation and the constraints on managerial action operating within profit maximising firms and the discipline of capital markets. FCG is unlikely to drive the structural and systemic changes in lifestyles and values, as well as production and consumption, that are urgently needed. Fundamentally, critics emphasise the contradictions inherent in expecting the stewards of capitalism to fix problems that arise from the system itself.

First, the financial interests of investors are not fully aligned with those of society, and this is clearly the case for climate change. While some investors and financial regulators are waking up to the systemic financial risk from climate change, action by individual companies is constrained by the large externalities associated with fossil fuels and the problems of collective action and free riding. At the firm level, climate change is often viewed as a long-term and rather abstract risk, especially if they do not face a substantial price on carbon emissions. In other words, 'win-win' climate opportunities can be more elusive than advocates sometimes claim. Companies can find profitable low-hanging fruit in areas such as energy efficiency and improving logistics, but moving towards 80% reductions or carbon neutrality is far more difficult, requiring a major structural shift in products and production processes, or a reliance on dubious carbon-offsets (Böhm and Dabhi 2009).

Moreover, the companies that will flourish in a zero-carbon economy are unlikely to be the same as those who will lose out—coal, oil, and gas companies have not fared well in clean energy and are likely to be replaced by those specialising in wind, solar, geothermal and energy storage. Traditional automobile companies will find it hard to compete with upstarts like Tesla that focus on advanced batteries and software. From a strategy perspective, the new low-carbon businesses have very different technologies, business models, and required competencies, making it difficult for incumbents to make the transition. A senior portfolio manager at Adams Fund, an investment company focused on the energy sector, commented after the successful activist campaign to appoint three new directors to Exxon's board that "[p]eople who are expecting substantive changes soon will likely be sorely disappointed

[...]. Repositioning XOM from a company focused on oil to one focused on climate change issues will take a long, long time."[6]

A second major limitation of FCG is that, in common with corporate social responsibility (CSR) and other sustainability efforts, it is open to 'greenwash', the disjuncture between corporate efforts to burnish their environmental reputation and actual outcomes (Berliner and Prakash 2015; Raghunandan and Rajgopal 2020). Institutional theorists refer to 'decoupling' along the implementation chain between public pronouncements, internal policies and targets, corporate practices, and actual emissions (Lyon and Montgomery 2013). It is true that FCG, as 'insider' corporate governance that demands more rigorous corporate disclosure of climate metrics, likely provides more credible verification of corporate practices than NGO-led initiatives such as the Global Reporting Initiative (GRI) or CDP. But companies can also game ESG reporting to satisfy external stakeholders. While investors themselves gain reputational value from signing on to initiatives such as Climate Action 100+, they do not have an incentive to press companies for emission cuts that are unprofitable, require reduction in sales, or even threaten continued viability. This may result in the institutionalisation of "organized hypocrisy" (Lim and Tsutsui 2012) that involve 'ceremonies' of corporate disclosure that are legitimised by investors, standard-setters and auditors. In one recent instance, Climate Action 100+ and Total's management issued a joint statement promoting a modest sustainability policy, which preempted a more aggressive resolution advanced by proxy activists (Mooney 2020). Indeed, it was the perception of such hypocrisy that helped drive the recent shareholder resolutions at Exxon and Shell.

A third source of caution regarding the potential of FCG is that it is incompatible with a transition to an economy and value system based on "sustainable lifestyles" (Levy and Spicer 2013) (as also highlighted in the chapters by Halme et al., North, Paterson and Sandover, this volume). Movements for sustainable consumption, localism, and more recently 'slowness' (Van Bommel and Spicer 2011) have been growing in recent years, inspired by visions of a simpler, less materialistic life that is more oriented toward leisure and community. It also envisages alternative

---

6   h t t p s : / / w w w . r e u t e r s . c o m / b u s i n e s s / e n e r g y / engine-no-1-win-third-seat-exxon-board-based-preliminary-results-2021-06-02/.

economic structures and market forms based on small-scale production, co-ops, widespread sharing and re-use of assets, and community-based services (Schor and White 2010). According to Jackson (2011: 35),

> [t]he prevailing vision of prosperity as a continually expanding economic paradise has come unraveled [...]. This chapter searches for a different kind of vision for prosperity: one in which it is possible for human beings to flourish, to achieve greater social cohesion, to find higher levels of well-being and yet still to reduce their material impact on the environment.

Such a radical transformation cannot easily be reconciled with investor demands for exponential economic growth and rising profits.

The fourth and final concern is that FCG shifts the balance of power in climate governance toward business and investors and away from environmental NGOs, activists, governments, and multilateral agencies. It is a continuation of the trend toward the privatisation of governance and self-regulation, with little inclusiveness or accountability (Bartley 2007; Levy and Kaplan 2008; Levy et al. 2010). Corporations have often pushed for self-regulation as a means to deflect external pressure, pre-empt governmental intervention (Malhotra et al. 2019), and increase business control over the political environment (Levy 1997; Sapinski 2015; Kaplan and Kinderman 2019; Kaplan and Lohmeyer 2020). The rise of FCG can be understood as an accommodation with the external pressures and financial risks of climate change but one that reaffirms the hegemony of capitalism and traditional modes of corporate governance by reasserting the confluence of corporate, investor and societal interests.

In conclusion, while the rise of FCG signals the mainstreaming of climate concerns in the business and investor communities, it also holds profound limitations that constrain its effectiveness in achieving the rapid transition to a low-carbon economy that is urgently needed. As Levy et al. (2016) observed in relation to the corporatisation of CSR, the paradox of FCG is that, while it accelerates incremental change in corporate practices, its inherent limitations will prevent the deeper systemic and structural shifts required in norms, corporate forms and governance, and patterns of production and consumption.

References

Asset Management Working Group, *The Materiality of Climate Change: How Finance Copes with the Ticking Clock* (New York: United Nations Environment Programme Finance Initiative, 2009).

Bartley, Tim, 'Institutional Emergence in an Era of Globalization: The Rise of Transnational Private Regulation of Labor and Environmental Conditions', *American Journal of Sociology*, 113(2) (2007), 297–351, https://doi.org/10.1086/518871.

Basar, Shanny, 'ESG Assets Have Grown 15% Annually' (Marketsmedia.org, 2020), https://www.marketsmedia.com/esg-assets-have-grown-15-annually/.

Berliner, Daniel, and Aseem Prakash, '"Bluewashing" the Firm? Voluntary Regulations, Program Design, and Member Compliance with the United Nations Global Compact', *Policy Studies Journal*, 43 (2015), 115–38, https://doi.org/10.1111/psj.12085.

BNP Paribas, Calpers, Clastrs, and others, 'Letter to CEOs on climate lobbying disclosure' (Paris: BNP Paribas, 2019).

Böhm, Steffen, and Siddhartha Dabhi, *Upsetting the Offset: The Political Economy of Carbon Markets* (London: MayFlyBooks, 2009), http://mayflybooks.org/?p=206.

British Petroleum, 'From International Oil Company to Integrated Energy Company: BP sets out strategy for decade of delivery towards net zero ambition' (British Petroleum, 2020), https://www.bp.com/en/global/corporate/news-and-insights/press-releases/from-international-oil-company-to-integrated-energy-company-bp-sets-out-strategy-for-decade-of-delivery-towards-net-zero-ambition.html.

Depoers, Florence, Thomas Jeanjean, and Tiphaine Jérôme, 'Voluntary Disclosure of Greenhouse Gas Emissions: Contrasting the Carbon Disclosure Project and Corporate Reports', *Journal of Business Ethics*, 134(3) (2016), 445–61, https://doi.org/10.1007/s10551-014-2432-0.

Eccles, Robert, and Svetlana Klimenko, 'The Investor Revolution', *Harvard Business Review*, 97(3) (2019), 106–16, https://hbr.org/2019/05/the-investor-revolution.

Herd, Emma, and Laura Hillis, *Climate Action 100+: Progress Report* (Climate Action 100+: 2019), https://www.climateaction100.org/wp-content/uploads/2020/10/English-Progress-Report-2019.pdf.

Jackson, Tim, *Prosperity without Growth: Economics for a Finite Planet* (New York: Routledge, 2011).

Kaplan, Rami, and Daniel Kinderman, 'The Business-class Case for Corporate Social Responsibility: Mobilization, Diffusion, and Institutionally

Transformative Strategy in Venezuela and Britain', *Theory & Society*, 48(1) (2019), 131–66, https://doi.org/10.1007/s11186-019-09340-w.

Kaplan, Rami, and Daniel Kinderman, 'The Business-led Globalization of CSR: Channels of Diffusion from the U.S. into Venezuela and Britain, 1962–1981', *Business & Society*, 59(3) (2020), 439–88, https://doi.org/10.1177/0007650317717958.

Kaplan, Rami, and Nora Lohmeyer, 'A Comparative Capitalism Approach to the Privatization of Governance: Business Power, Nonbusiness Resistance, and State Enforcement in Germany, 2000–2010', *Socio-Economic Review* (2020) mwaa001, https://doi.org/10.1093/ser/mwaa001.

Knox-Hayes, Janelle, and David L. Levy, 'The politics of carbon disclosure as climate governance', *Strategic Organization*, 9(1) (2011), 91–99, https://doi.org/10.1177/1476127010395066.

Levy, David L., 'Environmental Management as Political sustainability', *Organization and Environment*, 10(2) (1997), 126, https://doi.org/10.1177/0921810697102002.

Levy, David L., Halina S. Brown, and Martin de Jong, 'The Contested Politics of Corporate Governance: The Case of the Global Reporting Initiative', *Business and Society*, 49(1) (2010), 88–115, https://doi.org/10.1177/0007650309345420.

Levy, David L., and Rami Kaplan, 'CSR as Global Governance: Strategic Contestation in Global Issue Arenas', in *The Oxford Handbook of Corporate Social Responsibility*, ed. by Andrew Crane, Abagail McWilliams, Dirk Matten, Jeremy Moon, and Donald S. Siegel (Oxford: Oxford University Press, 2008), pp. 432–51 https://.doi.org/10.1093/oxfordhb/9780199211593.003.0019.

Levy, David L., Juliane Reinecke, and Stephan Manning, 'The Political Dynamics of Sustainable Coffee: Contested Value Regimes and the Transformation of Sustainability', *Journal of Management Studies*, 53(3) (2016), 364–401, https://doi.org/10.1111/joms.12144.

Levy, David L., and André Spicer, 'Contested Imaginaries and the Cultural Political Economy of Climate Change', *Organization* (20) (2013), 659–78, https://doi.org/10.1177/1350508413489816.

Lim, Alwyn, and Kiyoteru Tsutsui, 'Globalization and Commitment in Corporate Social Responsibility: Cross-National Analyses of Institutional and Political-Economy Effects', *American Sociological Review*, 77(1) (2012), 69–98, https://doi.org/10.1177/0003122411432701.

Lyon, Thomas, and Wren Montgomery, 'Tweetjacked: The Impact of Social Media on Corporate Greenwash', *Journal of Business Ethics*, 118 (2013), 747–57, https://doi.org/10.1007/s10551-013-1958-x.

Malhotra, Neil., Benoît Monin, and Michael Tomz, 'Does Private Regulation Preempt Public Regulation?', *American Political Science Review*, 113(1) (2019), 19–37, https://doi.org/10.1017/S0003055418000679.

Meyer, John W., Shawn Pope, and Andrew Isaacson, 'Legitimating the Transnational Corporation in a Stateless World Society', in *Corporate Social Responsibility in a Globalizing World*, ed. by Kiyoteru Tsutsui and Alwyn Lim (Cambridge: Cambridge University Press, 2015), pp. 27–72.

Mooney, Attracta, 'Corporate Eco-warriors Driving Change from Shell to Qantas' (Ft.com, 2020), https://www.ft.com/content/2db23ad7-da5c-4f1e-a100-a77a72226587.

Newell, Peter, and Matthew Paterson, '*Climate Capitalism: Global Warming and the Transformation of the Global Economy*' (Cambridge: Cambridge University Press, 2010).

Quinson, Tim, 'Fidelity, BlackRock reject Trump limits on 401(k) ESG investing' (Bloomberg.com, 2020), https://www.bloomberg.com/news/articles/2020-08-31/trump-plan-to-limit-esg-investing-by-401-k-s-opposed-by-funds.

Raghunandan, Aneesh, and Shivaram Rajgopal, 'Do Socially Responsible Firms Walk the Talk?', *SSRN* (2020), https://ssrn.com/abstract=3609056 or http://dx.doi.org/10.2139/ssrn.3609056.

Sapinski, Jean Philippe, 'Climate Capitalism and the Global Corporate Elite Network', *Environmental Sociology*, 1(4) (2015), 268–79, https://doi.org/10.1080/23251042.2015.1111490.

Schor, Juliet, and Karen E. White, *Plenitude: The New Economics of True Wealth* (London: Penguin Press, 2010).

Social Investment Forum Foundation, *Socially Responsible Investing Trends in the United States* (Washington, 2010), https://www.ussif.org/files/Publications/10_Trends_Exec_Summary.pdf.

Sullivan, Sian, 'Funds, Financiers and Values at the Impact Investing Edge in Environmental Conservation', in *Valuing Development, Environment and Conservation: Creating Values that Matter*, ed. by Sarah Bracking, Aurora Fredriksen, Sian Sullivan, and Philip Woodhouse (Abingdon: Routledge, 2018), pp. 101–21.

Umpierrez, Amanda, 'The Future for ESG Investing in Retirement Plans' (Plansponsor.com, 2020), https://www.plansponsor.com/in-depth/future-esg-investing-retirement-plans/.

UNEP-FI, 'The Asset Management Working Group: What, Why, Who?' (United Nations Environment Programme Finance Initiative, 2004), https://www.unepfi.org/fileadmin/documents/amwg_what_why_who_2004.pdf.

Van Bommel, Koen, and André Spicer, 'Hail the Snail: Hegemonic Struggles in the Slow Food Movement', *Organization Studies*, 32 (2011), 1717–44, https://doi.org/10.1177/0170840611425722.

Waddock, Sandra, 'Building a New Institutional Infrastructure for Corporate Responsibility', *Academy of Management Perspectives*, 22(3) (2008), 87–108, https://doi.org/10.5465/amp.2008.34587997.

Yan, Shipeng, Fabrizio Ferraro, and Juan Almandoz, 'The Rise of Socially Responsible Investment Funds: The Paradoxical Role of the Financial Logic', *Administrative Science Quarterly*, 64(2) (2019), 466–501, https://doi.org/10.1177/0001839218773324.

VII

ACTION(S)

22. What Is to Be Done to Save the Planet?

Peter North

This chapter uses the opportunity of the COP to take stock of the successes and failures of climate activism over the past decade. The COPs provide an opportunity for activists to meet, pressure COP delegates to take the action needed to avoid climate action, and discuss what a better world can look like. They can 'take stock' at a point in time about what they have done well, what did not work so well, and what still needs to be done. The chapter reviews mass and 'elite' communicative forms of direct action, and the longer-term programme of building community-based prefigurations of what could be. It argues that this taking stock and pressuring elites to act matters, but is not an alternative to building locally to transition to a world in which all, human and non-human, can live well.

Introduction

The "great acceleration" (McNeill and Engelke 2014) grows apace. Climate catastrophes intensify in the form of seemingly inexorable temperature and sea level rise, species extinction, ice sheet melting, and methane emission. Over the last couple of decades a wide-ranging set of social movements have emerged in a number of places globally to grapple with the politics of climate change, using a range of protest techniques, and with different conceptualisations about what to do. While to some extent put on pause as a result of COVID-19, this climate activism is a diverse space within which organisations, networks and

https://doi.org/10.11647/OBP.0265.22

activists act independently, coalesce, act together, disperse again, and emerge somewhere else later. Sometimes they organise in the streets— the classic protest march aimed at communicating a message to mass society and putting pressure on elites to act. Direct action is carried out by long-standing groups like Greenpeace or Earth First!, by anti-airport protesters such as 'Plane Stupid', by anti-coal protesters like 'Leave it in the ground' or anti-fracking groups, and more recently by Extinction Rebellion (XR). A third strategy, complementary to protesting 'against' catastrophic climate change, is that of community-based 'Transition Initiatives' that work at a grassroots level to develop fulfilling livelihoods based in more localised low-carbon economies (as also pointed towards by Sandover, this volume). They have created their own local currencies, local power and food initiatives and the like in an effort to prefigure the kind of low-carbon, localised and convivial economy they would like to see if dangerous climate change is to be avoided.

That many climate activists seem stereotypically 'middle class' means that the movement has its critics (as Gardham discusses, this volume). In contrast, I argue that globally-privileged citizens in high income, developed northern countries engaging with the geographies of their responsibility for the emissions that lead to anthropogenic climate change are to be applauded. There is nothing new about 'middle class radicalism'. What matters is how well the movement is doing, given the severity of the existential crisis humanity faces. This chapter aims to address this issue. While many of the examples below are based on what I know about activism in the UK, I hope my comments will be of wider interest to those with their eyes on the COP in Glasgow.

It Can All Come Together at the COPs

In their intensity and urgency, claims about the climate crisis echo concerns about the catastrophic nature and imminence of nuclear war in the early 1980s. Yet, while anti-nuclear, anti-war and anti-globalisation movements regularly mobilised upwards of 250,000 protesters, the numbers of protesters taking part in climate action marches, led by school and university students and XR, have not been at a level necessary to force the changes that the protesters (and I) feel are needed. The annual Conference of the Parties (COP) meetings provide a useful place and time to address that. At the COPs, a generally fissiparous 'movement' or series of 'convergence spaces' (Routledge 2003) join together or converge to reinforce and underline the existence of the existential threat of the climate in a world of competing issues for contestation.

The COPs enable climate activists to demand "meaningful, co-ordinated and urgent policy action" commensurate with the threat (Chatterton et al. 2012), take stock, meet like-minded people, discuss alternatives, and plan action. They can point to unequal geographies of responsibility for historic and contemporary emissions and environmental destruction, expose global inequalities and capacities to act in the face of this existential threat, demand global climate justice, and express solidarity. They enable local activists to focus on an issue of particular salience for them, for instance coal in Poland at the Katowice COP in 2019. They create a space where activists can lobby states, and spaces where corporate and business elites showcase technological solutions in line with neoliberal conceptions of how to live well (or cover up their nefarious activities, depending on how anti-capitalist or paranoid you are).

More resistant conceptions of how to live well in the Anthropocene are developed in the sometimes hidden, sometimes open autonomous Alternative Climate Forums, which act as spaces in which new knowledges (Melucci 1989) or grassroots innovations (Seyfang and Smith 2007) develop. The streets can be spaces for demonstrations where change can be demanded. Some activists believe that when they are, sometimes pre-emptively, attacked by local police forces this exposes the hidden violence of the seemingly liberal, democratic state supposedly committed to solving the climate crisis through a rhetorical commitment to the Sustainable Development Goals (SDGs). Given that the SDGs are simultaneously utopian yet insufficiently concrete, a rhetorical commitment to them is at best a cruel hoax, at worst a cover for slow violence, or even social murder, that the failure to avoid climate catastrophe represents.

Taking Stock

Activity at the COPs does not spring from nowhere—they provide a space in which this movement can emerge and converge, building on what has gone before. Social movement theorists Turner and Killian (1987) point to the emergence of new norms, timeliness and feasibility that help us move from a feeling that something is wrong to 'yes, we can' do something about it. This helps explain why an issue emerges in the first place, and then how it can be made to stay on the agenda. The COPs provide an opportunity to mark a time to restate a problem and come to a view about what is being done about it. It might therefore be useful at such a point to review what we know about how climate change has been contested, and how it has moved up and down agendas in competition with other issues, given that we live in a less than perfect world.

While climate change as an issue has been known in scientific circles for many years, a perception of its urgency, a feeling that 'something is wrong', emerged as the scale of problems associated with climate change began to be discerned in the early part of the twenty-first century. Global long-series temperature readings rose inexorably, culminating in a series of 'hottest ever' years and observable extreme weather events from the mass deaths from heatstroke in Europe (2003), Hurricane Katrina (2005) and Cyclone Nargis (2008), forest fires in Greece and California (2009), not to mention longer-lasting droughts in Sub-Saharan Africa and Australia. Al Gore's (2006) documentary *An Inconvenient Truth* communicated the issue to a wider audience, while the contemporaneous publication of IPCC's fourth report and the Stern Review, both in 2007, showed that global warming was accepted as happening by the overwhelming majority of climate scientists, and that something should be done. Climate activists used extreme weather events to suggest that global warming represented a clear danger to life itself. Mark Lynas's book *Six Degrees* (2007) constructed activist knowledge about what abstract phenomena like increasing global temperatures or atmospheric CO_2 levels mean in concrete, and increasingly apocalyptic, terms. Hot weather and extreme weather events suggested that 'something was wrong', and marches, spaces of grassroots innovation and direct action suggested that 'something could be done' and that action was 'timely' and 'feasible'.

Just as hot weather suggested that the planet was warming, a period of cold weather hit the northern hemisphere mid-latitudes during early 2010 and the years after were cooler. Newly confident climate denialist coverage in the media suggested that the need for 'something to be done' did not seem so pressing, and the severity of the issue was less clear cut.

Denialists argued the environmentalists were hysterical anticapitalist 'watermelons' (green on the outside, red on the inside) who, with the fall of communism, had lost the global battle for ideas and were now trying to re-impose their ideas in a new guise (Dellingpole 2012). Then, the global financial crisis hit in 2008, and in the UK at least the coalition government introduced austerity, and a long-term issue like the climate struggled to get visibility compared with other issues. Austerity led into Brexit. Jeremy Corbyn's Labour Party seemed to provide hope for many younger activists, promising action, including a Green New Deal. This showed that, wherever you are, other political issues and climate interact in complex ways that affect how activists can effectively mobilise against the climate crisis. In other places issues like environmental racism, struggles against right-wing populism, trade union struggles and organisations around gender are more prominent. For example, in contemporary Poland, environmentalists organise strongly against the climate crisis and against restrictions on women's reproductive rights, while the Solidarność trade union lobbies and marches in favour of coal.

On the other hand, the climate did not 'go away'. In the UK the focus moved on to airports and flying. Protests at Heathrow Airport in July 2015 and 2016 London City airport raised the issue of climate justice, pointing out that the victims of the climate catastrophe now, not in the future, are black and brown people in the majority of the world. But, on the other hand, extreme weather events seemed to be part of a 'new norm'—an unstable, changing climate that we could do little about and would have to adapt to. Methane continued to be released in the boreal high latitudes, ice melted, floods and fires continued, but not at an intensity that people believed that 'we could not go on like this', given that the world is imperfect and there are many problems to address.

Then, in the summer of 2018, one Swedish young woman with good communication skills, clarity of thought, and connected to people able and willing to get her message out, explained how angry she was at the situation. Many other young people agreed, took time out of school and university, and hit the streets in their thousands. Greta Thunberg's actions were presented in ways that mobilised others to believe again that action was 'timely', and not only 'feasible' but necessary—obligatory even. Many older people felt guilty enough to do something about it, and able to. The result was Extinction Rebellion (XR), which made a

Declaration of Rebellion and launched its protest in Parliament Square in October 2018. The 2018 IPCC reported the worsening situation in stark terms. The 2018 WWF annual Living Planet Report suggested a 60% decline of vertebrate species since 1979. This time, the driver was not urgency and optimism—if we recognise the problem we can use the creativity we used to build fossil fuel capitalism to build a more convivial alternative—but catastrophism and disaster. Young people were told they would probably not see their old age unless they rebelled (Doherty et al. 2020). They were understandably outraged about this.

XR's leaders had cut their teeth on climate and anti-austerity activism. Inspired by Chenoweth and Hayes's (2018) argument that 3.5% of the population engaged in non-violent direct action (NVDA) could force elites to change, they demanded that the government tell the 'truth' about the immediacy and potentially catastrophic nature of the climate crisis, commit to net zero carbon emissions by 2025, and create citizens' assemblies to make decisions about what should be done. They called on large numbers of people to take emergency action to compel politicians to act, including mass arrests to overwhelm the police. Many older people who felt that their complacency had led to this emergency believed that they should do something about it, recognising that retired people with time, money and no work or caring commitments can and should act for young people. For older people, guilt was a driver that fused with younger people's anger at what they believed was an awful fate.

For a time XR was successful. Mass actions in November 2018 and then in April and October 2019 saw mass NVDA and a large number of arrests in central London, with other less high-profile events around the world. Then, in November 2019, radical Islamists carried out another terrorist attack on London Bridge. And, in the spring of 2020, COVID hit, and seemingly the world was locked down. The 2020 COP planned for Glasgow in November was postponed for a year. At the time of writing (June 2021), it is not possible to know what opportunities for climate action will present themselves (although there were protests against the G7 meeting in Cornwall in June 2021), but we can take the opportunity to review what we know from this historical sketch of climate activism.

Reviewing Strategies and Tactics

Why do we go to the COPs? Coming together in convergence spaces at the COPs, particularly if this entails significant carbon emissions from long distance travel, might be seen as both unsustainable and ineffective politics, compared with locally- or community-based activism in which you work locally to prefigure the world you want to see (Taylor Aiken 2017). A focus on a 'once-in-a-lifetime deal' at key COP meetings might be ineffective if it is judged that global elites are not yet ready to make the fundamental changes in the global political order that activists claim are necessary (and they almost certainly are not). A focus on the annual merry-go-round of the COPs might distract from the hard work of grassroots activism, prefiguring the future that we want to see, building system change (see also Mannan et al., this volume). This is not to say that going to the COPs is a waste of time, but it might be that just raising the issue is not enough if nothing otherwise changes at the scale necessary to solve the issue.

Many anarchist-inspired ecoactivists have, for some time, had little faith in the capacity of demonstrations, even large ones, to make change by politely lobbying elites to change their mind (Wall 1999; Seel et al 2000). This perception was reinforced by the failure of the globally-coordinated demonstration of February 2003 to stop the war in Iraq. A wider range of activists began to feel that polite lobbies and attempts of persuasion are not enough—direct action to force change is necessary. This then suggests, 'what kind of direct action, by whom, and to what end?' Analytically, we can distinguish between openly organised or spontaneous acts of direct action involving all who wish to participate, and clandestinely-organised communicative direct action. What Barker et al. (2001: 21) call "exclusivist" direct action is planned and executed by an inner circle of activists, as distinct from the wider movement. Classic 'resource mobilisation' approaches to the organisation of social movements (McCarthy and Zald 1977) suggest that the role of an outer periphery is to support the core's decisions, providing material support and admiration. Those who undertake direct action lead by example, rather than by the interaction of persuasion. Organisations like Greenpeace have long organised stunts in which activists communicate to the wider populace through the media—for instance parachuting into a football stadium at the Euros, unfurling a banner on top of a power station chimney, or projecting a slogan on a building. They act for the passive masses who are framed as apathetic and self-interested.

Thus, we might distinguish between a small number of 'heroic' XR activists in London locking themselves on to an old boat painted in a pastel colour and named after a prominent environmental activist that has been clandestinely placed at a strategically important road junction early in the morning; and thousands of activists collectively blocking Westminster Bridge, getting arrested, filling the jails and declaring that this is 'not in my name'. One involves thousands in activity; the other communicates 'to' the passive majority. Another example is hundreds of young people in canoes stopping coal ships from leaving Newcastle, Australia. One is bodies on the line saying 'not in my name' and forcing change, the other is communicating the need for change by using a boat to block a junction, which does not require lots of people to produce a media stunt, and instead relies on the hope that elites will agree and act. Of course, they do not.

Individuals standing up to power, perhaps in heroic circumstances, matter—Tiananmen Square's 'Tank Man' comes to mind. Activists are also right to argue that an individual can march, take direct action, and engage in prefigurative politics at different times and in different spaces. But there are tensions. 'Muscular' forms of mass direct action and a refusal to negotiate with authorities within a political opportunity structure framed by the global 'war on terror' can bring down repression from the authorities on those they (wrongly, of course) label 'ecoterrorists'. Attempts by climate activists to temporarily shut down Kingsnorth Power Station in the UK were successfully thwarted by the police, and in uncompromisingly vigorous, if not violent, ways. This showed that the authorities can successfully defend a target named in advance, and control (repress) an activist camp in open countryside. Many people who are otherwise committed to low-carbon lifestyles might be put off from participating in an action that might involve significant levels of police harassment or even violence. There are complex trade-offs and debates about the extent to which radical disruptive direct action raises new issues, inspires and mobilises supporters, and creates new ways of understanding issues by social movements as 'knowledge producers' (Eyerman and Jamison 1991), or puts off potential supporters and provokes the authorities into taking measures that limit or close off their ability to organise and room for manoeuvre. There are consequently debates about the extent to which this is an effective tactic for social movements aiming at mass support (North 2011). Others argue that a 'radical flank' can open up spaces in which more moderate voices can make deals or advance policy goals in more pragmatic ways (Hains 2013; Mueller and Sullivan 2015).

On the other hand, media pictures of protesters being attacked can reveal the unsustainable and repressive face of the seemingly liberal state and of the slow climate violence, if not social murder, of ecocidal capitalism (White 2014): a key objective of the politics of anarchist-inspired direct action. Individual witness, saying that what is being done is 'not in my name', is important and has a long pedigree, especially in the peace movement. This is easier to achieve on Westminster Bridge than in a field far from the media. Many members of XR are older, middle-class, retired professional people from the south of the UK with the time, social capital and resources to take direct action that others—mainly younger people—lack. Negatively racialised male bodies will be treated more harshly than older, white, grandmotherly ones. But breaking the law, being arrested, charged, and prosecuted is stressful, time consuming, and expensive. The assumption that the costs of protest are undertaken by kindly older, generally white grandmothers suggests a rather liberal view that the police can be expected to act in a gentlemanly way that negatively racialised people can find problematic. The repression this form of activism can call down can also put off those less able or willing to put their bodies on the line through direct action, and in time the inconvenience caused to people trying to go about their everyday business, perhaps on minimum wages on a zero hours contract, will mean that sympathy for the aims of the protesters will wear thin, as in the case of the XR activists who stopped a Docklands train in the commuter rush hour. The media will lose interest.

The alternative to a march, which the media may cover but elites will ignore, and direct action, which lacks the capacity to force elites to change tack and which, in time, loses its efficacy is, of course, the slower work of movement building; the development of power to create the system change that we want to see rather than merely protest against the status quo. The problem here is that the prefigurative local activisms of Transition Towns and the like, Melucci's (1989) "nomads of the present", can be too small-scale, too hidden from view, and involve too few people promoting lifestyles that are not attractive enough to millions to trigger a systemic move to a low-carbon economy and society, avoiding catastrophic climate change and resource crunches. While much of this local activism is hidden from (the analysts') view, it must be remembered that activists happily work at a number of scales and use a variety of techniques utilising new communications technologies, to get their point across. Of course, festivals like the alternative COPs provide a space to do this, and this is massively important.

Conclusion

Avoiding dangerous climate change is not an issue that can be solved easily or quickly. No one demonstration at any one COP could ever be seen to 'succeed'. Adaptation to unavoidable climate change and mitigation of its worst effects requires a fundamental transformation of the way we organise human society. The real issue is to follow the effectiveness of these experiments, and use the spaces at the COP to come together to take stock of what has been done, how effective it has been, and what is still to be done. Adding COVID-19 to the mix suggests some possibilities for the development of a new politics of hope to be developed online rather than in convergence spaces and streets, to ask what the pandemic has stopped that we are happy to see stopped, and how we 'build back better' rather than succumbing to catastrophism. I look forward to watching this process unfold in Glasgow, online, or in person.

References

Barker, Colin, Alan Johnson, and Michael Lavalette (eds), *Leadership and Social Movements* (Manchester: Manchester University Press, 2001).

Chatterton, Paul, David Featherstone, and Paul Routledge, 'Articulating Climate Justice in Copenhagen: Antagonism, the Commons, and Solidarity', *Antipode*, 45 (2013), 602–20, https://doi.org/10.1111/j.1467-8330.2012.01025.x.

Chenoweth, Erica, and Maria J. Stephan, *How Civil Resistance Works: The Strategic Logic of Nonviolent Conflict* (New York: Columbia University Press, 2011).

Dellingpole, James, *Watermelons: How Environmentalists Are Killing the Planet, Destroying the Economy and Stealing your Children's Future* (London: Biteback Publishing, 2012).

Doherty, Brian, Clare Saunders, and Graeme A. Hayes, 'New Climate Movement? Extinction Rebellion's Activists in Profile', *CUSP Working Paper*, 25 (2020), https://www.cusp.ac.uk/themes/p/xr-study/.

Haines, Herbert H., 'Radical flank effects', in *The Wiley-Blackwell Encyclopaedia of Social and Political Movements*, ed. by David A. Snow, Donatella della Porta, Bert Klandermans, and Doug McAdam (Oxford: Wiley-Blackwell, 2013), pp. 1058–50.

Lynas, Mark, *Six Degrees: Our Future on a Hotter Planet* (London: Fourth Estate, 2007).

McNeill, J. R., and Peter Engelke, *The Great Acceleration: An Environmental History of the Anthropocene* (Cambridge, MA: The Belknap Press of Harvard University Press, 2014).

McCarthy, John D., and Mayer N. Zald, 'Resource Mobilisation and Social Movements: A Partial Theory', *American Journal of Sociology*, 82(6) (1977), 1212–41, https://doi.org/10.1086/226464.

Melucci, Alberto, *Nomads of the Present* (London: Hutchinson Radius, 1992).

Mueller, Tadzio, and Sian Sullivan, 'Making Other Worlds Possible? Riots, Movement and Counterglobalisation', in *Disturbing the Peace: Collective Action in Britain & France, 1381 to the Present*, ed. by Michael Davies (Basingstoke: Palgrave Macmillan, 2015), pp. 239–55.

North, Peter, 'The Politics of Climate Activism in the UK: A Social Movement Analysis', *Environment and Planning A*, 43(7) (2011), 1581–98, https://doi.org/10.1068/a43534.

Routledge, Paul, 'Convergence Space: Process Geographies of Grassroots Globalization Networks', *Transactions of the Institute of British Geographers*, 28(3) (2003), 333–49, https://doi.org/10.1111/1475-5661.00096.

Seel, Benjamin, Matthew Paterson, and Brian Doherty, *Direct Action in British Environmentalism* (London: Routledge, 2000).

Seyfang, Gill, and Adrian Smith, 'Grassroots Innovations for Sustainable Development: Towards a New Research and Policy Agenda', *Environmental Politics*, 16(4) (2007), 584–603, https://doi.org/10.1080/09644010701419121.

Taylor Aiken, Gerald, 'The Politics of Community: Togetherness, Transition and Post-politics', *Environment and Planning A*, 49 (2017), 2383–401, https://doi.org/10.1177/0308518x17724443.

Turner, Ralph, and Lewis M. Killian, *Collective Behaviour* (Brunswick, NJ: Prentice Hall, 1987).

Wall, Derek, *Earth First! And the Anti-roads Movement: Radical Environmentalism and Comparative Social Movements* (London: Routledge, 1999).

White, Rob, 'Climate Change, Ecocide and the Crimes of the Powerful', in *The Routledge International Handbook of the Crimes of the Powerful*, ed. by C. Barak (London: Routledge, 2014), pp. 211–22.

23. Climate Politics between Conflict and Complexity

Matthew Paterson

Climate politics needs both moments of sharp, highly politicising, even over-simplifying moves, to keep pressure up, but at the same time a sort of patient, careful attention to the complexity of socio-technical systems to work out how to generate radical shifts in infrastructure and practice. But these logics stand in quite a lot of tension—the post-political/agonistic logic can reduce to slogans and abstract from the details of how you actually decarbonise, while the complexity approaches can culminate in even more complex technocratic projects. This chapter navigates questions of how to keep both of these logics alive in climate politics.

On Climate Movement Rhetoric

An important undercurrent of recent climate movement rhetoric, echoed in sympathetic media, focuses on a specific number of global companies that are responsible for a particular percentage of global emissions. "Just 90 companies caused two-thirds of man-made global warming emissions", was the headline to one of the earliest media renditions of this argument (Goldenberg 2013). More recently, just to continue with material from The Guardian (where these claims are most prominently produced), we have had "[j]ust 100 companies responsible for 71% of global emissions, study says" (Riley 2017) and "[r]evealed: the 20 firms behind a third of all carbon emissions" (Taylor and Watts 2019). These claims have then circulated more broadly because of Extinction Rebellion (XR) strategies and particularly their use by philosopher Rupert Read

https://doi.org/10.11647/OBP.0265.23

within an XR context (as reported for example in Newsweek, see Mahmood 2020).

These claims are underpinned by pioneering research by Richard Heede, in particular in an article in *Climatic Change* (Heede 2013), and then maintained via the Carbon Majors Project of the Climate Accountability Initiative,[1] an organisation established by Heede, along with prominent analyst of climate denial, Naomi Oreskes (Oreskes and Conway 2011) and Greg Erwin.

Other similar narratives have been deployed. American anarchist, Utah Phillips, is often invoked: "[t]he earth is not dying, it is being killed, and those who are killing it have names and addresses" (see e.g. Climate and Capitalism 2009). Personalising the issue beyond the corporations to their chief executives, a world map has circulated widely with the "names and locations of the top 100 people killing the planet" on it, superimposed on a map with the country size representing cumulative emissions of that country from 1850 onwards.[2]

This sort of narrative represents a particular way that activists, and allied researchers, have sought to 'repoliticise' climate change in a specific way—to identify it as an existential struggle where specific organisations, even individuals, are the causal powers of climate collapse that need to be resisted and opposed. As Malm (2020: 15) succinctly and precisely puts it, "the enemy is fossil capital". This sort of repoliticisation, often entailing the identification of such a clear enemy, can be seen plainly in the school strikes for climate, XR, the Sunrise Movement, oil pipeline activism, and fossil fuel divestment activism (also see North, this volume). Analysis of divestment discourse shows that the dominant narrative is a war/enemy narrative, where fossil fuel producers are pitted against the rest of humanity (and occasionally beyond) in an existential struggle (Mangat et al. 2018; also Wright and Nyberg, this volume).

In academic debates about climate change politics, this is reflected in various literatures that have recently highlighted the conflictual qualities

---

1   See Climate Accountability Institute website (Climate Accountability Institute, no date), https://climateaccountability.org/carbonmajors.html.

2   See e.g. Decolonial Atlas, 'Names and Addresses of the Top 100 People Killing the Planet' (decolonialatlas.wordpress.com, 2019), https://decolonialatlas.wordpress.com/2019/04/27/names-and-locations-of-the-top-100-people-killing-the-planet/).

of climate politics, and specifically the conflicts between corporate/ fossil interests and the pursuit of climate policy. This is not entirely new—some analysts have made the power of fossil corporations central to their analyses of climate politics for a long time (Paterson 1996; Egan and Levy 1998; Newell and Paterson 1998, 2010; Newell 2000; Levy and Newell 2005). But there is a noticeable spread of this sort of focus in how academic analyses of climate change are conducted. This literature is various, encompassing: detailed empirical analyses of how corporations have blocked policy development in various countries, for example in the US (Stokes 2020; Mildenberger 2020), but also in Brazil and South Africa (Hochstetler 2020), as well as the large literature on corporate roles in climate denial, often with the frame of 'Exxon knew' (e.g. Supran and Oreskes 2017); broad attempts to theorise these empirical dynamics in general (Scoones et al. 2015; Breetz et al. 2018; Colgan et al. 2020); an argument derived theoretically from Chantal Mouffe's well-known arguments about democratic politics as intrinsically "agonistic" (Mouffe 2005; Machin 2013); and arguments that dominant forms of climate change response are 'post-political', that is, that they seek to take climate change decision-making out of the realm of democratic, public, decision-making and govern climate technocratically, while at the same time presenting climate change as a consensual issue in the interests of all humanity (see most notably Swyngedouw 2010; Kenis and Mathijs 2014; Macgregor 2014). In slightly less stark terms than the last of these claims, the argument that climate change is often depoliticised—actors seek to present responses as consensual, technocratic, in the interests of all—is widespread (e.g. Pepermans and Maeseele 2016; Mann and Wainwright 2018; Willis 2020).

Limits of the '100 Companies' Story

This return to an emphasis on the conflicts inherent in addressing climate change is welcome and has helped to mobilise climate activists in important ways—by articulating a sense of 'an enemy', enabling motivation for activists and highlighting key targets and goals. Even within the trajectory of movements, it has helped to direct action in more focused ways, as in the shifts in focus of XR actions towards corporate targets—banks, oil companies, for example—over time. So there is a good case for saying that this sort of repoliticisation of climate change is a crucial component in the search for more ambitious and adequate responses to climate change.

But at the same time, there is an important piece missing from the underlying narrative. We need to return to the '100 companies' story. A key component in the underlying analysis that has generated this frame is how emissions have been associated with these companies (and then, in the individualising version, to their CEOs). Specifically, what the analysis does is trace not only all emissions associated with the production activities of these companies, but also the consumption of all of their products over time, by individuals, other companies, governments and so on. This is what is called 'Scope 3' emissions in carbon accounting terms, 'Scope 1' being direct emissions by an entity, and 'Scope 2' being directly bought-in emissions as, most obviously, in electricity consumption. The '100 companies and their CEOs' analysis is an extreme version of a Scope 3 accounting procedure, which would normally include things like the commuting emissions of a company's workers, or travel emissions for work-related travel (in lots of organisations like universities, these latter emissions completely dominate overall organisational emissions). Inevitably there is all sorts of double-counting going on—the emissions of someone commuting by car are Scope 3 emissions for their employer, but Scope 1 emissions for themselves. But it is rare to associate downstream emissions from consumption with the producing organisation. The double-counting becomes even more complicated if we are also now saying that the commuter's emissions are Scope 3 emissions both for their employer and the oil company that sold them the fuel.

While we can clearly 'trace' those emissions from a car tailpipe back to Exxon, Shell, or whoever, as Heede and the Carbon Majors Project have done very effectively, it does not follow that there is a neat causal chain from Exxon to those emissions. There is a clearly overly simplistic causal narrative going on here to make this claim. Is Exxon really 'responsible' for the emissions by all the car drivers (individual and corporate) who buy their petrol? While it has been rhetorically really important for mobilising activists, and the basic claim about corporate power remains a powerful one, it is insufficient for thinking fully about what the politics of actually decarbonising the global economy entails (as also considered in Sullivan Chapter 11, this volume).

While activism focused on identifying key actors blocking policy change and undermining their political power is important, it needs to be supplemented with political action focused on the complexities of the large-scale socio-technical systems which constitute high carbon worlds, where the causal processes generated by emissions are not so neatly identifiable with specific agents, but are emergent properties arising out of the complex interactions between corporate strategy and power, ideology, technical change, social practices, and governance systems, irreducible to any one of those elements. These systems are also themselves quite heterogeneous—including food, transport, electricity, construction, raw materials extraction, and so on (also see Halme et al., this volume). These are all complex systems with their own specific sets of corporate structures and strategies, technical qualities, and daily practices that interact in specific ways. Interventions to shift them to get rid of fossil fuels and carbon will need to be differentiated accordingly. They may all be *capitalist* in important ways, but this does not therefore capture the specificities of their dynamics adequately. As a consequence, while agonistic activism and its associated rhetoric may shift one or two of these elements, it is implausible that it can shift the system as a whole, on its own. It may sometimes even get in the way of identifying ways forward by, in effect, mis-specifying the challenge.

Limits of Avoiding Conflict

On the other hand, there is plenty of reason to believe that the existing approaches that do take this socio-technical complexity seriously—most commonly going under the rubric of 'low carbon transitions'—fail to adequately incorporate the question of conflict and power relations. In Harriet Bulkeley's *Accomplishing Climate Governance* (2016), for example, the shift in focus to climate governance has a tendency (despite, I think, her intention) to present responses to climate change in depoliticised ways—with politics understood in terms of the interplay between agonistic conflict, power relations, and public democratic decision-making.[3] Bulkeley's account, arising out of a largely Foucauldian approach, is to think of climate politics in terms of the operations and

---

3   I develop this specific point, as well as some of the other arguments in this short piece, in more detail, in Paterson (2021).

effects of power—how it is exerted in climate governance, or how, in her terms, climate governance is "accomplished", at the expense of (if not total exclusion of) other dimensions of politics, notably public deliberation and conflict. She argues that climate politics "is not the politics of vested interests and decision points, but a slow burning, unfolding, enveloping and ongoing form of the working of power" (Bulkeley 2019: 14). But there does not need to be the choice that she presents here—rather, it is *both* the "politics of vested interests" (and therefore the contestation of those interests) *and* the "slow burning, unfolding" that helps us understand the dynamics of climate politics.

Bulkeley's is the most sophisticated of this sort of approach. Others collapse much more readily into a depoliticised, technocratic account of low-carbon transitions. There has been an undercurrent of critique of the transitions approach for underplaying questions of politics (Meadowcroft 2009), and clear attempts by leading transition scholars to respond to this critique and incorporate questions of politics (Geels 2014; Roberts et al. 2018). This has been mostly limited to thinking about the ability of incumbent actors to undermine transformational processes— "regime resistance" in Geels' (2014) terms (for a detailed analysis of this literature focused on how it thinks about incumbency, see Stirling 2019). But much of this literature is nevertheless dominated by a desire to elaborate models of complex systems, where the methodological devices of these modelling exercises obscure the ability to think fully about the political dynamics of such transitions. We are thus left with depoliticised accounts of path dependencies, lock-in processes, tipping points, niches and innovation, and so on.

Combining Conflict and Complexity

The challenge then seems to me, on the analytical or academic side, to work out how these two elements in climate politics—the detailed, focused, attention on governing and transforming large-scale, heterogeneous high-carbon systems, and the deeply contested questions of power, inequality and justice—interact. And on the practical politics side, to work out how the energy mobilised by the sharpened focus on 'fossil capital as the enemy' can enable not only continued pressure on politicians, corporations, and the like, but feed through into more concrete action to transform those high carbon systems (as discussed in the chapters by Halme et al., Sandover and Whitmarsh, this volume).

On the academic side, the implications of this argument are to generate a number of questions we might usefully focus our attention on. For example, we could focus more research attention on the conditions under which depoliticised governance 'works', and when it gets stuck because of incumbent resistance. Are there general lessons we can learn from these patterns? Do some aspects of the climate challenge lend themselves more readily to this sort of depoliticised governance than others? We could also ask, conversely, what types of repoliticisation actually shift the practices of governments and corporations? Or what types also generate novel initiatives that shift power relations in important ways and enable us to pursue more radical and rapid decarbonisation? For example, do community renewable energy or agro-ecology initiatives generate new sorts of social relations that undermine the power of fossil fuel corporations? What are the key moments in climate policy trajectories where activist pressure might have the most impacts? Finally, we could ask questions about whether, and how, novel institutional arrangements like citizens' climate assemblies, enable this sort of conflictual politics to be 'embedded' in formal climate policy and governance arrangements, and thus reshape the political landscape, more broadly favouring rapid change to accelerate decarbonisation?

I am not best placed to advise social movements on strategy, but it seems to me that the implications of this sort of argument are that we should work on activities that seek not only to put immediate pressure on governments and corporations, through the variety of well-known strategies we see in for example XR, divestment, or school strikes (and their analogies in earlier periods of climate activism), but also to generate initiatives that act more directly to shift power away from corporate and government actors—community energy or land ownership, community forestry or agriculture, and so on, that might both keep climate political in important ways but also start to build more long-term sustainable solutions. Many of these initiatives of course already exist, but perhaps need to be understood more deeply as political interventions in ways that they are often not.

References

Breetz, Hanna, Matto Mildenberger, and Leah Stokes, 'The Political Logics of Clean Energy Transitions', *Business and Politics*, 20(4) (2018), 492–522, https://doi.org/10.1017/bap.2018.14.

Bulkeley, Harriet, *Accomplishing Climate Governance* (Cambridge: Cambridge University Press, 2016).

Bulkeley, Harriet, 'Navigating Climate's Human Geographies: Exploring the Whereabouts of Climate Politics', *Dialogues in Human Geography*, 9(1) (2019), 3–17, https://doi.org/10.1177/2043820619829920.

Climate Accountability Institute (Climateaccountability.org, no date), https://climateaccountability.org/carbonmajors.html.

Climate and Capitalism, 'Utah Phillips: Who Is Responsible?' (*ClimateandCapitalism.com*, 2009), https://climateandcapitalism.com/2009/05/12/quotable/.

Decolonial Atlas, 'Names and Addresses of the Top 100 People Killing the Planet' (Decolonialatlas.wordpress.com, 2019), https://decolonialatlas.wordpress.com/2019/04/27/names-and-locations-of-the-top-100-people-killing-the-planet/.

Egan, Daniel, and David Levy, 'Capital Contests: National and Transnational Channels of Corporate Influence on the Climate Change Negotiations', *Politics and Society*, 26(3), 1998, 337–61, https://doi.org/10.1177/0032329298026003003.

Geels, Frank W., 'Regime Resistance against Low-Carbon Transitions: Introducing Politics and Power into the Multi-Level Perspective', *Theory, Culture & Society*, 31(5) (2014), 21–40, https://doi.org/10.1177/0263276414531627.

Goldenberg, Suzanne, 'Just 90 Companies Caused Two-Thirds of Man-Made Global Warming Emissions' (Theguardian.com, 2013), http://www.theguardian.com/environment/2013/nov/20/90-companies-man-made-global-warming-emissions-climate-change.

Heede, Richard, 'Tracing Anthropogenic Carbon Dioxide and Methane Emissions to Fossil Fuel and Cement Producers, 1854–2010', *Climatic Change*, 122(1–2) (2013), 229–41, https://doi.org/10.1007/s10584-013-0986-y.

Hochstetler, Kathryn, *Political Economies of Energy Transition: Wind and Solar Power in Brazil and South Africa* (Cambridge: Cambridge University Press, 2020).

Kenis, Anneleen and Erik Mathijs, 'Climate Change and Post-Politics: Repoliticizing the Present by Imagining the Future?', *Geoforum*, 52 (2014), 148–56, https://doi.org/10.1016/j.geoforum.2014.01.009.

Levy, David, and Peter Newell (eds), *The Business of Global Environmental Governance* (Cambridge MA: MIT Press, 2005).

MacGregor, Sherilyn, 'Only Resist: Feminist Ecological Citizenship and the Post-Politics of Climate Change', *Hypatia*, 29(3) (2014), 617–33, https://doi.org/10.1111/hypa.12065.

Machin, Amanda, *Negotiating Climate Change: Radical Democracy and the Illusion of Consensus* (London: Zed Books, 2013).

Mahmood, Basit, 'There Are 100 Companies Responsible for Climate Change, Activist Says' (Newsweek.com, 2020), https://www.newsweek.com/climate-change-xr-extinction-rebellion-fossil-fuels-climate-greenhouse-gasses-emissions-1530084.

Malm, Andreas, *Corona, Climate, Chronic Emergency: War Communism in the Twenty-First Century* (London: Verso Books, 2020).

Mangat, Rupinder, Simon Dalby, and Matthew Paterson, 'Divestment Discourse: War, Justice, Morality and Money', *Environmental Politics*, 27(2) (2018), 187–208, https://doi.org/10.1080/09644016.2017.1413725.

Mann, Geoff, and Joel Wainwright, *Climate Leviathan: A Political Theory of Our Planetary Future* (London: Verso, 2018).

Meadowcroft, James, 'What about the Politics? Sustainable Development, Transition Management, and Long Term Energy Transitions', *Policy Sciences*, 42(4) (2009), 323–40, https://doi.org/10.1007/s11077-009-9097-z.

Mildenberger, Matto, *Carbon Captured: How Business and Labor Control Climate Politics* (Cambridge, MA: MIT Press, 2020).

Mouffe, Chantal, *On the Political*, New Ed edition (London; New York: Routledge, 2005).

Newell, Peter, *Climate for Change: Non-State Actors and the Global Politics of the Greenhouse* (Cambridge: Cambridge University Press, 2000).

Newell, Peter, and Matthew Paterson, 'A Climate for Business: Global Warming, the State and Capital', *Review of International Political Economy*, 5(4) (1998), 679–703, https://doi.org/10.1080/096922998347426.

Newell, Peter, and Matthew Paterson, *Climate Capitalism: Global Warming and the Transformation of the Global Economy* (Cambridge: Cambridge University Press, 2010).

Oreskes, Naomi, and Erik M. Conway, *Merchants of Doubt: How a Handful of Scientists Obscured the Truth on Issues from Tobacco Smoke to Global Warming* (New York: Bloomsbury Publishing USA, 2011).

Paterson, Matthew, *Global Warming and Global Politics* (London: Routledge, 1996).

Pepermans, Yves, and Pieter Maeseele, 'The Politicization of Climate Change: Problem or Solution?', *Wiley Interdisciplinary Reviews: Climate Change*, 7(4) (2016), 478–85, https://doi.org/10.1002/wcc.405.

Riley, Tess, 'Just 100 Companies Responsible for 71% of Global Emissions, Study Says' (Theguardian.com, 2017), https://www.theguardian.com/sustainable-business/2017/jul/10/100-fossil-fuel-companies-investors-responsible-71-global-emissions-cdp-study-climate-change.

Roberts, Cameron, Frank W. Geels, Matthew Lockwood, Peter Newell, Hubert Schmitz, Bruno Turnheim, and others, 'The Politics of Accelerating Low-Carbon Transitions: Towards a New Research Agenda', *Energy Research & Social Science*, 44 (2018), 304–11, https://doi.org/10.1016/j.erss.2018.06.001.

Scoones, Ian, Melissa Leach, and Peter Newell, *The Politics of Green Transformations* (London: Routledge, 2015).

Stirling, Andy, 'How Deep Is Incumbency? A "Configuring Fields" Approach to Redistributing and Reorienting Power in Socio-Material Change', *Energy Research & Social Science*, 58 (2019), 101239, https://doi.org/10.1016/j.erss.2019.101239.

Stokes, Leah Cardamore, *Short Circuiting Policy: Interest Groups and the Battle Over Clean Energy and Climate Policy in the American States* (Oxford: Oxford University Press, 2020).

Supran, Geoffrey, and Naomi Oreskes, 'Assessing ExxonMobil's Climate Change Communications (1977–2014)', *Environmental Research Letters*, 12(8) (2017), 084019, https://doi.org/10.1088/1748-9326/aa815f.

Swyngedouw, Erik, 'Apocalypse Forever? Post-Political Populism and the Spectre of Climate Change', *Theory, Culture & Society*, 27(2–3) (2010), 213–32, https://doi.org/10.1177/0263276409358728.

Taylor, Matthew, and Jonathan Watts, 'Revealed: The 20 Firms behind a Third of All Carbon Emissions', Theguardian.com, 2019, https://www.theguardian.com/environment/2019/oct/09/revealed-20-firms-third-carbon-emissions.

Willis, Rebecca, *Too Hot to Handle? The Democratic Challenge of Climate Change* (Bristol: Policy Press, 2020).

24. Sustainable Foodscapes: Hybrid Food Networks Creating Food Change

Rebecca Sandover

Food matters, from modes of production to global supply chains, what we eat and how we address food waste. Food practices shape not only climate and ecological breakdown but also human health and well-being including within our food producing communities, unequal access to food, food justice, animal welfare and more. Agriculture, Forestry and Other Land Use (AFOLU) activities account for 21–37% of total net anthropogenic GHG emissions (IPCC 2019). Considering these 'wicked issues' in the UK, and how to work for more sustainable food systems, centres debates on intersecting issues of land use, food distribution, community-based innovation and social justice amongst others. Within the present food policy vacuum in England, place-based community groups have been self-organising and connecting with different national organisations whose campaigns overlap to form hybrid food networks. Hybrid food networks focus on central food issues, such as sustainable local food supply chains, access to sustainable local food, household food insecurity and more. These networks intersect at a place-based scale where locally acting communities take forward programmes of work to enact sustainable food change, whilst also linking to the campaigns of national and translocal networks and frameworks. This essay will explore the dynamic potential of these hybrid networks in working towards place-based sustainable food solutions, via a case study of Devon.

https://doi.org/10.11647/OBP.0265.24

Introduction

Food has become an organising principle through which we measure the impacts of crises on our lives in a time of multiple emergencies. From issues of effective food supply chains and access to food during the UK COVID-19 pandemic lockdown, to contestations around how to eat for a sustainable planet. Food, how it is produced, and how we get access to it as consumers, is of central concern when considering a time of multiple emergencies. Action to address these critical issues has been seen at a place-based scale, from climate assemblies and juries[1] to local civil society food organisations collaborating to effect food change, including forging partnerships with local authorities to support place-based food assistance programmes.[2]

Local food initiatives are now collaborating at a range of scales across the UK, from city level, to borough or countywide projects, as exemplified by The Sustainable Food Places, Food Power Alliances and Feeding Britain projects. Intersecting actions of civil society organisations via hybrid food networks act as an increasingly important mechanism to link food actors, community-based organisations and policymakers in addressing the critical food issues faced by communities in the UK today. Civil society food organisations are concerned with issues of boosting sustainable food production, household food insecurity, access to fresh food, diet-related ill-health, promoting sustainable diets, boosting community food resilience, and more (Blake 2019; Sandover 2020a) (also see Halme et al., this volume). Self-organising local food initiatives connect place-based organisations with nationwide bodies to form hybrid networks working for food change, linking grassroots community organisations with policymakers (Moragues-Faus and Sonnino 2019; Santo and Moragues-Faus 2019). By being comprised of actors representing a range of community groups, hybrid food networks are able to generate community-based knowledge and work to effect policy change (Moragues-Faus and Morgan 2015; Sonnino et

---

1   Recent and current examples here include Leeds Climate Change Citizens' Jury, Adur and Worthing Climate Assembly, Kendal Climate Jury, Devon Net Zero Citizens' Assembly and others.

2   See the work of Milan Urban Food Policy Pact, La Via Campesina, EAT Nordic Cities Initiative, African Food Security Urban Network for examples of trans-local and regional food policy organisations.

al. 2016; Sandover 2020a). At a wider scale, place-based food networks are common features of a number of countries' food policy landscapes, including USA and Canada, and they often intersect with trans-local food networks, enabling international knowledge sharing and target setting for place-based sustainable food action (Santo and Moragues-Faus 2019).

This essay explores how food traverses debates focused on taking action on climate change, specifically issues of governance and the role of civil society organisations in shaping sustainability agendas. In particular, it focuses on hybrid food networks and the dynamic potential of these hybrid networks in working towards place-based sustainable food solutions that also consider imperatives of food justice.

Hybrid Food Networks

City-regions, and other local administrative areas, have been recognised as being ideally placed to promote localised food strategies and to join up the disparate actors working towards similar ends (Morgan 2013). Recent scholarship has focused on the development of effective food governance frameworks in the cities of London, Toronto and New York, city-regions of Bristol, Brighton and Hove, Glasgow, and trans-local frameworks such as The Milan Urban Food Policy Pact and the C40 Cities network, who work for more just and sustainable policies (Morgan and Sonnino 2010; Sonnino et al. 2016; Santo and Moragues-Faus 2019). A more integrative approach to food policy thinking is being progressed via city-based and place-based initiatives that enable policymakers to work with civil society actors and trans-local networks on common issues (Betsill and Bulkeley 2007; Morgan and Sonnino 2010). Place-based and trans-local networks of civil society food actors are emerging as coherent voices for a reconfigured food system (Sonnino et al. 2016; Moragues-Faus and Morgan 2015; Santo and Moragues-Faus 2019).

Hybrid food networks focus on central food issues such as sustainable local food supply chains, access to sustainable local food, household food insecurity and more. Many food-focused activist and campaigning networks such as Sustainable Food Places (SFP), Landworkers' Alliance, Food Power, La Via Campesina, and others intersect at a place-based scale where locally acting self-organising communities take forward programmes of work to enact local food change, whilst also linking to the campaigns of national and translocal networks and frameworks. National UK networks, like SFP and Food Power, assist place-based civil society organisations' action for food change by offering support, limited funding and sharing tools such as evidence-based reports (Sonnino et al. 2016; Santo and Moragues-Faus 2019). In the UK, SFP have been working since 2013 with localised food partnerships to work towards transforming food cultures and food systems. The network has grown from the first six cities in 2015 to the over fifty-five places that are working today towards common goals of implementing and supporting sustainable and fair food systems that meet the needs of local communities (Moragues-Faus and Morgan 2015). SFP has the potential to promote access to sustainable and healthy food by influencing policy makers, local communities and local businesses. There are examples of enduring and impactful work by local food networks across the UK, which in England have partly arisen in response to the policy vacuum, although there is hope that this will change via the work of the National Food Strategy (2020), plus progressive food policy action in the devolved nations.[3] Examples include Bristol Food Network, Brighton and Hove Food Partnership, Manchester Food Board, Sheffood, Food Durham, Food Cardiff, Glasgow Food Policy Partnership and others

In recent years, Sustainable Food Places have also linked to other national movements for food change. In particular, the Food Power and Feeding Britain movements have connected with sustainable food cities, and independent self-organising communities to work together on the rising issue of UK household food insecurity. These networks share commonalities of objectives but have differences in terms of their wider aims and ambitions. By working together on specific objectives and in the absence of effective government policies in England, these organisations are creating hybrid food networks that address complex concerns that require a multi-partnership, multi-issue response.

---

3   See https://www.nourishscotland.org/ and https://www.foodsensewales.org.uk/
     for insights into different political approaches within the UK.

Action on Food at a Place-Based Scale: Food Exeter

The formation and operation of hybrid food networks are visible in the recent history of Food Exeter, which was established as a sustainable food city in 2014 (previously known as Exeter Food Network) to work towards sustainable and healthy food for all in the city. In 2018 Food Exeter also became a 'Food Poverty Alliance' after securing funding from Food Power and began a dialogue process to explore a cross-city approach to addressing household food insecurity. In a time of COVID-19, Food Power funding enabled Food Exeter to begin working with emergency food providers, on their 'Signposting Project' to begin first steps in exploring ways of reducing disjointed operations across the city. Food Power funding has also assisted Food Exeter work with other regional organisations working to address household food insecurity. The South West Food Power Alliance supports the sharing of best practices and exploring common experienced challenges for organisations working on access to food and food equity across the region.

Local governance structures in Devon have shaped Food Exeter's independence as a civil society organisation working without formal links to local authorities. In the county of Devon, public health responsibilities around food and health sit with the county council, reducing pathways to engagement with Exeter City Council. However, the COVID-19 crisis and mass climate emergency protests in 2019 have impacted the urgency with which food issues are now perceived by local authorities, with both Devon County Council and Exeter City Council taking steps forward in establishing food strategies or food partnerships in late 2020 and early 2021. Cascading government funding on food support, Devon County Council have funded organisations at each district council level to run short programmes to join up emergency food providers and trial programmes in order to prevent household food insecurity. In Exeter, Food Exeter are working with Exeter Community Initiatives to run this programme in collaboration with emergency food providers and agencies.

Devon County Council also supports a multi-stakeholder partnership, Devon Climate Emergency, to run a Carbon Plan process in the wake of all councils in Devon declaring a climate emergency in 2019. The Carbon Plan process is a multi-faceted approach with the Devon

Climate Emergency Response Group, made up of all local authorities in Devon plus other key environmental, land-use and business membership groups, managing the process. Alongside them, The Devon Climate Emergency Taskforce operates as an independent group of experts weighing up the evidence sourced from expert hearings and public consultations. Between them they have agreed an Interim Devon Carbon Plan, with key controversial issues, such as sustainable food production and consumption issues being deliberated by a Climate Assembly in summer 2021.

Alongside their focus on access to food and household food insecurity, Food Exeter are focused on action to build capacity for sustainable food in the city. In 2020, Food Exeter supported the establishment of a new community benefit enterprise, Good Food Exeter, to set up neighbourhood farmers' markets in communities where good-quality, fresh produce was less available. With the uncertainties of lockdown and with the support of 'Veg Cities' funding from The Sustainable Places, Food Exeter decided to turn this into an online farmers' market where customers made online orders based on what produce was available and then collected from a designated collection point on a set day or received their delivery by e-cargo bike. With forty (and growing) Greater Exeter local and micro-producers supplying them with affordable, high quality produce, the market has won loyal support from local customers. A difference between Good Food Exeter and other online food suppliers is that Good Food Exeter's producers include micro-producer startups who may have a limited weekly stock. This enables micro-producers to sell as little or as much as they can and so support their first steps as a sustainable food producer.

The Networks Powering Local Food Initiatives

COVID-19 has shown that local food producers, distributors and shops have proved to be "small and nimble and people-powered" (Tom Steele of the Kentish Town Box Scheme), highlighting that their ability to provide access to food relies on their adaptability and their community embeddedness. Independent, local food shops source produce from a diverse range of suppliers including wholesalers, local producers and micro-producers. Food initiatives enact community embeddedness that forge strong relationships both with their suppliers and with their customers. An innovative local food market, Good Food Exeter draws on the enduring relationships of their sister organisation, Food Exeter with the UK-wide Sustainable Places network, who host a range of channels for sharing best practices and learning from other place-based food policy organisations. Enduring relationships with local food producers, shops, charities and community centres in Exeter also created the local knowledge networks needed for Good Food Exeter to be formed. The networks underpinning the responsiveness and adaptability of the local food economy may be place-based and contingent, however sustainable food networks, producer membership organisations and associated charities enable the building of capacity within the local food economy. This in turn builds the potential to invigorate food security at regional and local scales via relocalised supply chains. Local food initiatives have proven themselves to be innovative, adaptable and creative in a time of crisis. However, a longer-term flourishing of the local food economy will require more than the dedication and ingenuity of local food leaders.

In her new book, *Sitopia*, Carolyn Steel (2020) calls for a redesigning of the local food market in collaboration with local authorities, who have the ability to support the local food economy via policy and planning. Multi-purpose covered markets, rate reductions for local food shops, supporting pop-up micro food businesses, exploring access to land for new entrants, and more, would enable the visibility of local food producers within our high streets. By providing access to popular shopping spaces, policies that support local food producers would also have the potential to boost the vibrancy of these spaces, in line with thinking on the experience economy (Poulsson and Kale 2004). A decentralised approach to food policy and redesigning local food retail spaces needs to go hand-in-hand with national policies that support this sector. The COVID-19 crisis shows that an over-reliance on supermarkets for the nation's food needs has created fragile agri-food supply chains that are not only vulnerable to disruption but also hamper the placing of local, sustainable and regional fresh produce at the heart of our communities (Sandover 2020b).

Conclusion

In 2019/2020 national and local authorities made declarations of a climate emergency and as outlined here at a regional level, Devon is taking steps to address the role of food in the climate change emergency. Alongside this, the COVID-19 crisis has accelerated action to address household food insecurity and promote sustainable food supply chains. Drawing on its history of work funded by SFP and Food Power, Food Exeter found itself at the centre of this place-based action for sustainable and equitable food change. National linkages provided Food Exeter with essential opportunities to learn from other place-based food networks. Local linkages and its history of operation enabled it to move swiftly to innovate new programme areas of work, such as the new emergency food project and the establishment of Good Food Exeter.

Taking a civic-led and bottom-up approach, local and regional food and health programmes can bring together networks of organisations to work collaboratively on pressing food issues within their localities. Working with local authorities, key decision-makers and regional bodies, hybrid food networks support place-based food networks to realise the goal of transforming local and regional food systems whilst working to address climate change. Developing a regional, sustainable food plan would support these hybrid food networks to build capacity in the supply and sale of local, sustainable food, including increasing the procurement of local, sustainable food by public bodies and anchor institutions.

Spatially, these models increasingly act as nested partnerships, where place-based networks engage with partners within regional and national frameworks based on agreed aims and objectives. Through these actions of network formation and co-producing knowledges, hybrid food networks can influence and implement locally adapted programmes and policies that enact sustainable food change.

References

Betsill, Michele, and Harriet Bulkeley, 'Looking Back and Thinking Ahead: A Decade of Cities and Climate Change Research', *Local Environment*, 12(5) (2007), 447–56, https://doi.org/10.1080/13549830701659683.

Blake, Megan K., 'More than Just Food: Food Insecurity and Resilient Place Making through Community Self-Organising', *Sustainability*, 11(10) (2019), 2942, https://doi.org/10.3390/su11102942.

Devon Community Foundation, 'Community Food Programme' (2020), https://devoncf.com/apply/community-food-programme-2/.

Devon Climate Emergency, 'Devon Climate Emergency—Creating a resilient, net-zero carbon Devon—where people and nature thrive' (2020), https://www.devonclimateemergency.org.uk/.

Good Food Exeter (2021), https://goodfoodexeter.co.uk.

IPCC, 'Climate Change and Land', IPCC (2019) https://www.ipcc.ch/site/assets/uploads/2019/08/4.-SPM_Approved_Microsite_FINAL.pdf.

Moragues-Faus, Ana, and Kevin Morgan, 'Reframing the Foodscape: The Emergent World of Urban Food Policy', *Environment and Planning A*, 47(7) (2015), 1558–73, https://doi.org/10.1177/0308518X15595754.

Moragues-Faus, Ana, and Roberta Sonnino, 'Re-Assembling Sustainable Food Cities: An Exploration of Translocal Governance and Its Multiple Agencies', *Urban Studies*, 56(4) (2019), https://doi.org/10.1177/0042098018763038.

Morgan, Kevin, 'The Rise of Urban Food Planning', *International Planning Studies*, 18(1) (2013), 1–4, https://doi.org/10.1080/13563475.2012.752189.

Morgan, Kevin, and Roberta Sonnino, 'The Urban Foodscape: World Cities and the New Food Equation', Cambridge Journal of Regions, *Economy and Society*, 3(2) (2010), 209–24, https://doi.org/10.1093/cjres/rsq007.

Poulsson, Susanne H. G., and Sudhir H Kale, 'The Experience Economy and Commercial Experiences', *The Marketing Review*, 4 (2004), 267–77.

Pullman, Nina, 'Support local food after Covid, shoppers urged' (2020), https://wickedleeks.riverford.co.uk/news/local-sourcing-ethical-business/support-local-food-after-covid-shoppers-urged.

Sandover, Rebecca, 'Participatory Food Cities: Scholar Activism and the Co-Production of Food Knowledge', *Sustainability*, 12(9) (2020a), 3548, https://doi.org/10.3390/su12093548.

Sandover, Rebecca, 'Covid-19 Reset: Place-based Food Initiatives in a Time of Crisis' (2020b), http://www.ccri.ac.uk/sandoverblog/.

Santo, Raychel, and Ana Moragues-Faus, 'Towards a Trans-Local Food Governance: Exploring the Transformative Capacity of Food Policy

Assemblages in the US and UK', *Geoforum*, 98 (2019), 75–87, https://doi.org/10.1016/j.geoforum.2018.10.002.

Sonnino, Roberta, Ana Moragues-Faus, and Terry Marsden, 'Relationalities and Convergences in Food Security Narratives: Towards a Place-Based Approach', *Transactions of the Institute of British Geographers*, 41(4) (2016), 477–89, https://doi.org/10.1111/tran.12137.

Steel, Carolyn, *Sitopia: How Food Can Save the World* (London: Chatto & Windus, 2020).

The National Food Strategy (2020), https://www.nationalfoodstrategy.org.

UK Gov, 'Coronavirus (COVID-19): Local authority Emergency Assistance Grant for Food and Essential Supplies' (2020), https://www.gov.uk/government/publications/coronavirus-covid-19-local-authority-emergency-assistance-grant-for-food-and-essential-supplies/coronavirus-covid-19-local-authority-emergency-assistance-grant-for-food-and-essential-supplies.

25. Telling the 'Truth': Communication of the Climate Protest Agenda in the UK Legacy Media

Sharon Gardham

This essay draws on the results of a thematic discourse analysis of UK media coverage of climate strike actions that took place in 2019, and reflects on the importance of the framing of protester claims-making and identity for wider adoption of climate protest messages. It revisits a key question for the organisers of such protests regarding how they can overcome the potential conflict between ensuring their actions pass the test of newsworthiness required to ensure media attention, without failing the tests of claims-making legitimisation necessary for an issue to become accepted as a societal problem that requires urgent resolution.

Introduction

The momentum of climate protests grew exponentially throughout 2018 and 2019, culminating in two key climate protest actions in the autumn of 2019: the Fridays for Future (FFF) Climate Strike, which was the biggest climate strike ever held (Laville and Watts 2019); and the Extinction Rebellion (XR) International Rebellion, which saw acts of civil disobedience take place in cities around the world over a two-week period.

https://doi.org/10.11647/OBP.0265.25

Despite the size and scale of these protests, in the UK at least, issues surrounding the environment failed to make any obvious political headway at the extraordinary general election held in December of that year. Instead, the Conservative Party enjoyed a landslide victory, despite the relatively scant coverage of environmental issues in their campaigning or manifesto (Richards 2019). This situation was reflected in the media, whose coverage in the run up to the election was dominated by Brexit and leadership personalities, with the environment way down the list of news priorities (Loughborough University 2019).

This gap in reporting raises the question of where and how the protest momentum stalled, and what role the press has in setting the environmental agenda for their readership.

For the protest organisations involved, this issue begs the question of how they can strike a balance between creating an "image event" (Cox and Schwarze 2015: 76) sufficient to ensure it earns media coverage, whilst gaining enough claims-making legitimisation to promote environmental issues to recognised social problems (Hansen 2015). How can protesters "command attention, gain legitimacy and invoke action" (Hansen 2015: 30) against the backdrop of a media who "more often than not, prefer to maintain and reproduce the dominant mainstream frames and cultural codes" (Hannigan 2014: 137), and who are naturally wedded to the reporting of newsworthy content in rapid news cycles, which the relatively slow-moving issue of environmental degradation does not seem to support (Carvalho 2010)?

This chapter presents the key findings of a thematic discourse analysis carried out on UK newspaper reporting of the two climate protest actions mentioned above that took place in the autumn of 2019. More than 4000 excerpts were coded using a six-stage method (after Braun and Clarke 2006), examining the balance of event reporting in comparison to underlying communication of protester claims-making, the presentation of protester motivation and identity and the potential politicisation of protest news reporting.

Analysis was carried out on legacy media[1] publications from across the political spectrum in the UK (*The Guardian, BBC Online, The Daily Telegraph* and *Daily Mail*), and on protester organisation press

---

1   I.e. traditional print or television media news organisations that predate online formats.

releases, covering four events: FFF's Climate Strike, XR's International Rebellion opening actions, XR's Tube action and the outcome of XR's legal challenge to the Metropolitan Police on their London-wide ban on protesters meeting in groups.

Key Findings

The analysis showed that media coverage of the events was highly politicised, demonstrated by the proportion of coverage given to voices in support of, or opposition to, protest, which was directly relatable to the political bias of the various publications. For example, for one coded item 'Protest or protester opposition directly quoted', 77% of excerpts came from the right-leaning publications studied (*The Telegraph* and *Daily Mail*). This propensity towards confirmation bias was evident across publications and across events, with various methods employed to promote established positions, and reinforce the views and expectations of the paper's readership.

In fact, all publications appeared to stay firmly within a preferred narrative, whether they were reporting on FFF or XR actions. For example, *The Guardian* stuck to its depiction of protests as peaceable and non-violent, even when violence broke out during the XR Tube action (detailed below), instead focusing the majority of its coverage on the peaceable actions that took place as opposed to the assault perpetuated on protesters by members of the travelling public (Gayle 2019). *The Daily Telegraph*, on the other hand, focused its coverage of protests on issues relating to law and order, with even legal protest actions being likened to criminality via comments that protests were taking police away from dealing with 'serious crime' (Sawer and Roberts 2019). Whilst a focus on matters of law and order may be understandable when reporting on XR actions for which disruption and arrest are deliberate tactics (Taylor 2020), it is perhaps more telling of a commitment to a preferred narrative when it is used to report on the FFF action. This action saw very few arrests, with a Metropolitan Police commander quoted as saying that "overall the day ran smoothly" and any real disruption was attributable to a "tiny minority" of protesters (*BBC Online* 2019). A focus on law-and-order issues at this protest would therefore seem incongruous, unless it is deployed in order to support a predefined position.

The protest organisations themselves were no less inclined to stick to the script when it came to their own publications. FFF displayed a strong bias towards an environmental injustice narrative, enough so to frighten *The Daily Telegraph* into branding their claims as "anti-science" and "hugely dangerous" (Sawer and Roberts 2019). XR on the other hand avoided any direct challenge to the status quo (at least in the texts I studied), relying instead on what Hannigan (2014) terms the Arcadian narrative, rallying supporters around a sense of history and a love of land, as well as self-sacrifice for the greater good (Extinction Rebellion 2019).

These favoured narratives fail to make much impression on news reporting, however, instead becoming either lost in translation or else eclipsed by protest event-driven news coverage. Aside from *The Daily Telegraph*'s slightly alarmist reporting on the demands of the UK Student Climate Network, communication of underlying protester claims-making was far less evident than reporting on other so-called newsworthy items. In fact, if we exclude the protester press releases, just 3% of the analysed excerpts related to the communication of underlying protest messages. Given that media framing is so important to the recognition of an issue, particularly where there is little perceived direct experience of it (Happer and Philo 2013; Hansen 2015; Whitmarsh 2015), this gap should be considered a problem for these organisations.

On this note, analysis of mention of the specifics of environmental problems showed that scientific explanation or legitimation of environmental issues accounted for fewer than 1% of all excerpts, including the statements of the protester organisations. General comments about the issues were more prevalent, but specifics were thin on the ground. The reasons for this absence are not clear: perhaps the British public are already considered familiar enough with the specifics of the issues so as not to need further explanation, as Jukneviciute et al.'s (2011) comparative study of the Swedish and Lithuanian media suggests, or perhaps the science narrative is not favoured due to a perceived mistrust of science, even though 'empirically [...] blanket mistrust of scientists is rare in most countries' (Fairbrother 2017: 3).

In terms of who or what was responsible for either the cause or resolution of environmental issues, the government were most commonly blamed, with the economy and personal responsibility coming a distant second and third. Very occasional mention was made of corporate responsibility. Considering that corporations are in fact responsible for so much environmental damage (Riley 2017), and that they wield considerably more power in some cases than nation states (Rodionova 2016), this is both surprising and worrying.

Stahel (2016) suggests that whilst state responsibility is key to environmental agreements being made, the state-centred approach to change is less relevant in today's global markets (as also alluded to in Bracking's chapter, this volume). If FFF's environmental injustice narrative is to be credible, however, surely corporate responsibility should feature more prominently?

There were also distinct differences in the amount of coverage given over to questions of responsibility in the different sections of the press, with *The Guardian* and *BBC Online* together accounting for 42% of excerpts on this issue, compared to 7% for *The Daily Telegraph* and *Daily Mail*.

Another consideration is the propensity towards a binary journalism; a presentation of us and them where journalist and reader combine in juxtaposition to an enemy other, with one single view claimed as the only worldview; a style of journalism which, according to Sonwalker (2005: 262) "came out of the closet after 9/11" and which is evident today, not only in the media, but in popular politics, not least during the 2020 US presidential election. The identification of protesters as other was strongly evident, particularly in the right-leaning press. Various tactics were used to perpetuate this view, with FFF protesters largely othered on the grounds that they were naïve truants out for an educational day trip, despite the fact that the September FFF Climate Strike was an inter-generational effort. This fact is in line with the depiction of FFF protesters found in the media of both Germany (Bergmann and Ossewaarde 2019) and Sweden (Jacobssen 2019).

XR protesters are similarly othered in some publications, on occasion via direct name calling. The *Daily Mail*, for example, described XR protesters as an "eco-mob" of "nose-ringed crusties" (Sinmaz et al. 2019). *The Daily Telegraph* described XR protesters as engaged in various activities (e.g. yoga, drumming and chanting, lighting incense (Dixon and Lyons 2019)) that might be juxtaposed against the hard-working parents, commuters, hospital patients and gig-economy workers presented at odds with the protesters: the "people like us" whose lives and livelihoods are disrupted by the actions of protesters. Alternatively, protesters of all types are at times depicted as deviant or criminal. In fact, reporting of protesters as either naïve children or figures of fun often sits alongside depictions of protesters as being complicit in violence or being likened to criminality without the slightest irony or sense of contradiction.

Even when protesters were the victims of criminal acts there was little sympathy expressed for them either on a personal level, or in terms of their cause. The XR Tube action on 17 October 2019 saw protesters dragged from the roof of a train at Canning Town and beaten and kicked on the platform (Gardner 2019). Despite this reality, all publications mentioned the arrest of protesters rather than their assailants. The messages of protest were all but drowned out, with communication of underlying claims-making at this event barely registering.

Similarly, the successful XR legal challenge did not garner much sympathy for protesters from some corners of the press, who, despite their recent vociferous defence of democracy when threatened by direct XR action in the autumn of 2020 (*The Daily Telegraph* 2020), failed to defend the democratic right of peaceable protest. It instead quoted Scotland Yard as saying that it required "new powers to help it shut down future green protests", making much of the fact that protesters who were wrongly arrested could now sue the Metropolitan Police, thereby placing additional burdens on taxpayers who "already face a bill of at least £24 million" as a result of the International Rebellion (Ledwith 2019: online). That the overreach of police powers might, in itself, pose a threat to democratic rights was only mentioned in *The Guardian*.

Whilst the right-leaning press seem to resort to scare-tactics or othering of protesters in some cases in order to disguise or delegitimate protester claims-making, more sympathetic treatment was evident in the left-leaning publication. *The Guardian*'s coverage of the FFF march was both enthusiastic and extensive and whilst their reporting on XR's actions was less so, it was certainly more measured in tone than reporting found elsewhere. *BBC Online* coverage, however, was very event-driven. Whilst extensive, it appeared to be the most neutral, perhaps reflecting its wish to avoid accusations of bias to which, as a public broadcaster, it is particularly sensitive, so much so that its new Director General has introduced new stringent rules on impartiality amongst staff, that even extend to their personal lives (Waterson 2020).

Given that the right-leaning publications with their potentially greater readership reach are more vociferous in opposition to protest than other media outlets,[2] the implications are that the clearest statement to the largest audience is one of protester message de-legitimation and incitement of opposition to protest, drowning out the less strident voices of support. *BBC Online*'s greater reach[3] is negated by its focus on neutrality. Whether the recent XR blockade of Murdoch-owned publications, with its subsequent chorus of approbation, was the right way to redress this balance is debatable, however, since it allowed the right-leaning press the chance to play the victim and, perhaps justifiably, accuse the protesters of being actively anti-democratic.

Conclusion

There will always be competing items clamouring for media coverage, with the next big story always just around the corner. After five years of Brexit dominating the UK news and political debate, we might have longed for an alternative story. That it comes along in the shape of a deadly, life-limiting global pandemic is something that few would have anticipated and that even fewer would have desired. Against this fierce competition for column inches, the environment often does not just play second fiddle, but instead barely makes the line-up at all.

The efforts of FFF and XR to redress this balance, to give voice to the often voiceless other-than-human world, should therefore be admired (see also North and Paterson, this volume). As this research demonstrates however, it is debatable how much this voice is heard amongst the noise of other aspects of protests—those "image events" (Cox and Schwarze 2015) that may capture headlines, but do not necessarily win the hearts and minds of the public, or at least not enough to represent an electoral threat to government.

---

2   *The Daily Telegraph* and *Daily Mail* had a 1.6 million circulation between them during 2019, compared to *The Guardian's* 141,000 (Audit Bureau of Circulations 2020).

3   In 2015 its news app was used by 51% of UK smartphone users (National Union of Journalists 2015).

In addition, whilst the deployment of ever more disruptive and spectacular tactics may gain column inches, the greater shock value may mean that the fewer of those inches are dedicated to communicating a sympathetic message of protester claims, even when protesters have been attacked or legally wronged. These events may instead distract from the underlying message, meaning that even the determined reader might struggle to discover any real details of the issues, let alone suggestions for their resolution. Reporting in effect becomes less focused on messages that may aid the legitimation of protester claims-making, and more on the reporting of the sensational aspects of those events, or else on direct denigration of protesters and their messages in support of the status quo. This observation, however, is not intended to detract from the equally important experiential and community-making dimensions of participation in protest events, which are critical elements of building 'social movement potency' regarding environmental justice concerns (cf. Salter and Sullivan 2008; Mueller and Sullivan 2015).

XR and FFF both request that a citizens' assembly is formed with a focus on the environment, showing their recognition that protest can only be a short-term strategy in striving for urgent and meaningful change. That we have been discussing climate change for more than thirty years, that the main NGOs have already traversed the path from protest to a seat at the table, and that despite all this we have moved during that time from anticipated problems with climate change to a real and present climate crisis, should be a warning to those who see the chance for direct political participation as an outcome in itself.

In majoritarian democracies such as the UK, political change can be brought about by the perception of an electoral threat (Vliengenthart et al. 2016). Winning sufficient public support from all sections of society is necessary if we are to signal this electoral threat to whichever incumbent government is in power, such that they do indeed make decisions on behalf of the environment that override purely economic considerations. For corporations that continue to perpetuate environmental degradation and social injustice, consumer and shareholder pressure, combined with widespread public support for globally-agreed governmental controls and legislation, are necessary to bring about a change in attitude and strategy.

The fact that transnational corporations control such a high proportion of the media (Stahel 2016), the media's propensity to remain within hegemonic norms (Sonwalker 2005) and readers' preference for affirmation of their pre-established views rather than seeking a challenge to them (Happer and Philo 2013), all suggest that garnering media coverage is currently of limited use in the urgent fight to reverse our increasingly disastrous environmental trajectory. Media from across the political spectrum will deliver the news that it thinks reflects the priorities and views of its readership and its corporate owners. Circumvention of the legacy media via a restoration of faith in the democratising influence of the Internet and the citizen journalist may provide one way forward. Following this logic, if the environment can be made to be a priority for readers, voters and consumers, then the news, corporate behaviour and the political will to make fundamental changes should follow.

References

BBC Online, 'Climate Strike: Thousands Protest Across the UK' (BBC.co.uk, 2019), https://www.bbc.co.uk/news/uk-49767327.

Bergmann, Zoe, and Ringo Ossewaarde, 'Youth Climate Activists Meet Environmental Governance: Ageist Depictions of the FFF Movement and Greta Thunberg in German Newspaper Coverage', *Journal of Multicultural Discourses*, 15(3) (2019), 1–24, https://doi.org/10.1080/17447143.2020.1745 211.

Braun, Virginia, and Victoria Clarke, 'Using Thematic Analysis in Psychology', *Qualitative Research in Psychology*, 3(2) (2006), 77–101, https://doi.org/10.1191/1478088706qp063oa.

Carvalho, Anabela, 'Reporting the Climate Change Crisis', in *The Routledge Companion to News and Journalism*, ed. by Stuart Allan (London: Routledge, 2010), pp. 485–95.

Cox, Robert, and Steve Schwarze, 'The Media/communication Strategies of Environmental Pressure Groups and NGOs', *The Routledge Handbook of Environment and Communication*, ed. by Anders Hansen and Robert Cox (London: Routledge, 2015), pp. 73–85.

Dixon, Hayley, and Izzy Lyons, 'Police Chiefs Criticise Extinction Rebellion: More Than 270 Activists Arrested as Force Takes Hard Line Against Protests That Bring Capital to a Standstill', *Daily Telegraph*, 8 October

2019, p. 7. *Gale OneFile: News,* link.gale.com/apps/doc/A601999236/STND?u=bsuc&sid=bookmark-STND&xid=d091163c.

Extinction Rebellion, 'The Moment of Truth Arrives—International Rebellion Begins' (Extinctionrebellion.uk, 2019), https://extinctionrebellion.uk/2019/10/07/the-moment-of-truth-arrives-international-rebellion-begins/.

Fairbrother, Malcolm, 'Environmental Attitudes and the Politics of Distrust', *Sociology Compass,* 11(5) (2017), 1–10, https://doi.org/10.1111/soc4.12482.

Gardner, Bill, 'Extinction Rebellion Admits Tube Protest was "Own Goal": Rush-hour Activists Attacked by Angry Commuters After climbing on Train', *The Daily Telegraph,* 18 October 2019, p. 13, *Gale OneFile: News,* https://link-gale-com.bathspa.idm.oclc.org/apps/doc/A603038252/STND?u=bsuc&sid=STND&xid=e5722c22.

Gayle, Damien, 'Protester Dragged from Roof of Train as Extinction Rebellion Targets Underground; Commuters Clash With Demonstrators During Morning Rush Hour Disruption in London', *The Guardian,* 17 October 2019, *Gale OneFile: News,* https://link-gale-com.bathspa.idm.oclc.org/apps/doc/A602956963/STND?u=bsuc&sid=STND&xid=d4309c87.

Hannigan, John, *Environmental Sociology* Third edition (London: Routledge, 2014).

Hansen, Anders, 'Communication, Media and the Social construction of the Environment', in *The Routledge Handbook of Environment and Communication,* ed. by Anders Hansen and Robert Cox (London: Routledge, 2015), pp. 26–38.

Happer, Catherine, and Greg Philo, 'The Role of the Media in the Construction of Public Belief and Social Change', *Journal of Social and Political Psychology,* 1(1) (2013), 321–26, https://doi.org/10.5964/jspp.v1i1.96.

Jacobsson Diana, 'Young vs Old? Truancy or Radical New Politics? Journalistic Discourses About Social Protests in Relation to the Climate Crisis', *Critical Discourse Studies,* 18(4) (2019), 481–97, https://doi.org/10.1080/17405904.2020.1752758.

Jukneviciute, Laura, Vilmantè Liubinienè, and Daniel Persson-Thunqvist, 'The Role of Media in Transforming Outlooks on Environmental Issues. A Comparative study of Lithuania and Sweden', *Social Sciences,* 73(3) (2011), 23–33, https://doi.org/10.5755/j01.ss.73.3.789.

Laville, Sandra, and Jonathan Watts, 'Across the Globe, Millions Join Biggest Climate Protest Ever' (Theguardian.com, 2019), https://www.theguardian.com/environment/2019/sep/21/across-the-globe-millions-join-biggest-climate-protest-ever.

Ledwith, Mario, 'Eco Rebels: Now we Will Sue Over Arrests; Police Facing £2m Bill After Demo Ban is Ruled Illegal', *Daily Mail,* 7 November 2019, *Gale OneFile: News,* https://link-gale-com.bathspa.idm.oclc.org/apps/doc/A605002036/STND?u=bsuc&sid=STND&xid=dc3baf82.

Loughborough University, 'News: What About the Environment?' (Lboro. ac.uk, 2019), https://www.lboro.ac.uk/news-events/news/2019/november/what-about-the-environment/.

Mueller, Tadzio, and Sian Sullivan, 'Making Other Worlds Possible? Riots, Movement and Counterglobalisation', in *Disturbing the Peace: Collective Action in Britain & France, 1381 to the Present*, ed. by Michael Davies (Basingstoke: Palgrave Macmillan, 2015), pp. 239–55.

National Union of Journalists, 'BBC Stats and Facts' (Nuj.org.uk, 2015), https://www.nuj.org.uk/news/bbc-facts-and-stats/.

Richards, Phil, 'Labour's Plans for Climate and Nature Score Twice as High as the Conservatives, According to Election Manifesto Ranking' (Greenpeace. org.uk, 2019), https://www.greenpeace.org.uk/news/labours-plans-for-climate-and-nature-score-twice-as-high-as-the-conservatives-according-to-election-manifesto-ranking/.

Riley, Tess, 'Just 100 Companies Responsible for 71% of Global Emissions, Study Says' (Theguardian.com, 2017), https://www.theguardian.com/sustainable-business/2017/jul/10/100-fossil-fuel-companies-investors-responsible-71-global-emissions-cdp-study-climate-change.

Rodinova, Zlata, 'World's Largest Corporations Make More Money Than Most Countries on Earth Combined' (Independent.co.uk, 2016), https://www.independent.co.uk/news/business/news/world-s-largest-corporations-more-money-countries-world-combined-apple-walmart-shell-global-justice-now-report-a7245991.html.

Salter, Kat, and Sian Sullivan, 'Shell to Sea' in Ireland: Building Social Movement Potency, *Non-Governmental Public Action (NGPA) Working Paper Series* 5 (London: London School of Economics, 2008).

Sawer, Patrick, and Lizzie Roberts, 'Police Complain Climate Strikes Take Officers Away From the Beat; Politics Schoolchildren Join Thousands Marching Across Country on Day of Global Protests', *The Daily Telegraph*, 21 September 2019, p. 4, *Gale OneFile: News*, https://link-gale-com.bathspa.idm.oclc.org/apps/doc/A600264767/STND?u=bsuc&sid=STND&xid=477db748.

Sinmaz, Emine, Jim Norton, and Mario Ledwith, 'Nose-ringed Crusties and Their Hemp Smelling Tents'; PM Blasts Eco Mob: Capital at Standstill: Patients Stranded', *The Daily Mail*, 8 October 2019, *Gale OneFile: News*, https://link-gale-com.bathspa.idm.oclc.org/apps/doc/A601998106/STND?u=bsuc&sid=STND&xid=155c18d8.

Sonwalker, Prasun, 'Banal Journalism: The Centrality of the 'Us-them' Binary in News Discourse', in *Journalism: Critical issues*, ed. by Stuart Allan (Maidenhead: Open University Press, 2005), pp. 260–73.

Stahel, Richard, 'Climate Change and Social Conflicts', *Perspectives on Global Development & Technology*, 15(5) (2016), 480–97, https://doi.org/10.1163/15691497-12341403.

Taylor, Matthew, 'Sign of the Cross: Extinction Rebellion', *The Guardian Weekly*, 14 August 2020, 34–39.

Telegraph Reporters, 'Extinction Rebellion: Printworks Protest "Completely Unacceptable" says Boris Johnson' (Telegraph. co.uk, 2020), https://www.telegraph.co.uk/news/2020/09/05/extinction-rebellion-blockade-rupert-murdochs-printing-presses/.

Vliegenthart, Rens, et al., 'The Media as a Dual-Mediator of the Political Agenda-setting Effect of a Protest. A Longitudinal Study in Six Western European Countries', *Social Forces*, 95(2) (2016), 837–60, https://doi.org/10.1093/sf/sow075.

Waterson, Jim, 'BBC "No Bias" Rules Prevent Staff Joining LGBT Pride Marches' (Theguardian.com, 2020), https://www.theguardian.com/media/2020/oct/29/bbc-no-bias-rules-prevent-staff-joining-lgbt-pride-protests.

Whitmarsh, Lorraine, 'Analysing Public Perceptions, Understanding and Images of Environmental Change', in *The Routledge Handbook of Environment and Communication*, ed. by Anders Hansen and Robert Cox (London: Routledge, 2015), pp. 339–53.

Wikipedia, 'List of newspapers in the United Kingdom by circulation' (En. wikipedia.org, 2020), https://en.wikipedia.org/wiki/List_of_newspapers_in_the_United_Kingdom_by_circulation#Daily_newspapers.

26. Climate Justice Advocacy: Strategic Choices for Glasgow and Beyond

Patrick Bond

The Paris Climate Agreement and subsequent United Nations follow-up conferences have not taken seriously the ecological crisis now unfolding. Not only does prominent scientist James Hansen describe its tokenistic measures in scathing terms, but those seeking climate justice have long despaired of multilateral climate policymaking dominated by imperial and sub-imperial elite negotiators from high-emitting economies. Mid-2021 negotiations confirmed the lack of UN progress. Instead, there are two strategies worth considering: *delegitimisation* of elites, and *'Blockadia'* of high-carbon projects. Both are proceeding but both need more clarity in strategic approaches—as in the 'Glasgow Agreement' promoted by leading civil society activist groups— that apply to the 2021 climate summit and many other struggles beyond.

Introduction

In June 2019, at the first Climate Justice Forum dedicated to scholars now embracing the field, I had the opportunity to speak following Mary Robinson's opening plenary address to the Glasgow Caledonian University Centre for Climate Justice (2019). The former Irish president and UN Human Rights Commissioner was as eloquent as ever. Her most powerful advice to the group, with regard to a strategic advocacy agenda, was that since the United Nations Framework Convention on Climate Change (UNFCCC) 2015 Paris Climate Agreement was a useful start to decarbonising the world economy, the critical next step was to relegitimise Paris by compelling national governments to shift its 'non-binding' provisions to binding.

Robinson's approach would entail returning to an essential principle of UN treaties dealing with global ecological crises: for example, the 1987 Montreal Protocol that banned CFCs to prevent catastrophic ozone hole growth, or the Kyoto Protocol's 1997 binding conditions. She proposed transcending the sleazy back-room deal permitting 'bottom-up' voluntary emissions commitments made in December 2009 by leaders of the United States, Brazil, China, India and South Africa— i.e., a "league of super-polluters and would-be super-polluters", as Bill McKibben (2009: online) of 350.org put it—at the fifteenth Copenhagen Conference of the Parties. (From then on, the nickname Conference of Polluters would often be used by critics to describe the UNFCCC's annual gathering.) Nevertheless, insisted Robinson, such a reform to ensure binding non-voluntary adherence to Paris should be the orientation we adopt as scholar-activists, so as to incrementally strengthen the case that the planet can be saved, top down.

Against this approach, I pointed out, were dilemmas associated with implementation mechanisms implied at Paris, such as ongoing emissions trading and offsets to maximise Northern emissions' efficiency (no matter the speculative bubbles forever roiling their price), or sequestering CO_2 through dubious "carbon neutrality" gambits (see chapters by Hannis and Dyke et al., this volume). These strategies she has supported in the past under the rubric of climate justice (CJ), even though the CJ movement universally opposed carbon markets and so-called "false solutions" (Bond 2012a). She did not acknowledge that the mere act of signing the Paris Climate Agreement meant acknowledging no accountability mechanisms or penalties (such as "border adjustment taxes" on climate scofflaws), as Donald Trump showed in June 2017 when he pulled the US out of the agreement. Robinson was not concerned that when countries signed the Paris deal that meant they legally forgave the West and BRICS for what is their historic "climate debt" (i.e., ecological reparations to the victims of the correlated "loss and damage"). She did not grapple with the three missing sectors conveniently left out of the Paris Climate Agreement: military, shipping and air transport. Nor was the failure of Paris to include a Just Transition for workers in carbon-intensive sectors to find alternative employment in a greener economy worth mentioning. Nor did Paris mention the urgent need to force fossil fuel firms into accepting that there is vast "unburnable carbon" in their portfolios, that in a sane world would be adjusted radically downward in valuation accounts (as "stranded assets"). The divestment pressures that were building up in civil society——removing funds from firms and financiers that refuse these logical capitalist self-correction mechanisms—were not considered, nor did Paris negotiators pay due respect to activists, especially those in grassroots, indigenous, anti-extractivist struggles and especially the youth.

Although the university's Centre for Climate Justice has firmly defined its field on the progressive end of the spectrum, some of the gathered intellectuals seemed quite content with Robinson's approach. It would allow them ongoing participation within the mainstream of global climate policy, and hence sustained potentials for receipt of research and education grants, more rapid academic publications and membership in the UN Intergovernmental Panel on Climate Change. As a result, without properly interrogating the politics of Paris, some intellectuals proceeded to take debates into the standard explorations of justice applied to climate: procedural, recognition, distributive, compensatory, restitutive and corrective. To be fair, some scholars also acknowledge the dangers that "neoliberal justice" would become a potentially dangerous trajectory (see Khan et al. 2020, for a review applied to climate finance; also see Harris, this volume). But there was a solid bloc of academics who were satisfied with the prevailing wisdom that the Paris Climate Agreement is essentially sound, and if the ambition is ratcheted up in quinquennial revisions of Nationally Determined Contributions, the central goal of reducing greenhouse gas emissions and maintaining temperatures below a rise of 1.5 degrees above pre-industrial levels during this century, is achievable.

Not everyone sees the framing in this way. If the Paris parameters, instead, offer a profoundly unsound basis for making climate policy— from global to local scales—then a very different set of principles, analyses, strategies, tactics and alliances (PASTA) should present themselves. And if the presumption that global climate policy does far more harm than good is correct, Greta Thunberg (2020) put her finger on the problem: "we are still in a state of complete denial, as we waste our time, creating new loopholes with empty words and creative accounting." As she accused the United Nations in 2019, "[w]e are in the beginning of a mass extinction and all you can talk about is money and fairytales of eternal economic growth. How dare you."

A change is needed. For if the flaws in global climate policy processes and content identified above (as well as others), then no matter how much debate proceeds on injecting various justice framings into the UNFCCC, it will be impossible to generate an outcome worthy of human endeavour, and planetary survival will be moot. That outcome appears, in mid-2021, far more likely than any other, so a Plan B is needed based on an entirely different strategy to Robinson's: i.e., a strategy to delegitimise Paris and its elite negotiators, and instead turn to immediate direct actions, more flexible scales of international engagement, and more creative strategies for bottom-up activism. The challenge is simple: how to most rapidly overturn what can be considered climate-policy mal-governance. Is one of the approaches to delegitimise the UNFCCC and especially the COPs? If so, what to put in its place?

The Case of the Glasgow Agreement

Climate Justice (CJ) is typically the alternative to "Climate Action" of the sort the UNFCCC promotes. Three of the most famous activist-based statements on CJ came from meetings of the Durban Group for Climate Justice (hosted in South Africa) in 2004, the Bali (Indonesia) COP in 2007, and the Cochabamba (Bolivia) alternative climate summit in 2010 (Bond 2012a). They were ambitious. The Cochabamba statement, for instance, made concrete demands for reparations, emissions-cutting targets and institutional mechanisms such as ecocide courts, amplifying Indigenous People's power, and formal Rights of Mother Earth.

In subsequent years, less systematic approaches by the movement were taken at various COPs and occasional meetings in between. There was a systemic failure in the CJ movement to generate the kind of global coordination achieved by, for example, La Via Campesina whose main force was the Brazilian Landless Workers Movement but which successfully moved the federated network's global headquarters around affiliates. However, despite CJ movement complaints that the UNFCCC should no longer be a central focus of global organising initiatives, that was the terrain of struggle from Bali in 2007 until at least Warsaw COP in 2013. Then in September 2014, a march of 400,000 climate activists in New York coincided with the UN General Assembly heads of state meeting, and while the November-December period was invariably one of global days of action and critique, September 2019 became the most active month of global climate action yet, thanks to the campaigning of Fridays for Future. Unlike other CJ local actions which failed to generate global-scale coordination, the youth were successfully catalysed by Greta Thunberg's weekly sit-in at the Swedish parliament from mid-2018.

In late 2020, as COVID-19 continued to disrupt the potential for wide-scale, coordinated and increasingly radical climate activism, a "Glasgow Agreement" was offered by leading forces driven especially from within southern Europe's CJ movement. It caught on internationally, with participation and Agreement sign-on from 170 mainly grassroots environmental movements across the world. Several of the agreement's features help to define what we can consider—following the French sociologist Andre Gorz (1967)—the distinct terrains of "reformist" and "non-reformist" reforms:

The People's Climate Commitment: The Glasgow Agreement (main excerpts)

The purpose of the Glasgow Agreement is to reclaim the initiative from governments and international institutions and create an alternative tool for action and collaboration, for the climate justice movement...

The institutional framework used by governments, international organisations and the whole economic system to address the climate crisis is failing in keeping global warming below 1.5 or 2°C by 2100. From its onset, developed countries and polluting corporations like the fossil fuel industry have orchestrated the repeated failure of this institutional framework.

Instead, an illusion of climate action was created while decisive steps were delayed and greenhouse gas emissions were allowed to continue rising. As a result of decades of interference by these actors, weak commitments have been continually dishonoured, and thus the main institutional arrangements on climate change, namely the Kyoto Protocol and the Paris Agreement, have not produced the reduction in global greenhouse gas emissions required to halt the worst impacts of climate change.

The Paris Agreement is only a procedure, and will not be able to achieve its stated goal of preventing the worst consequences of climate change.

Hundreds of governments, municipalities and organisations have declared a climate emergency. Massive protests in streets all around the world have repeatedly called for decisive action for climate justice inside the deadline of 2030, with scientific consensus on the need for a minimum cut by 50% of global greenhouse gas emissions within this period. To achieve any measure of these objectives, no new fossil fuel (coal, oil and gas) projects or infrastructure can be developed. A powerful climate justice movement needs new and enhanced tools to address these fundamental contradictions and to reverse the global narrative from institutional impotence into social power that brings about lasting change.

As such, the undersigned organisations and social movements assume:

1. The political framework for the required cuts and climate action will be that of climate justice, which is defined as a social and political demand that advocates for the redistribution of power, knowledge and wellbeing. It proposes a new notion of prosperity within natural limits and just resource distribution, advocating for a true connection between traditional and westernised knowledge systems. It calls for a public and participatory science to address the needs of humanity and of the earth, principally to stop the climate crisis.

In this respect:

- It recognises the interdependence between all species and affirms the need to reduce, with an aim to eliminate, the production of greenhouse gases and associated local pollutants;

- It acknowledges and integrates the care economy into daily life, with the shared responsibility of persons, regardless of their gender identity, for care and maintenance activities, both inside homes and within society—climate justice puts life at the centre;

- It supports the structural changes in society to redress centuries of systemic racism, colonialism and imperialism— climate justice is racial justice;

- It perceives the economy to be under the rules of the environment, and not the other way around, defending democratic planning based on real needs, replacing oppression, imposition and appropriation for cooperation, solidarity and mutual aid;

- It defends a just transition for workers currently employed in the sectors that need to be dismantled, reconfigured or downsized, providing support to these workers in different economies and societies, introducing energy sovereignty and energy sufficiency. This transition must be just and equitable, redressing past harms and securing the future livelihoods of workers and communities, approaching the necessary shift from an extractive economy into a climate-safe society, to build economic and political power for a regenerative economy;

- It means to recover knowledge from indigenous communities, promoting the pragmatic human activity that has beneficial effects on life cycles and ecosystems;

- It defends the introduction of reparation for communities and peoples at the frontlines of colonialism, globalisation and exploitation, acknowledging that there is a historical and ecological debt that must be paid to the Global South, and that the origins of said debts need to be stopped;

- It recognises that the effects of climate breakdown are here and now. The poorest communities in the world are experiencing loss of their homes and livelihoods, damage to their lands and culture, and are in urgent need of funding. Global solidarity and pressure is needed, to shine a light on the corporations and governments responsible for loss and damage, and to uplift the voices of the people and places most affected;

- It defends the full protection, freedom of movement, and civil, political, and economic rights of migrants;

- It defends food sovereignty as the peoples' right to define their agricultural and food policies, without any dumping vis-à-vis third countries;

- It opposes exponential and unbound economic growth—contemporarily reflected in the sovereignty of capital—understanding capitalism as incompatible with the principles of life systems;

- It refuses green capitalism and its proposed "solutions" (whether "nature based," geo-engineering, carbon trading, carbon markets or others), as well as extractivism.

2. Taking into their own hands the need to collectively cut greenhouse gas emissions and keep fossil fuels in the ground. While participating in the Glasgow Agreement, organisations will maintain their main focus away from institutional struggle—namely from negotiations with governments and the United Nations;

3. The production of an inventory of the main sectors, infrastructures and future projects responsible for the emissions of greenhouse gases in each territory, that will be nationally and internationally publicised. There will be a technical working group to support and follow-up the creation of this inventory;

4. The production of a territorial climate agenda based on the inventory. The climate agenda is an action plan, designed by communities, movements, and organisations working on the ground, that is informed by the inventory of the biggest greenhouse gas emissions sources (existing and planned) in its area of concern. It aims to set us on track for staying below 1.5ºC global warming by 2100 inside a clear framework of climate justice;

5. That political and economic noncooperation, as well as nonviolent intervention, in particular civil disobedience, are the main tools for the fulfilment of the Glasgow Agreement. At the same time, we recognise that for oppressed groups and those living in more oppressive societies, it is much more difficult to partake directly in civil disobedience. The tactic of civil disobedience is only one of the tactics through which the Glasgow Agreement's objectives can be fulfilled. Additionally, we acknowledge that the strategy of civil disobedience has long been used, under various names, by many before us, particularly in marginalised communities and in the Global South, and we would not be able to join this struggle without these historical and contemporary sacrifices, and continuous action against climate change through struggles to keep fossil fuels underground and resistance to other industrial causes of global warming;

6. Support each other and coordinate to define their own local and national strategies and tactics on how to enact the climate agenda, and to call for the support of other member organisations of the Glasgow Agreement (nationally and internationally). The organisations from the Global North underline their commitment to support those in the Global South, through solidarity with existing struggles and by directly addressing projects led by governments, corporations, banks and financial institutions based in the Global North...

Glasgow Agreement Gaps

The Glasgow Agreement is a profound, eloquent input into global climate politics, one that various strains of progressives and radicals right through to eco-socialists could warm to. However, the emphasis on leaving fossil fuels underground—absolutely essential as a first priority—means that, like the Paris Climate Agreement, some critical areas (e.g. cutting emissions that emanate from militaries, air and maritime transport) are left out. For example, there is no gender analysis, which is a huge flaw.

Below, however, let me address four other central points that are vital for future drafters: the balance of forces represented by Washington's return to COP leadership; intergenerational equity; tactics; and the need for alignment with growing anti-extractivist movements. In taking up the latter four shortcomings, a broader concern arises, associated with a warning from the militant eco-feminist group Accion Ecologica from Quito, Ecuador. Its founder expressed frustration at the agreement's prioritisation of an 'emissions inventory' that distracted from root capitalist causes of the climate crisis (Yanez 2021).

First, the agreement could better alert readers to the current *balance of forces*—and how to change that array of power. After all, there is a dangerous new factor that became apparent in January 2021: the US corporate-neoliberal re-entry to the UNFCCC, led by Joe Biden and his climate envoy John Kerry (former Secretary of State in 2015 at Paris) (Bond 2021a). One result of the shift from Trump climate denialism to this new regime is renewed emphasis on market strategies and 'net zero' accounting gimmickry. Such "green capitalism" and associated false solutions are noted in the agreement's final statement of principles—and flagged in much more detail by, among others, Corporate Accountability, Global Forest Coalition and Friends of the Earth International (2021).

Second, the Agreement does not address *rights of future generations*, notwithstanding rising youth rage. This is an absolutely critical new factor in climate politics, so it represents a surprising gap given Fridays for Future's potential and the clarity with which Thunberg and her allies continue to express exceptionally tough critique. Thunberg's successful approach, based on speaking truth to power at elite events that gain her unprecedented publicity for the climate cause, has thus far focused on delegitimising the corporate and multilateral establishment. To illustrate, when in mid-2021 Kerry was quoted endorsing mythical technofix strategies—"I am told by scientists, not by anybody in politics, but by scientists, that 50% of the reductions we have to make are going to come from technology that we don't yet have"—she tweeted, "Great news! I spoke to Harry Potter and he said he will team up with Gandalf, Sherlock Holmes & The Avengers and get started right away!" The anger and sense of urgency that leading youth activists can generate stunned the world since her Stockholm sit-ins began mid-2018, especially in September 2019 when seven million protesters coordinated international events over the course of a week. No one can doubt how desperately we need a post-COVID revival of that spirit, especially given internal divisions in the US Sunrise Movement on the one hand, but on the other, a rising network of Global South youth preparing to take greater leadership once COVID-19 threats to unified international actions recede.

Third, in relation to *tactics*, the agreement's framing is unsatisfyingly narrow. The authors do not acknowledge that, unfortunately, there's a long-standing style of *tokenistic* climate-related civil disobedience (CD): set-piece, pre-negotiated arrests that are mainly publicity enhancing. Such predictable, non-disruptive CD characterises leading currents within climate-action politics and also some strains within Climate Justice. It needs rethinking since the approach is so readily assimilated, with accompanying platitudes, by those wielding power (also see chapters by Gardham and Paterson, this volume). Indeed, CD as practiced in this way provides diminishing public-educational opportunity, much less the capability to actively threaten *status quo* polluting activities (Malm 2021).

So, on the one hand, the agreement certainly recognises that many activists in vulnerable situations cannot take steps toward CD for fear of extreme repression. But, on the other, the agreement is not quite brave enough to openly address a different, more militant approach: *blocking and even sabotaging extraction, transport, refining, combustion and financing of fossil fuels and other sources of greenhouse gas emissions.*

This is not terribly unusual activism against fossil fuel corporations, as in the Global South such uncivil disobedience was pioneered against oil extraction during the early 1990s by Ken Saro-Wiwa's Movement for the Emancipation of the Ogoni People in the Niger Delta (before his execution in 1995). Disruptive CD is increasingly being practiced by many others, for example XR in countless sites of corporate power, the Standing Rock Sioux Tribe against the Dakota Access Pipeline, or Ende Gelände in Germany's coal fields. For Naomi Klein (2014), this spirit deserves the term "blockadia," and the Environmental Justice Atlas at http://ejatlas.org documents hundreds of such cases.

When it comes to this contradiction, there is a need to rebalance the always-uncomfortable division of labour between rigorous *tree-shakers*—hard-core activists who are ready to disrupt power and face jail time in the process—and *jam-makers* on the inside of the COPs, doing more polite advocacy. Of the many civil society COP attendees, several prominent Glasgow Agreement signatories are typically leaders. Yet notwithstanding all their passion and strategic insight, they rarely attempt to actively empower the tree-shakers by paying tribute to *their most radical actions.*

The COP17 People's Space in Durban was a good case site to understand these flaws. Our comrades and I (as a university-based host of the People's Space) (Bond 2012b) failed miserably along these lines. Although our South African and African CJ forces possessed powerful principles and sound analyses, the team was distracted when it came to establishing effective strategies, tactics and alliances. Counter-summitry and protests were impotent, in part because distinctions between tree-shakers in the People's Space, and jam-makers inside the Durban International Convention Centre, were never clearly established by the C17 network, one that sought unity over clarity. Most subsequent COP outside-protest and inside-advocacy scenes reflected the same failure, leading in Paris to confusing stances within the "climate movement," reflecting uncivil society militantly promoting CJ on the one hand, and on the other, civilised society groups begging for mere climate action (Bond 2018). The problem has persisted to this day, in Africa generally and South Africa specifically (Mwenda and Bond 2020).

Fourth, there is a profound challenge from Accion Ecologica, a signatory whose April 2021 letter from the eco-feminist organisation's co-founder Ivonne Yanez (2021) warns that by lacking clarity on broader ideology, the agreement risks "colliding with the anti-extractivist movements in the world." These include many struggles

Yanez herself supports across the Andes, especially Ecuador. The specific contradictions relate to how "minimally-necessary mining" might be defined, and whether some of the ingredients necessary for a decarbonised economy—lithium for batteries, titanium dioxide for highly-reflective white paint, palladium and rhodium for so-called "green hydrogen" fuel cells, and other rare-earth minerals—themselves are being contested in sites like the Andes and several South Africa anti-mining conflicts (also see Dunlap, this volume). For Yanez (2021), "asking anti-extractivist social movements—mainly in the South—to 'make inventories of emissions' is like asking us to take inventories of future forms of dispossession and exploitation."

Like many who soon tired of COP-oriented advocacy work, Yanez (2021) instead adopts—and amplifies—the tradition of delegitimisation:

> As for the Paris Agreement, and its predecessors, they were designed precisely to confuse. And they succeeded. They were conceived so that organisations, instead of talking about how to confront extractivism, how to end injustices and inequalities, would be busy talking about degrees of temperature, and calculating tons of CO2. The Paris Agreement and the absurd and malevolent proposals it entails divert attention from what is important: confronting patriarchal, neo-colonial and racist capitalism. They have succeeded for almost 25 years in distracting attention. And so, we end up thinking that first come the IPCC numbers with an army of experts counting molecules and in second place come the anti-capitalist extractivist resistances.
>
> To confront climate change we have to confront the capitalist system that is institutionalised (for example, through the Paris Agreement) and global. But who are the anti-capitalist movements? The main ones in the world today are anti-extractivist movements, anti-capitalist labour movements, territorial and community-based feminist movements, anti-white supremacist movements, anti-colonial movements, movements fighting for water, anti-debt movements, anti-agribusiness movements... A *movement to reduce emissions* falls short among this tide of struggles, and I doubt it will make much difference in the struggle against capitalism. And while the Glasgow Agreement takes up many ideas, the anti-capitalist, concrete and territorial struggles that are also global are more important. Learning and listening from these frontline climate movements is a task.

As a final point, although Accion Ecologica does not advocate overly-technicist work such as the agreement's proposed census of emissions, there are nevertheless two rationales for doing so *if conjoined with anti-extractivist struggles.* The first is to identify whether a given country's activists have been maximising their potential to link up and challenge their economy's most egregious polluters, in the form of an accompanying inventory of anti-emissions campaigning. This is something that autonomist-style blockadia strategies require better networking to achieve: linkage of their local organic (and sometimes atomised) struggles for maximum impact, including tackling various national state subsidies, regulatory fora, legislation, and more generally, politicians' (and often police or even army) support for extractive industries.

The second rationale is one that appeals to eco-socialists, namely *the planned reduction of emissions*—a process which would otherwise be accomplished erratically and unreliably through either protest (rarely) or market forces. The danger of relying on the latter was evident in April 2020 when there was great cheering by climate activists at the collapse in fossil fuel prices, but disillusionment when they very quickly recovered.

Conclusion: A Routing from Climate Injustice to Eco-Socialism

The UNFCCC continued to disappoint reformers into mid-2021 as COVID-19 dragged on. After eighteen months of no negotiations, the Bonn intersessional was conducted via Microsoft Teams. "Progress is pretty slow if not non-existent at this session, but I wouldn't just blame it on the virtual format," one analyst told *Climate Brief* (2021). (But the distanced format, worsened by time zone difficulties, did reduce the impact of some crucial Global South negotiators who suffered communication interruptions). As the US West witnessed record heat-waves and another terrible fire season loomed, the leaders remained hesitant to tackle critical problems of adaption and finance, leaving "nothing substantive" to agree on in the Glasgow COP26, according to Bangladeshi negotiator Mizan Khan. The "vast majority" of poor countries voiced objections to Western sabotage of the talks, given that the latest climate loss and damage accounts (from 2019) showed that when the Global North suffered, 60% of the damage was commercially insured, in contrast to 4% in the Global South. And as Carbon Brief (2021) reported, "it is universally assumed that climate finance is currently falling short of the $100 billion goal" for annual disbursements especially if grant (not loan) finance is considered independently of prevailing aid. In sum, the insider strategy had met its limits.

A revealing French working-class strategic choice in earlier (mid-1960s) battles—as articulated by Gorz (1967)—was whether activists could identify opportunities for *non-reformist, transformative reforms,* or instead settle for 'reformist reforms' that in turn strengthen the assimilationist power of the *status quo.* Most climate activists working at global scale have only achieved reformist reforms to date, and the cost—legitimising the counterproductive Paris Climate Agreement—is enormous. But when it comes to the UNFCCC, or even micro-campaigning against specific emitters, 'fix it or nix it' choices, and resulting openings for more radical reforms, i.e. to break not polish the chains of oppression, sometimes arise when least expected.

Typically there are two contrary directions for framing campaigns. First, reformist reforms

- strengthen the internal logic of the system, by smoothing rough edges,
- allow the system to relegitimise,
- give confidence to status quo ideas and forces,
- leave activists disempowered or coopted, and
- confirm society's fear of power, apathy and cynicism about activism.

But second, in contrast, non-reformist reforms (or 'transformative reforms')

- counteract the internal logic of the system, by confronting core dynamics,
- continue to delegitimise the system of oppression,
- give confidence to critical ideas and social forces,
- leave activists empowered with momentum for the next struggle, and
- replace social apathy with confidence in activist integrity and leadership.

We have seen this in South Africa on occasion, such as in the defeat of apartheid. In 1983, as economic crisis began to worry the country's white leaders, several wide-ranging reformist reforms were offered by the apartheid regime to black voters: assimilationist seats offered in second-tier sites of representation (segregated parliamentary bodies, satellite municipalities and Bantustan pseudo-countries). Black liberation activists rejected these, for as Archbishop Desmond Tutu put it, these reforms represented "polishing the chains of apartheid," when the chains needed to be *broken*. Principled activists campaigned for a non-reformist principle: one person, one vote in a unitary state. In 1994, with Nelson Mandela by then free from his 1963–1990 jail term and leading the broad-based anti-apartheid movement, they changed the balance of forces sufficiently to win democracy. Since the early 2000s there have been similar battles and victories. When South African activists waged struggles against state and capital to gain free anti-retroviral AIDS medicines in the early 2000s or free tertiary education for the working class in 2015–2017, these entailed successful national coordinations of localised grievances (Ngwane and Bond 2020).

With this in mind, my own sense is that the Glasgow Agreement *principles* are very appealing. Yet there is a vagueness when it comes to *analysis, strategies, tactics and alliances*, beginning with the very obvious question of whether the COP26 and future UNFCCC events will be sites of clarity—or instead confusion—over legitimation or delegitimation. This difficult choice is shared by virtually all the climate movements I have seen working towards some form of influence over the Glasgow COP26 in 2020–21. The groups involved in the agreement are generally the most admirable from the perspective of CJ, but all remain unclear on whether and how to pursue the delegitimation strategy Thunberg has embodied so eloquently.

The alignments of this PASTA framework are vital in the cases I have seen in South Africa—against both apartheid and post-apartheid socio-economic oppression—and are parallel to what is now needed for global and local CJ movements, given the UNFCCC's failures. No matter how much 2021 propaganda is offered about bandaging the Paris deal at Glasgow COP26 or subsequent COPS, the power relations remain terribly adverse. In this context, the PASTA framing for climate justice takes two forms, one based on past activist practice, including limitations; and the other based on the contradictions between CJ and "ecological modernisation" strategies, in which a dialectical resolution in eco-socialism can be theorised (Bond 2021b).

Without the space here to address how difficult a process that is (e.g. in technological choices or use of ecological valuation techniques), it should nevertheless be obvious that a major problem confronts CJ and efforts like the Glasgow Agreement. The arguments above presume increasing clarity over the major differences between what CJ advocates historically insisted upon, by way of non-reformist reforms that can end the climate crisis in a manner that is just both globally and locally, and the UNFCCC COP26 agenda of reformist reforms based on market and technological strategies. But the latter, even when articulated by the most enlightened elites (like Mary Robinson), are "designed precisely to confuse," to recall Yanez.

So to arrive at such far-reaching reforms—parallel to South Africans ending apartheid and then decommodifying essential state services using an anti-neoliberal, proto-socialist "commons" approach—the activists must first confront and defeat the reformist reforms put in their way. Delegitimation of the elites, as Thunberg and Glasgow Agreement authors agree, should both embrace and transcend personal insults, and from there, rapidly address the full set of divergent principles, analyses, strategies, tactics and alliances that distinguish CJ from the elites' self-proclaimed climate action, which in reality is so passive that the future of humanity and all other species is, increasingly, in question.

References

Bond, Patrick, *Politics of Climate Justice* (Pietermaritzburg: University of KwaZulu-Natal Press, 2012a).

Bond, Patrick, 'Durban's Conference of Polluters, Market Failure and Critic Failure', *ephemera*, 12 (2012b), 42–69, http://www.ephemerajournal.org/contribution/durban%E2%80%99s-conference-polluters-market-failure-and-critic-failure.

Bond, Patrick, 'Social Movements for Climate Justice during the Decline of Global Governance', in *Rethinking Environmentalism*, ed. by S. Lele, E. Brondizio, J. Byrne, G. M. Mace, and J. Martinez-Alier (Cambridge: Massachusetts Institute of Technology Press, 2018), pp. 153–82, https://doi.org/10.7551/mitpress/11961.003.0013.

Bond, Patrick, 'Biden-Kerry International Climate Politricks', CounterPunch, 2 February (2021a), https://www.counterpunch.org/2021/02/01/biden-kerry-international-climate-politricks.

Bond, Patrick, 'As Climate Crisis Worsens, the Case for Eco-socialism Strengthens', *Science & Society* (2021b, forthcoming).

Carbon Brief, 'UN climate talks: Key outcomes from the June 2021 virtual conference', 18 June (2021), https://www.carbonbrief.org/un-climate-talks-key-outcomes-from-the-june-2021-virtual-conference.

Corporate Accountability, Global Forest Coalition and Friends of the Earth International, *The Big Con* (2021), https://www.corporateaccountability.org/wp-content/uploads/2021/06/The-Big-Con_EN.pdf.

Glasgow Agreement (2020), https://glasgowagreement.net/en/.

Glasgow Caledonian University Centre for Climate Justice, *Proceedings of the World Forum on Climate Justice*, 19–21 June (2019), Glasgow, https://researchonline.gcu.ac.uk/ws/portalfiles/portal/39771260/Proceedings_of_the_World_Forum_on_Climate_Justice_Online.pdf.

Gorz, Andre, *Strategy for Labor* (Boston: Beacon Press, 1967).

Khan, Mizan, Stacy-Ann Robinson, Romain Weikmans, David Ciplet, and J. Timmons Roberts, 'Twenty-five Years of Adaptation Finance Through a Climate Justice Lens', *Climatic Change*, 161 (2020), 251–69, https://doi.org/10.1007/s10584-019-02563-x.

Klein, Naomi, *This Changes Everything* (Toronto: Alfred A. Knopf, 2014).

Malm, Andreas, *How to Blow up a Pipeline* (London: Verso, 2021).

McKibben, Bill, 'With Climate Agreement, Obama guts Progressive Values', *Grist*, 19 December (2009), https://grist.org/article/2009-12-18-with-climate-agreement-obama-guts-progressive-values.

Milman, Oliver, 'James Hansen, Father of Climate Change Awareness, calls Paris Talks "A Fraud"', *The Guardian*, 12 December (2015), https://www.theguardian.com/environment/2015/dec/12/james-hansen-climate-change-paris-talks-fraud.

Mwenda, Mithika and Patrick Bond, 'African Climate Justice Articulations and Activism', in *Climate Justice and Community Renewal: Resistance and Grassroots Solutions*, ed by B. Tokar and T. Gilbertson (London: Routledge, 2020), 108–28, https://doi.org/10.4324/9780429277146-8.

Ngwane, Trevor, and Patrick Bond, 'South Africa's Shrinking Sovereignty: Economic Crises, Ecological Damage, Sub-Imperialism and Social Resistances', *Vestnik RUDN. International Relations*, 20(1) (2020), 67–83, https://doi.org/10.22363/2313-0660-2020-20-1-67-83.

Thunberg, Greta, 'If World Leaders Choose to Fail Us, My Generation Will Never Forgive Them.' *The Guardian*, 23 September (2019),

https://www.theguardian.com/commentisfree/2019/sep/23/
world-leaders-generation-climate-breakdown-greta-thunberg.

Thunberg, Greta, 'We are Speeding in the Wrong Direction on
Climate Crisis', *The Guardian*, 10 December (2020), https://
www.theguardian.com/environment/2020/dec/10/
greta-thunberg-we-are-speeding-in-the-wrong-direction-on-climate-crisis.

Thunberg, Greta, Tweet. Stockholm, 16 May (2021), https://twitter.com/
GretaThunberg/status/1393974674867036164?s=20.

Yanez, Ivonne, 'Para el Grupo del Acuerdo de Glasgow', Letter to the Glasgow
Agreement, Quito, 15 April (2021).

27. Public Engagement with Radical Climate Change Action

Lorraine Whitmarsh

The role of people in addressing climate change is often relegated to merely consumers. While adopting electric vehicles and heat pumps, for example, will indeed be critical for reaching climate targets, people will also need to engage as political, social and professional actors to achieve the scale of societal transformation needed. This includes actively engaging in both decision-making and in delivery in respect of climate action. Here, I discuss the varied roles that the public can play in decision-making and in taking rapid and radical climate action, their current levels of engagement with climate change, and how to foster further public action. I argue that we have a unique opportunity as we build back society post-COVID-19 to lock in low-carbon habits created during the pandemic, and to build on the growing social mandate for bold policy action to support sustainable lifestyles.

Introduction

In this chapter, I discuss the multiple roles the public can play in climate action—as consumers, citizens, parents, community members, employees and professional decision-makers—and how we can better engage the public in decision-making and action to achieve rapid and significant emission cuts to mitigate climate change. I argue that public engagement is critical both for building a public mandate for radical social change, but also for achieving profound lifestyle, community, organisational and policy transformation. Public engagement is

https://doi.org/10.11647/OBP.0265.27

potentially a very broad concept that captures: (a) engagement in *decision-making* (including policy-making) about how to reach net zero; and (b) engagement in *delivery* of action to reach net zero (i.e., 'behaviour change' in its broadest sense, including lifestyle change, technology adoption/use, policy support, activism, and awareness raising). These two forms of engagement are interlinked—if we have joined-up thinking and a national conversation on these issues (i.e., engagement with decision-making) that would also help with the delivery of net zero since it provides the context and rationale for specific behavioural and structural interventions; and fosters collective efficacy and trust (Dietz and Stern 2009; Capstick et al. 2019).

Evidence shows a key predictor of policy acceptance by the public is perceived *fairness*, including procedural fairness (i.e., involving people in decisions that affect them; Dreyer and Walker 2013; Schmocker et al. 2012). This means that we cannot have a net zero transition without the public 'noticing' (i.e., via supply-side change and consumer nudges); indeed, social/behavioural change is required for the *majority* of measures to reach net zero (CCC 2019). Reconfiguration of urban environments, food and transport systems, energy technologies, and provision of goods and services, are hugely disruptive to lifestyles and may require changes in values and norms. Developments that may be less disruptive to lifestyles but still pose risks and costs to society (e.g., supply-side and negative emissions technologies) also require public buy-in (RCUK 2010). Thus, the inevitably visible, disruptive and risky transformation to net zero requires a *public mandate*—hence active engagement with publics to co-produce net zero futures and pathways, including collectively assessing their risks and benefits, and to achieve buy-in to their delivery (also see Halme et al., this volume).

How Engaged Is the Public?

So, how 'engaged' is the public with climate change at the moment? The last few years have seen a significant rise in public concern about climate change: polling in the UK and elsewhere showed unprecedented worry about climate change during 2019, which has been maintained into 2020 despite competing concerns over COVID-19 (BEIS 2019; Leiserowitz et al. 2020; Ipsos MORI 2020). In fact, one UK survey (Whitmarsh et al. 2020) found that the perceived urgency of tackling climate change was higher during the pandemic (May 2020) than in August the previous year (74% up from 62% seeing it as an 'extremely high' or 'high' level of urgency). Furthermore, support for climate change mitigation policies, including measures to decrease meat consumption and flying, was higher during the pandemic (67% and 85%, respectively) than in 2019 (53% and 67%, respectively). This apparent support for ambitious action to address climate change has been reflected in (and strengthened by) high-profile public protests and 'school strikes' around the world (Thackeray et al. 2020), as well as a shift in media language and societal discourse to reconceptualise the issue as a 'climate emergency' (Carrington 2019; Zhou 2019).

There has also been a growth in deliberative democracy activities that provide a stronger voice for the public in national and local policy-making on climate change, notably the Climate Assembly UK, the French Citizens' Convention for Climate, and various regional and city-level citizens' assemblies and juries (Capstick et al. 2020). These engagement activities seek to elicit informed public opinion on low-carbon visions and pathways, and have shown strong support for ambitious climate action (e.g., CAUK 2020; Citizens Convention on Climate 2020). Yet, it is widely acknowledged that stated preferences (via polls, interviews, deliberative discussions, etc.) often diverge from actual behaviour—the so-called 'value-action' gap (Blake 1999). Indeed, despite people's good intentions, there remain significant structural and social barriers to engagement with climate change at the behavioural level (e.g., Lorenzoni et al. 2007). Demand for material goods, car travel and aviation, for example, have grown rapidly in recent years; while low-carbon behaviours—such as walking and cycling, eating a plant-based diet, and reduced consumption—often remain inconvenient, inaccessible, socially and/or economically costly (CAUK 2020; Whittle et al. 2019).

Here, I provide some concrete suggestions for how to build public engagement with climate change, both in terms of providing a stronger role for the public in decision-making, and in public participation in the delivery of action to achieve the UK's net zero goal. These suggestions are grounded in psychological, sociological, economics and political science literatures which provide insights on public participation and behaviour change in its broadest sense.

Engagement in Decision-Making

In the case of engagement in decision-making, this would ideally involve local deliberative processes to identify tailored solutions and build community participation, as well as an overarching 'national conversation' on options for reaching net zero, including *supply-side and demand-side* changes. This would require being:

a. Joined up across sectors and scales (i.e., consistent messaging and policies embedded across government departments, devolved governments, local authorities, etc.). This could involve co-development of a shared vision and 'branding' around net zero (similar to the 'Energiewende' in Germany; Moss et al. 2015) that provides the coordinated, overarching, joined-up vision demanded by citizens (CAUK 2020), tying together the variety of changes people see and are asked to make, and giving a sense of collective efficacy and ownership; and

b. Timely in influencing decision-making—i.e., upstream engagement in policy-making at national and local levels (using deliberative approaches, such as citizens' assemblies/juries, online deliberative polling, etc.; Dietz and Stern 2009).

Engagement in Delivery

Behaviour change is a central element of delivering net zero, so public engagement is key for realising this goal (CCC 2019). Behaviour change, though, is not only required for consumer-citizens, but also other individuals and groups across a range of contexts (e.g., parents, communities, employees, employers, political actors; Whitmarsh et al. 2010). Behaviour change is not only about adoption of net zero technology—though consumer behaviour is important. It more fully encompasses:

c. Adoption of low/no-carbon technology and products (in personal or professional contexts);

   d. Use and disposal of low/no-carbon technology and products (in personal or professional contexts);

   e. Acceptance of (or demand for) large-scale low/no-carbon infrastructure, including supply-side and greenhouse gas removal technologies;

   f. Political action to support or demand climate change action (voting, protesting, boycotting);

   g. Community and voluntary action to promote low-carbon choices (hosting or owning low/no-carbon developments, volunteering for climate causes, etc.); and

   h. Creating and disseminating climate change narratives/discourses that normalise and promote low-carbon lifestyles, call out inaction (by people, businesses, policy-makers, schools, family members, etc.), and raise awareness through conversations, as well as modelling change through action.

There is a vast literature on how to change behaviour, and much can be learnt from historical and international examples of transformation (e.g., tobacco control, urban sustainable transport). Key findings from this evidence base include:

   a. Change is required across *multiple levels* and *using various levers* (information and incentives alone will not be sufficient; broader social, infrastructural, technical, and regulatory interventions are also required; Lorenzoni et al. 2007; Corner et al. 2019);

   b. Interventions should exploit and be framed around *co-benefits* or *win-wins* (e.g., wellbeing/health, equity, cost-saving/profit; Bain et al. 2016; Maibach et al. 2010; Whitmarsh and Corner 2017);

   c. Interventions should be *timely* at the point of decision-making (e.g., buying a car or appliance; renovating a house) and when habits are disrupted/malleable (see below; Graham-Rowe et al. 2011; Wilson et al. 2015);

d. Changing social norms through *leadership,* exemplifying/
    disseminating innovations and *good practice* through
    networks, and using *trusted messengers* to communicate, are
    important (Clayton et al. 2015; Corner et al. 2019; Pettifor
    et al. 2017); and

e. Building *public support* is key to leveraging government
    action for behavioural interventions (particularly if there is
    industry resistance; Willis 2017; Corner et al. 2019).

Moments of Change and COVID-19

A growing literature points to the importance not only of *how* to
intervene to achieve social and lifestyle change, but also *when.* Much
of our behaviour is habitual—unconscious routines triggered by
contextual cues (e.g., 'it's 8am, time to drive to work') rather than
conscious deliberation of alternatives (e.g., 'which mode of transport
would be best today?'; Kurz et al. 2015). Habits are one of the
strongest impediments to lifestyle change, acting to 'lock in' behaviour
(Marechal and Lazaric 2011). Many interventions (e.g., information
campaigns) are ineffective because they are not strong enough to
disrupt habits (Verplanken et al. 1997). But since habits are cued by
stable contexts (i.e., the same time, place and/or social group; Wood
et al. 2005), change in context disrupts habits (Verplanken et al. 2008).
Consistent with this observation, 'moments of change'—defined as
"occasions where the circumstances of an individual's life change
considerably within a relatively short timeframe" (Thompson et al.
2011)—have been identified as one of the most important levers for
lifestyle change (House of Lords 2011; Capstick et al. 2014). Research
shows that disruptions—whether concerning a person's life-course
(e.g., moving home) or structural (e.g., economic downturn, extreme
weather events)—can provide opportunities to recraft social practices
in new directions (Verplanken et al. 2018; Birkmann et al. 2010), for
example shifting from commuting by car to home-working (Marsden
et al. 2020). Furthermore, interventions targeted to moments of
change are more effective than at other times (Verplanken et al. 2018).
Several studies show that mobility interventions are more effective
when targeted to relocation (Thøgersen 2012; Ralph and Brown 2017;

Bamberg 2006). Other low-carbon behaviours, such as energy efficiency and wasted reduction measures, have also been shown to be more effectively changed using low-cost interventions in the twelve weeks following relocation (Verplanken and Roy 2016; Maréchal 2010), as well as at other moments of change, such as buying an electric vehicle (Nicolson et al. 2017). Other such opportunities to intervene include temporal milestones (e.g., New Year, becoming an adult), having a child, retiring, infrastructure disruption (e.g., road closures), and COVID-19 (e.g., Verplanken et al. 2018; Burningham & Venn 2020).

COVID-19 and measures to respond to it may be the most significant disruption to lifestyles since World War II. Citizens are working, consuming and interacting in new ways, some of which may be more desirable both personally and environmentally (e.g., commuting less). For example, one UK study (Whitmarsh et al. 2020) found that during lockdown: online food shopping more than doubled; food waste and consumption of energy and goods reduced; working from home rose significantly and most people found this a positive experience. In line with this, around a third said they intend to increase the amount they work from home (compared to pre-lockdown) once restrictions are removed, and even more plan to socialise more online (43%) and to fly less on holidays (47%). Importantly, of course, intentions do not always manifest in behaviour change (Whitmarsh 2009). Since new habits take two to three months to form (Lally et al. 2010), lockdown periods in most countries have been long enough to establish new routines. However, when lockdowns are lifted, there is a risk of recidivism into pre-existing habits (Carden and Wood 2018), particularly if economic stimulus measures promote unfettered, high-carbon consumption (Peters 2020). So, while COVID-19 may represent a unique window of opportunity to promote low-carbon lifestyles, this is only likely to occur with appropriate infrastructure, incentives, and norms to encourage and lock in new low-carbon routines. Fortunately, there is strong public support for net zero policies (e.g., shifting to low-carbon transport; reducing red meat consumption) and a green recovery (CAUK 2020; Whitmarsh et al. 2020), which provides a mandate for policy-makers to take bold climate change measures to establish and lock in low-carbon habits.

Conclusion

Public engagement in decision-making and action is essential for radical societal transformation to address climate change. While public support for climate action has grown in recent years, demand-side emission reductions lag far behind supply-side reductions, highlighting the need to focus efforts on achieving society-wide behavioural change (CCC 2019). Much can be learnt on how to engage the public with climate change from COVID-19; however, there are unique challenges associated with climate change that make it a "different kind of crisis" (Howarth et al. 2020). Although the pandemic has shown that measures to change behaviour and society can be taken rapidly, we require a social mandate for such radical interventions to be implemented for the longer term, for example via further deliberative democratic opportunities and a coherent national conversation on climate change. COVID-19 as a 'moment of change' has also created many low-carbon habits that could be locked in with the right policy measures, such as reallocating road space from cars to active and public modes, economic (dis)incentives to promote consumption of low-carbon products and services, and support for businesses to encourage more flexible working and teleconferencing (e.g., Cairns et al. 2002; Henderson and Mokhtarian 1996; Capstick et al. 2014; CAUK 2020). Embarking on a green economic recovery from COVID-19 requires using the insights outlined here on how to engage the public to achieve a low-carbon societal transformation.

References

Bain, Paul, Taciano Milfont, Yoshi Kashima, et al., 'Co-benefits of Addressing Climate Change Can Motivate Action Around the World', *Nature Climate Change*, 6 (2016), 154–57, https://doi.org/10.1038/nclimate2814.

Bamberg, Sebastian, 'Is a Residential Relocation a Good Opportunity to Change People's Travel Behaviour? Results From a Theory-driven Intervention Study', *Environment & Behavior*, 38 (2006), 820–40, https://doi.org/10.1177/0013916505285091.

BEIS, *BEIS Public Attitudes Tracker: Wave 29—Key Findings* (2019), https://www.gov.uk/government/statistics/beis-public-attitudes-tracker-wave-29.

Birkmann, Joern, P. Buckle, Jill Jaeger, et al., 'Extreme Events and Disasters: A Window of Opportunity for Change? Analysis of Organizational,

Institutional and Political Changes, Formal and Informal Responses After Mega-disasters', *Natural Hazards*, 55 (2010), 637–55, https://doi.org/10.1007/ s11069-008-9319-2.

Blake, James, 'Overcoming the "Value-action Gap" in Environmental Policy: Tensions Between National Policy and Local Experience', *Local Environment*, 4 (1999), 257–78, https://doi.org/10.1080/13549839908725599.

Burningham, Kate, and Susan Venn, 'Are Lifecourse Transitions Opportunities for Moving to More Sustainable Consumption?', *Journal of Consumer Culture*, 20 (2020), 102–21, https://doi.org/10.1177/1469540517729010.

Cairns, Sally, Stephen Atkins, and Phil Goodwin, 'Disappearing Traffic? The Story so Far' (Nacto.org, 2002), https://nacto.org/docs/usdg/disappearing_ traffic_cairns.pdf.

Capstick, Stuart, Irene Lorenzoni, Adam Corner, and Lorraine Whitmarsh, 'Social Science Prospects for Radical Emissions Reduction', *Carbon Management*, 4 (2014), 429–45, https://doi.org/10.1080/17583004.2015.1020011.

Capstick, Stuart, Christina Demski, Catherine Cherry, Caroline Verfuerth, and Katharine Steentjes, *Climate Change Citizens' Assemblies*, CAST Briefing Paper 03 (2020), https://orca.cardiff.ac.uk/131693/1/CAST-Briefing-03-Climate-Change-Citizens-Assemblies.pdf.

Carden, Lucas, and Wendy Wood, 'Habit Formation and Change', *Current Opinion in Behavioral Sciences*, 20 (2018), 117–22, https://doi.org/10.1016/j. cobeha.2017.12.009.

Carrington, Damien, 'Why the Guardian is Changing the Language it Uses About the Environment' (Guardian.com, 2020), https://www.theguardian. com/environment/2019/may/17/why-the-guardian-is-changing-the-language-it-uses-about-the-environment.

CAUK, 'The Path to Net Zero: Climate Assembly UK full report' (Climateassembly. uk, 2020), https://www.climateassembly.uk/.

CCC, *Net Zero: The UK's Contribution to Stopping Global Warming* (UK Committee on Climate Change, London, 2019), https://www.theccc.org.uk/ wp-content/uploads/2019/05/Net-Zero-The-UKs-contribution-to-stopping-global-warming.pdf.

Citizens Convention on Climate, 'Citizens Convention on Climate Final report' (conventioncitoyennepourleclimat.fr, 2020), https://www. conventioncitoyennepourleclimat.fr/wp-content/uploads/2020/07/062020-CCC-propositions-synthese-EN.pdf.

Clayton, Susan, Patrick Devine-Wright, Paul Stern, Lorraine Whitmarsh, Amanda Carrico, Linda Steg, Janet Swim, and Mirilia Bonnes, 'Psychological Research and Global Climate Change', *Nature Climate Change*, 5 (2015), 640–46, https://doi.org/10.1038/nclimate2622.

Corner, Adam, Hilary Graham, and Lorraine Whitmarsh, 'Engaging the Public on Low-carbon Lifestyle Change', CAST Briefing Paper 01 (Cast.ac.uk, 2019), http://cast.ac.uk/wp-content/uploads/2020/01/CAST-briefing-01-Engaging-the-public-on-low-carbon-lifestyle-change-min.pdf.

Dietz, Tom, and Paul Stern (eds), *Public Participation in Environmental Assessment and Decision-Making* (Washington DC: National Academies Press, 2009).

Dreyer, Stacia, and Iain Walker, 'Acceptance and Support of the Australian Carbon Policy', *Social Justice Research*, 26 (2013), 343–62, https://doi.org/10.1007/s11211-013-0191-1.

Graham-Rowe, Ella, Stephen Skippon, Benjamin Gardner, and Charles Abraham, 'Can We Reduce Car Use and, if so, How? A Review of Available Evidence', *Transportation Research Part A: Policy & Practice*, 45 (2011), 401–18, http://dx.doi.org/10.1016/j.tra.2011.02.001.

Henderson, Dennis, and Patricia Mokhtarian, 'Impacts of Center-based Telecommuting on Travel and Emissions: Analysis of the Puget Sound Demonstration Project', *Transportation Research Part D: Transport & Environment*, 1 (1996), 29–45, https://doi.org/10.1016/S1361-9209(96)00009-0.

House of Lords, *Behaviour Change* (House of Lords Select Committee on Science & Technology, London, 2011).

Howarth, Candice, Peter Bryant, Adam Corner, et al., 'Building a Social Mandate for Climate Action: Lessons from COVID-19', *Environmental & Resource Economics*, 76 (2020), 1107–15, https://doi.org/10.1007/s10640-020-00446-9.

Ipsos MORI, *Two Thirds of Britons Believe Climate Change as Serious as Coronavirus and Majority Want Climate Prioritised in Economic Recovery* (Ipsos.com, 2020), https://www.ipsos.com/ipsos-mori/en-uk/two-thirds-britons-believe-climate-change-serious-coronavirus-and-majority-want-climate-prioritised.

Kurz, Tim, Benjamin Gardner, Bas Verplanken, and Charles Abraham, 'Habitual Behaviors or Patterns of Practice? Explaining and Changing Repetitive Climate-relevant Actions', *WIREs Climate Change*, 6 (2015), 113–28, https://doi.org/10.1002/wcc.327.

Lally, Phillippa, Cornelia van Jaarsveld, Henry Potts, and Jane Wardle, 'How Are Habits Formed: Modelling Habit Formation in the Real World', *European Journal of Social Psychology*, 40 (2010), 998–1009, https://doi.org/10.1002/ejsp.674.

Leiserowitz, Anthony, et al., 'Climate Change in the American Mind' (Climatecommunication.yale.edu, 2020), https://climatecommunication.yale.edu/publications/climate-change-in-the-american-mind-april-2020.

Lorenzoni, Irene, Sophie Nicholson-Cole, and Lorraine Whitmarsh, 'Barriers Perceived to Engaging with Climate Change Among the UK Public and their Policy Implications', *Global Environmental Change*, 17 (2007), 445–59, https://doi.org/10.1016/j.gloenvcha.2007.01.004.

Maibach, Edward, Matthew Nisbet, Paula Baldwin, et al., 'Reframing Climate Change as a Public Health Issue: An Exploratory Study of Public Reactions', *BMC Public Health*, 10 (2010), 299, https://doi.org/10.1186/1471-2458-10-299.

Maréchal, Kevin, and Nathalie Lazaric, 'Overcoming Inertia: Insights from Evolutionary Economics into Improved Energy and Climate Policies', *Climate Policy*, 10 (2011), 103–19, https://doi.org/10.3763/cpol.2008.0601.

Maréchal, Kevin, 'Not Irrational but Habitual: The Importance of 'Behavioural Lock-in' in Energy Consumption', *Ecological Economics*, 69 (2010), 1104–14, https://doi.org/10.1016/j.ecolecon.2009.12.004.

Marsden, Greg, Jillian Anable, Tim Chatterton, Iain Docherty, James Faulconbridge, Lesley Murray, Helen Roby, and Jeremy Shires, 'Studying disruptive events: innovations in behaviour, opportunities for lower carbon transport policy?' *Transport Policy*, 94 (2020), 89–101, https://doi.org/10.1016/j.tranpol.2020.04.008.

Moss, Timothy, Sören Becker, and Matthias Naumann, 'Whose Energy Transition is it, Anyway? Organisation and Ownership of the Energiewende in Villages, Cities and Regions', *Local Environment*, 20 (2015), 1547–63, https://doi.org/10.1080/13549839.2014.915799.

Nicolson, Moira, Gesche Huebner, David Shipworth, and Simon Elam, 'Tailored Emails Prompt Electric Vehicle Owners to Engage with Tariff Switching Information', *Nature Energy*, 2 (2017), 17073, https://doi.org/10.1038/nenergy.2017.73.

Peters, Glen, 'How Changes Brought on by Coronavirus Could Help Tackle Climate Change', *The Conversation* (Theconversation.com, 2020), https://theconversation.com/how-changes-brought-on-by-coronavirus-could-help-tackle-climate-change-133509.

Pettifor, Hazel, Charlie Wilson, David McCollum, and Oriane Edelenbosch, 'Modelling Social Influence and Cultural Variation in Global Low-carbon Vehicle Transitions', *Global Environmental Change*, 47 (2017), 76–87, https://doi.org/10.1016/j.gloenvcha.2017.09.008.

Ralph, Kelcie, and Anne Brown, 'The Role of Habit and Residential Location in Travel Behavior Change Programs, a Field Experiment', *Transportation*, 46 (2019), 719–34, https://doi.org/10.1007/s11116-017-9842-7.

RCUK, *Progressing UK Energy Research for a Coherent Structure with Impact Report of the International Panel for the RCUK Review of Energy* (Rcuk.ac.uk, 2010), www.rcuk.ac.uk/documents/reviews/reviewpanelreport-pdf.

Schmocker, Jan-Dirk, Pierre Pettersson, and Satoshi Fujii, 'Comparative Analysis of Proximal and Distal Determinants for the Acceptance of Coercive Charging Policies in the UK and Japan', *International Journal of Sustainable Transportation*, 6 (2012), 156–73, https://doi.org/10.1080/15568318.2011.570856.

Thackeray, Stephen, Sharon Robinson, Pete Smith, et al., 'Civil Disobedience Movements Such as School Strike for the Climate are Raising Public Awareness of the Climate Change Emergency', *Global Change Biology*, 26 (2020), 1042–44, https://doi.org/10.1111/gcb.14978.

Thøgersen, John, 'The Importance of Timing for Breaking Commuters' Car Driving Habits', in *The Habits of Consumption*, ed. by A. Warde and D. Southerton (Helsinki: Helsinki Collegium for Advanced Studies, 2010), pp. 130–40.

Thompson, Sam, Juliet Michaelson, Saamah Abdallah, et al., *'Moments of change' as Opportunities for Influencing Behaviour. A Report to the Department for Environment, Food and Rural Affairs* (London: NEF / Defra, 2011).

Verplanken, Bas, and Deborah Roy, 'Empowering Interventions to Promote Sustainable Lifestyles: Testing the Habit Discontinuity Hypothesis in a Field Experiment', *Journal of Environmental Psychology*, 45 (2016), 127–34, https://doi.org/10.1016/j.jenvp.2015.11.008.

Verplanken, Bas, Deborah Roy, and Lorraine Whitmarsh, 'Cracks in the Wall: Habit Discontinuities as Vehicles for Behavior Change', in *The Psychology of Habit*, ed. by B. Verplanken (Dordrecht: Springer, 2018), pp. 189–205.

Verplanken, Bas, Henk Aarts, and Ad van Knippenberg, 'Habit, Information Acquisition, and the Process of Making Travel Mode Choices', *European Journal of Social Psychology*, 27 (1997), 539–60, https://doi.org/10.1002/(SICI)1099–0992(199709/10)27:5<539::AID-EJSP831>3.0.CO;2-A.

Verplanken, Bas, Ian Walker, Adrian Davis, and Michaela Jurasek, 'Context Change and Travel Mode Choice: Combining the habit discontinuity and self-activation hypotheses', *Journal of Environmental Psychology*, 28 (2008), 121–27, https://doi.org/10.1016/j.jenvp.2007.10.005.

Whitmarsh, Lorraine, Claire Hoolohan, Olivia Larner, Carly McLachlan, and Wouter Poortinga, *How Has COVID-19 Impacted Low-Carbon Lifestyles and Attitudes towards Climate Action?* CAST Briefing Paper 04 (2020), https://cast.ac.uk/wp-content/uploads/2020/08/CAST-Briefing-04-Covid-low-carbon-choices-1.pdf.

Whitmarsh, Lorraine, and Adam Corner, 'Tools for a New Climate Conversation: A Mixed-methods Study of Language for Public Engagement Across the Political Spectrum', *Global Environmental Change*, 42 (2017), 122–35, https://doi.org/10.1016/j.gloenvcha.2016.12.008.

Whitmarsh, Lorraine, Saffron O'Neill, and Irene Lorenzoni (eds), *Engaging the Public with Climate Change: Behaviour Change and Communication* (London: Earthscan, 2010).

Whitmarsh, Lorraine, 'Behavioural Responses to Climate Change: Asymmetry of Intentions and Impacts', *Journal of Environmental Psychology*, 29 (2009), 13–23, https://doi.org/10.1016/j.jenvp.2008.05.003.

Whittle, Colin, Paul Haggar, Lorraine Whitmarsh, Phil Morgan, and Dimitrios Xenias, *Decision-Making in the UK Transport System. Future of Mobility: Evidence Review*, Foresight, Government Office for Science (2019), https://assets.publishing.service.gov.uk/government/uploads/system/uploads/attachment_data/file/773667/decisionmaking.pdf.

Willis, Rebecca, 'How Members of Parliament understand and respond to climate change', *The Sociological Review*, 66 (2017), 475–91, https://doi.org/10.1177/0038026117731658.

Wilson, Charlie, L. Crane, and George Chryssochoidis, 'Why do homeowners renovate energy efficiently? Contrasting perspectives and implications for policy', *Energy Research & Social Science*, 7 (2015), 12–22, https://doi.org/10.1016/j.erss.2015.03.002.

Wood, Wendy, Leona Tam, and Melissa Guerrero Wit, 'Changing circumstances, disrupting habits', *Journal of Personality and Social Psychology*, 88 (2005), 918–33, https://doi.org/10.1037/0022-3514.88.6.918.

Zhou, Naaman, 'Oxford Dictionaries declares 'climate emergency' the word of 2019' (Theguardian.com, 2019), https://www.theguardian.com/environment/2019/nov/21/oxford-dictionaries-declares-climate-emergency-the-word-of-2019.

28. Five Questions whilst Walking: For Those that Decided to Participate in *Agir Pour le Vivant*

Isabelle Fremeaux and Jay Jordan

This chapter republishes an intervention to clarify our choice to ask participants to desert a big festival of ideas, *Agir Pour le Vivant* (Action for the Living), that took place in Arles in France in August 2020. We felt that the festival's intention of 'action for the living' was dissonant with the event's sponsorship by a series of toxic corporations. Our demand precipitated a series of public responses, ending with this final letter by us that asked a series of questions, our intention being to foreground the sorts of difficult choices that need to be made if we are collectively to walk away from the forces propelling global ecological crisis.

I Am a Boycotter

I am a boycotter. I am and always have been for some worlds and not others.
     If ever there were a time for life-affirming anti-capitalism it is NOW (Donna Haraway).[1]

This chapter republishes an intervention to clarify our choice to ask participants to boycott an event, *Agir Pour le Vivant* (Action for the

---

1   Personal correspondence between Donna Haraway, Isabelle Fremeaux and Jay Jordan, August 2020, quoted with permission.

 https://doi.org/10.11647/OBP.0265.28

Living) that took place in Arles in France in August 2020. The event's description and intention were advertised as follows:

> A large festival open to all and rooted in its territory, AGIR POUR LE VIVANT creates a new space for reflection and discussion beyond ideologies. For a week, it combines approaches, crosses the skills and proposals of writers, philosophers, scientists, gardeners, botanists, agronomists, herbalists, entrepreneurs and environmental activists who are trying to renew the great history of man's relationship with nature. They redefine the place of rivers in the world; claim royalty-free and reproducible organic seeds; campaign for the recognition of herbalism, for social and climatic justice or for a decolonial ecology; imagine resaving humanity; support the transition of companies, territories... [2]

We felt this statement to be dissonant with the event's sponsorship by a series of toxic corporations and financiers. Our open letter demanding people not to participate was published in terrestres.org as 'Choosing which Culture to Feed: An Open Letter about Friendships and a Call to Desert"[3] (Fremeaux and Jordan 2020). It became something of a *cause celèbre*, being shared widely on social media and precipitating further published letters between ourselves and participants who decided to attend the event but managed to eject one of the funders (for example, Fremeaux and Jordan 2020; Morizot and Zhong Mengual 2020). We share here the final letter of the exchange that poses a series of questions to foreground the sorts of difficult choices that need to be made if we are collectively to walk away from the forces propelling the global ecological crisis.

Question 1: What about the Forest?

I'm lost in a forest
All alone
The girl was never there
It's always the same
I'm running towards nothing
(Again and again and again and again)[4]

---

2   https://www.agirpourlevivant.fr/copie-de-programme-2.
3   An   English   version   is   here: https://www.terrestres.org/wp-content/uploads/2020/08/Choosing-which-culture-to-feed.pdf.
4   The Cure, 'A forest' (1980), ppm 337.

Let us begin with celebration and joy. Joy that words have led to action as they always should. The action being that one of the sponsors of *Agir Pour le Vivant* has had to retreat and has thus liberated the forum from one of its toxic ties. BNP Paribas's logo has been taken off the website and its money will be returned. "We would like to thank them here for their commitment to the living", says the forum's page, covered in logos.

BNP Paribas's "commitment to the living" would have cost them 20,000 euros—a little under 10% of the 270,000 total budget of the festival we learn from the article in the Arlesian local paper about this controversy (L'Arlesienne 2020). For a company whose 2019 revenue was 44.6 billion euros, and profits 8.17 billion, their support is a drop in the ocean, but their retreat is significant. What is just as significant to us is that their staff will not be present at the forum, nor speaking at the public events, nor in the closed-door workshops such as "L'empreinte naturelle des entreprises" ("The natural imprint of companies"), where they would have met with the other staff and CEOs of corporations for what is called an 'atelier de travaille'[5] (a 'working work-shop'). Of course, this event does not appear on the website's programme, and is not accessible to the public, even to those who have payed fifty euros for their special access pass, but it is perhaps the place where the real *work* of the greenwashers takes place and the false suicidal solutions to this omnicidal crisis are dreamt up and planned.

Nonetheless, this is an historic victory. It joins the growing list of cultural institutions that have liberated themselves from the funders and drivers of this culture of extinction over the last few years. In the UK alone both the Tate Museum and the Royal Shakespeare Company have freed themselves from British Petroleum's sponsorship, London's Science Museum, National Theatre and National Gallery have ended their relationship with Shell, the Edinburgh Science Festival has severed ties with ExxonMobil and Total. In the Netherlands, the Dutch art museum the Mauritshuis, the science and culture museum Museon, and—close to the heart of Arlesians—Amsterdam's Van Gogh Museum, will no longer accept Shell's sponsorship money.

---

5   See   https://www.eterritoire.fr/detail/activites-touristiques/agir-pour-le-vivant-jour-3/666132924/provence-alpes-cote-d-azur,bouches-du-rhone,arles(13200).

Of course, none of these institutions did this voluntarily. They changed their behaviour and let go of their sponsors because of uncomfortable words written to them, and most importantly because people acted on their ideas and put their vulnerable bodies on the line, often with stunningly beautiful performative protests[6] and creative disruptions. Many of these disobedient bodies belonged to artists, intellectuals and researchers who, by entering into conflict with these institutions, were biting the hand that fed them. But they had decided that their individual cultural capital was less important than being part of a culture of resistance against those who, as Donna Haraway writes, "greenwash the exterminators".[7]

The other thing that brings us joy is that some participants have chosen to desert, to walk away, including AfroEuropean anthropologist Dénètem Touam Bona and landscape architect Giles Clément. We say joy in contrast to the neoliberal duty of happiness, because as Silvia Federici says, joy is

> not satisfaction with things as they are. It's part of feeling power's capacities growing in you and growing in the people around you. It's a feeling, a passion, that comes from a process of transformation [...] You feel that you have the power to change and you feel yourself changing with what you're doing, together with other people. It's not a form of acquiescence to what exists (Federici et al. 2017).

For us this feeling of power to change our lives and circumstances is at the core of collective resistance and the construction of forms of culture and life that affirm the living.

One of the other conditions that Baptiste Morizot, Estelle Zhong Mengual and their friends (including Rob Hopkins, Cyril Dion and Vinciane Dépres), set to the organisers of the festival in their open letter —*Quel trouble voulons-nous habiter?* (*Which Trouble do we Want to Inhabit?*) (Morizot and Zhong Mengual 2020)—was that all the corporate logos must be taken off communications. We are writing this nearly a week later, and not only are the other logos still on the website, but there are now thirty-three of them, as opposed to the twenty-six that were visible

---

6   See, for example, https://www.liberatetate.org.uk/ and https://www. fossilfreeculture.nl/.

7   Personal correspondence between Isabelle Fremeaux, John Jordan and Donna Haraway, 13 August 2020, quoted with permission.

when we wrote the first letter. What is surprising and somewhat absurd is that it seems as though as one bank left another one came in, not even through the back door, but right on the front page of the forum's website. Amongst these new additions is the logo of Crédit du Nord, which is entirely owned by Société Générale, by far the biggest funders of North American shale gas. Since the signing of the UNFCCC COP21 Paris Agreement in December 2015 they have pumped over 11 billion euros into this death-dealing industry (Chocron and Wakim 2020). What is the difference between Société Générale and BNP Paribas ("The bank for a changing world"[8])? (also see chapters by Wright and Nyberg, and Bracking, this volume).

We do not want to bore anyone with another cartography of poisonous sponsors. But to change something you need to know the texture of that thing. For us, this means being attuned and deeply sensitive to the specific details of situations and particular relationships in which we are enmeshed. The philosopher Spinoza, who we must never forget was despised by most of his contemporaries, taught us that such situated understanding enables us to move along in accordance with what is required in that moment. Surely this is the key to ethics. We are not interested in those old forms of rigid radicalism which try to control things, but in response-ability, in building our capacities to remain responsive to specific changing situations and opening up common spaces that support, rather than control, mutual transformation. The key is surely that we feel more alive together.

And we certainly do not feel such joy when we see that all the other corporations remain and three of the new logos include Faber and Novel, a talent and technology company whose clients include Total.[9] Fondation Yves Rocher, who expose low-paid workers to pesticides and recently sacked 132 Turkish women workers because they joined a union (Billette 2019). And last but not least, the great polluters of public space and our imaginaries, dealers of the dangerous drug of endless consumption, the world's largest outdoor advertising corporation, JC Decaux.

Was the felling of BNP Paribas the tree that is hiding the forest?

---

8   https://group.bnpparibas/en/.
9   https://www.fabernovel.com/fr/clients/cases.

Question 2: Is It Just about Fossil Fuels?

You have stolen my dreams and my childhood with your empty words. And yet I'm one of the lucky ones. People are suffering. People are dying. Entire ecosystems are collapsing. We are in the beginning of a mass extinction, and all you can talk about is money and fairy tales of eternal economic growth. How dare you! (Greta Thunberg speaking at the *UN Climate Action Summit in New York City*, September 2019).

We disagree with the assertion in *Quel trouble voulons-nous habiter?* that "after analysis, the other sponsors do not seem to have the same degree of seriousness at all" (Morizot and Zhong Mengual 2020). Does this suggest that, by removing the most obvious 'exterminators', it is OK for you to keep cooperating with the others by attending the forum? Is the designing of airports and supermarkets and the creation of new financial markets in water, air, soil and forests—and thus the effective privatisation of nature—really less serious? Is this not about wielding the great magical rootless tool of the new spirit of 'green' capitalism: offsetting?

We have been involved in the climate justice movement for a quarter of a century. When we were setting up climate camps over a decade ago (Fremeaux and Jordan 2011), merging the yes and the no, entangling the creation of alternatives with resistance, demonstrating forms of non-hierarchical ecological life, and simultaneously taking action against airport expansion and coal-fired power stations, we still had to convince people that climate change existed: 'keep the oil in the soil' was seen as a radical statement. Now such words are commonly heard in board rooms and chanted by the biggest youth movement in world history on our streets, calling for "system change not climate change". We can only celebrate the fact that fossil fuel corporations and their funders are rapidly losing their social acceptance and a fossil fuel-free future is no longer just the dream of rebels. But there is a blind spot. When those in power talk of 'anthropogenic' climate it would be infinitely more accurate to refer to it as capitalist climate change (Tanuro 2014). As one of the beautiful pink and green banners at climate camp proclaimed, "capitalism is crisis".

Whether capitalism comes in red, pink or green, it is its cancer-like logic of limitless growth that is at the heart of the problem. In *This Changes Everything: Capitalism vs. the Climate*, which brilliantly details how the economy is at war against life, Naomi Klein (2015: 21) wrote "[w]hat the climate needs to avoid collapse is a contraction in humanity's use of resources; what our economic model demands to avoid collapse is unfettered growth". This contradictory, suicidal logic of capitalism, a legacy of colonialism, patriarchy and dispossession, cannot be smoothed over by words that demand us to "inhabit incoherence". This contradiction is rendering this world uninhabitable.

Scientists everywhere tell us there are limits and key planetary boundaries that must be respected to avoid triggering collapse, but we should no longer fear, because a new panacea has been found, namely, 'green growth'. This buzz word is now the core tenet of the UN Sustainable Development Goals, and since 2012 has been promoted by institutions such as the World Bank and the OECD. The goal is to achieve "absolute decoupling" of GDP from the total use of natural resources. The trouble is that three recent empirical studies (Hickel 2018) (including one by cheerleaders of green growth, the UN Environment Programme), show that this seemingly elegant solution to the catastrophe is a pipe dream. Even under the best conditions—including state-of-the-art, government-supported technological innovation to develop absolute energy efficiency, massive taxation raising the price of carbon from $50 to $600 per metric tonne, and taxing resource extraction—every computer model of the figures pushed us way over the planetary limits. As Sian Sullivan (2013) writes:

> [t]he utopian vision here is that capitalism will thus become better aligned with 'nature', so as to generate the multiple wins of a 'green economy' wherein economic growth is maintained and 'natural capital' is too.

In the new documentary, *Fairytales of Growth*, sixteen-year-old Tokatawin Iron Eyes, President of Standing Rock Youth Council, looks into the camera.[10] Her life-giving land is threatened by the Dakota Access Pipeline, initially funded in part by international financial services firm Natixis, owners of Mirova, the sponsors whose '#naturalcapital' belief system could not be further from her world and her community of "water defenders" who risked everything to keep life flourishing on their land (Earthjustice 2020). The belief that humans will only

---

10   https://www.fairytalesofgrowth.com.

protect nature if it is measured, valued and becomes integrated into a profit-making market accounting system, could not be further from her life-affirming culture, where people did *"did not own land individually, but instead believed in the importance of honoring the earth as our common home and sharing its resources responsibly"* (Ignatian Solidarity Network 2019). "One of the biggest things that anybody can learn from this youth climate movement right now, being built on the work of indigenous, black and brown communities is the fact that it is an issue of priorities", she gesticulates with calm rage, "[b]ecause when we want to talk about economic growth over people having clean water and the right to a livable future and planet that is a sign that something is wrong".

Question 3: Who Is Contaminating Who?

Friendship will be the soil from which a new politics will emerge (Ivan Illich n.d., quoted in bergman and Montgomery 2017: online).

The mechanism to gain social licence to operate in an event such as *Agir pour le Vivant* seems not to have been understood. It is neither a question of an indirect contact with an economic actor being turned into a sort of recruitment operation, nor of participants becoming spokespeople despite themselves being 'contaminated' and losing their critical intelligence and lucidity. What we are pointing to is actually the reverse: it is your critical intelligence, your dazzling analyses, your innovative proposals that positively spill over on to them. Simply by association, they repair their often shaky reputation.

The very notion of social licence to operate is not an activist concept, infused with 'ideological' or 'theological unconscious': it was born in corporate offices. For instance, Henderson and Williams (respectively Shell's Project Director for External Affairs, and Chair of corporate PR firm Fishburn Hedges) described it thus when they were in charge of "a global reputation management programme to 'build, maintain and defend Shell's capital'", after the Brent Spar debacle:[11] "[i]t is opinion formers that grant the licence to operate and often set the tone for how

---

11 In which in the mid-1990s Shell controversially proposed to decommission the Brent Spar oil rig in the North Sea by simply sinking the platform into deep water in the North Atlantic, causing an outcry amongst environmental campaigners, see https://en.wikipedia.org/wiki/Brent_Spar.

the general public hears about and assesses companies" (Henderson and Williams 2002: 12, quoted in Evans 2015: 79). Putting their strategy into action, Shell went on to sponsor a large number of cultural institutions and high-profile cultural events in the fifteen years that ensued.

Allowing ourselves an analogy with COVID-19, the problem is not becoming representatives of the virus but finding oneself aiding its spread. We need to stop the infection of all corners of life with capitalist logic. What is being called for here is some 'social distancing', so as to not unwittingly become 'spreaders', even if one can feel proud of being asymptomatic.

Question 4: Which Friendships Are Fertile for Whom?

It was never for us a question of issuing an *ultimatum* about friendship, a sort of emotional blackmail. To imagine that such a thing could be a real political lever would have been rather presumptuous. For us, friendship is not the neoliberal "banal affair of private preferences [...] with those who are already like us, [with whom] we keep each other comfortable, rather than becoming different and more capable together", as bergman and Montgomery (2017: 96) sum up so sharply: friendship is a "relationship crucial to life, worth fighting for".

That said, we do not subscribe to the Bush-like logic that seems to be attributed to us—"you are either with us or against us"—because we are not confused as to whom the real enemies are. The aim of our call to desert the event was *not* to sort out friends from foes, allies from traitors; it was to defuse the nefarious organisations' strategy of gaining a social licence to operate.

As Dénètem Touam Bona, the first deserter, underlined to us in his reading of *Which Trouble Do we Want to Inhabit?* (Morizot and Zhong Mengual 2020),

[t]here is an assumption here that 'attachments', bonds are good in themselves, and that out of their proliferation, salvation will inevitably be born. As far as I am concerned, my conception of the lyannaj [coalition] cannot be dissociated from what I call 'maroon secession'. The maroonnage that I conceive of as 'running away', as forms of life and resistance in a minor mode, is an operation of subtraction, similar to that La Boétie already praised in his *Discourse on voluntary servitude*, or to that Foucault evoked when he linked becoming-fascist with falling

in love with power (and recognition, prestige, honours... are part of the attributes of power) (Personal communication).

In the end, we are actually in complete agreement with Donna Haraway who wrote to us about the call out and its response:

> I am a boycotter. I am and always have been for some worlds and not others. If ever there were a time for life-affirming anti-capitalism it is NOW [...] I also affirm the ongoing possibility of future alliances with people who did not boycott, and who disagree, but not on just any terms. Coming together is always finite, fragile, open to change. It is not easy not to demonize after fierce disagreement, but it is crucial. But sympoiesis is not a grand neoliberal festival of co-becoming (Personal communication, 13 August 2020).

Question 5: What Are We Capable of?

First of all, to clarify, as it is one of several reversals of our arguments: we did not ask for coherence from our addressees. We explained that what has often motivated our numerous non-collaboration decisions was a need for coherence. Not to alleviate guilt, but as care for mental health (which has little to do with 'psychological comfort'). George Orwell, who knew what it meant to embody words and ideas and was prepared to die for them on the anti-fascist front of the 1936 Spanish Revolution, coined the term "doublethink", in his dystopian novel *1984* (1949). An imposed practice at the heart of maintaining a totalitarian regime founded on inequality, 'doublethink' was "the power of holding two contradictory beliefs in one's mind simultaneously, and accepting both of them" (Orwell 1949: 244; also see Sullivan's Chapter 11, this volume). For Orwell, with 'doublethink' came the mental state necessary to make sure a society of equality could never be put in place: he called this managed authoritarian deferral, "controlled insanity."

There is no doubt that the moralistic hunt for daily incoherences is absurd at best, most often noxious. We certainly also strive for a world where contradictions can be "melting pots and sources of creative tensions": yet, and as Dénètem Touam Bona also remarks, "[t]he praise of trouble must not serve the nihilistic mechanics of general equivalence of capital" (personal communication); it cannot be a handless concept, as Baptiste Morizot would say.

It is indeed crucial and urgent to embrace 'an art of consequences': we are not calling for much more. And maybe the question at the core of such an art would no longer be 'what should one do?' but 'what is one capable of?'

Bibliography

Bergman, carla, and Nick Montgomery, *Joyful Militancy: Building Thriving Resistance in Toxic Times* (London: AK Press, 2017).

Billette, Alexandre, 'Turkey: The Union Struggle of Women Sacked by a Subsidiary of Yves Rocher' (Rfi.fr, 2019), https://www.rfi.fr/fr/europe/201903009-turquie-lutte-syndicale-femmes-licenciees-une-filiale-yves-rocher.

Chocron, Véronique, and Nabil Wakim, 'Oil: French Banks Stuck in American Shale' (Lemonde.fr, 2020), https://www.lemonde.fr/economie/article/2020/05/13/petrole-les-banques-francaises-engluees-dans-le-schiste-americain_6039530_3234.html.

Earthjustice 2020, 'Judge Orders Dakota Access Pipeline to Shut Down' (Earthjustice.org, 2020), https://earthjustice.org/news/press/2020/judge-orders-dakota-access-pipeline-to-shut-down.

Evans, Mel, *Artwash* (London: Pluto Press, 2015).

Federici, Silvia, Carla Bergman, and Nick Montgomery, 'Feeling Powers Growing: An Interview with Silvia Federici', in *Joyful Militancy: Building Thriving Resistance in Toxic Times*, ed. by carla bergman and Nick Montgomery (London: AK Press, 2017), https://joyfulmilitancy.com/2018/06/03/feeling-powers-growing-an-interview-with-silvia-federici/.

Fremeaux, Isabelle, and John Jordan, *Les Sentiers de l'utopie* (Paris: Zones, 2011).

Fremeaux, Isabelle, and Jay Jordan, 'What Culture do we Want to Nurture?' (Terrestres.org, 2020), https://www.terrestres.org/2020/08/04/quelle-culture-voulons-nous-nourrir/2020.

Henderson, Tom, and John Williams, 'Shell: Managing a Corporate Reputation', in *Public Relations Cases*, ed. by Barbara DeSanto and Daniel Moss (London: Routledge, 2002).

Hickel, Jason, 'Why Growth Can't Be Green' (Foreignpolicy.com, 2018), https://foreignpolicy.com/2018/09/12/why-growth-cant-be-green/.

Ignatian Solidarity Network, 'Tokatawin Iron Eyes and Greta Thunberg Join Youth Climate Crisis Panel at Red Cloud Indian School' (Ignationsolidarity.net, 2019), https://ignatiansolidarity.net/blog/2019/10/07/tokata-greta-climate-crisis-red-cloud/.

Illich, Ivan, to Madhu Suri Prakash, "Friendship," n.d., in *Joyful Militancy: Building Thriving Resistance in Toxic Times*, ed. by carla bergman and Nick Montgomery (London: AK Press, 2017), https://theanarchistlibrary.org/library/joyful-militancy-bergman-montgomery#fn59.

L'Arlesienne, 'Actes Sud: "toxic" partnerships at the Agir pour le vivant festival' (Larlesienne.info, 2020), https://larlesienne.info/2020/08/14/actes-sud-partenariats-toxiques-au-festival-agir-pour-le-vivant/.

Morizot, Baptiste, and Estelle Zhong Mengual, 'What Disorder do we Want to Live in? Response to Isabelle Fremeaux and John Jordan' (Terrestres.org, 2020), https://www.terrestres.org/2020/08/12/quel-trouble-voulons-nous-habiter-reponse-a-isabelle-fremeaux-et-john-jordan/.

Sullivan, Sian, 'At the Edinburgh Forums on Natural Capital and Natural Commons, 2013' (The-natural-capital-myth.net, 2013), https://the-natural-capital-myth.net/2013/11/21/at-the-edinburgh-forums-on-natural-capital-and-natural-commons-from-disavowal-to-plutonomy-via-natural-capital/.

Tanuro, Daniel, *Green Capitalism: Why It Can't Work* (Winnipeg: Fernwood Publishing, 2014).

Index

About the Team

Alessandra Tosi was the managing editor for this book.

Melissa Purkiss performed the copy-editing, proofreading and typesetting.

Anna Gatti designed the cover. The cover was produced in InDesign using the Fontin font.

Luca Baffa produced the paperback and hardback editions. The text font is Tex Gyre Pagella; the heading font is Californian FB. Luca produced the EPUB, MOBI, PDF, HTML, and XML editions—the conversion is performed with open source software freely available on our GitHub page (https://github.com/OpenBookPublishers).

This book need not end here...

Share

All our books — including the one you have just read — are free to access online so that students, researchers and members of the public who can't afford a printed edition will have access to the same ideas. This title will be accessed online by hundreds of readers each month across the globe: why not share the link so that someone you know is one of them?

This book and additional content is available at:

https://doi.org/10.11647/OBP.0265

Customise

Personalise your copy of this book or design new books using OBP and third-party material. Take chapters or whole books from our published list and make a special edition, a new anthology or an illuminating coursepack. Each customised edition will be produced as a paperback and a downloadable PDF.

Find out more at:

https://www.openbookpublishers.com/section/59/1

Like Open Book Publishers

Follow @OpenBookPublish

Read more at the Open Book Publishers **BLOG**

You may also be interested in:

Right Research
Modelling Sustainable Research Practices in the Anthropocene
Chelsea Miya, Oliver Rossier and Geoffrey Rockwell (eds)

https://doi.org/10.11647/OBP.0213

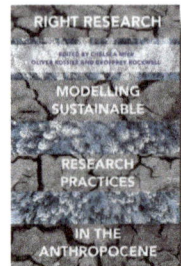

Global Warming in Local Discourses
How Communities around the World Make Sense of Climate Change
Michael Brüggemann and Simone Rödder (eds)

https://doi.org/10.11647/OBP.0212

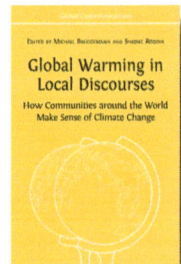

Living Earth Community
Multiple Ways of Being and Knowing
Sam Mickey, Mary Evelyn Tucker, and John Grim (eds)

https://doi.org/10.11647/OBP.0186

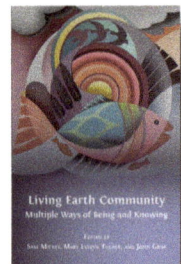